Lecture Notes in Physics

Volume 1051

The series Lecture Notes in Physics (LNP), founded in 1969, reports new developments in physics research and teaching - quickly and informally, but with a high quality and the explicit aim to summarize and communicate current knowledge in an accessible way. Books published in this series are conceived as bridging material between advanced graduate textbooks and the forefront of research and to serve three purposes:

- to be a compact and modern up-to-date source of reference on a well-defined topic;
- to serve as an accessible introduction to the field to postgraduate students and non-specialist researchers from related areas;
- to be a source of advanced teaching material for specialized seminars, courses and schools.

Both monographs and multi-author volumes will be considered for publication. Edited volumes should however consist of a very limited number of contributions only. Proceedings will not be considered for LNP.

Volumes published in LNP are disseminated both in print and in electronic formats, the electronic archive being available at springerlink.com. The series content is indexed, abstracted and referenced by many abstracting and information services, bibliographic networks, subscription agencies, library networks, and consortia.

Proposals should be sent to a member of the Editorial Board, or directly to the responsible editor at Springer:

Dr Lisa Scalone
lisa.scalone@springernature.com

Rishabh Jha

Theory of Strongly Correlated Systems

An Introduction to Sachdev-Ye-Kitaev Model

Rishabh Jha
Institute for Theoretical Physics
Georg-August-Universität Göttingen
Göttingen, Germany

ISSN 0075-8450 ISSN 1616-6361 (electronic)
Lecture Notes in Physics
ISBN 978-3-032-20909-2 ISBN 978-3-032-20910-8 (eBook)
https://doi.org/10.1007/978-3-032-20910-8

This Springer imprint is published by the registered company Springer Nature Switzerland AG
The registered company address is: Gewerbestrasse 11, 6330 Cham, Switzerland

Acknowledgement I gratefully acknowledge financial support from the Deutsche Forschungsgemeinschaft (DFG, German Research Foundation) through Grant No. 217133147 as part of SFB 1073 (Project B03). I have benefited immensely from discussions about the SYK physics with researchers and collaborators, in particular Prof. Dr. Stefan Kehrein and Dr. Jan C. Louw. Any errors or shortcomings remain solely my responsibility.

Introduction

The Sachdev-Ye-Kitaev (SYK) model provides an analytically tractable framework for exotic strongly correlated phases where conventional paradigms like Landau's Fermi liquid theory collapse. This review offers a pedagogical introduction to the SYK physics, highlighting its unique capacity to model *strange metals*—systems exhibiting linear-in-temperature resistivity, Planckian dissipation, and quasiparticle breakdown. We systematically construct both Majorana and complex fermion variants, transforming them into training grounds for modern many-body physics techniques, for instance, (1) large-N formulations via disorder averaging and replica symmetry, (2) Schwinger-Dyson and Kadanoff-Baym equations, (3) imaginary time Matsubara formulation, (4) real-time dynamics via Keldysh formalism, and the associated (5) non-perturbative Keldysh contour deformations. These tools lay the foundation for equilibrium thermodynamics, quantum chaos, quench dynamics, and transport in the thermodynamic limit, all within a solvable, chaotic quantum system. Intended as a self-contained resource, the review bridges advanced technical machinery to physical insights, with computational implementations provided. Though principally treating the SYK model as a condensed matter laboratory, we also highlight its profound connection to quantum gravity, woven throughout this work, underscoring how this solvable chaotic fermionic model serves as a lens onto black hole thermodynamics and holographic duality.

Contents

List of Figures

Invitation to the Sachdev-Ye-Kitaev Paradigm 1

Abstract

This chapter introduces the Sachdev-Ye-Kitaev (SYK) paradigm within the broader context of quantum many body physics. It begins with a review of Landau's Fermi liquid theory as the conventional framework for metallic systems, emphasizing both its experimental successes, such as the Wiedemann-Franz law and quadratic temperature dependence of resistivity, and its fundamental limitations. The discussion then turns to strange metals, namely materials that display linear-in-temperature resistivity and other anomalous properties which violate Fermi liquid predictions, with high temperature cuprate superconductors providing key examples. Against this backdrop, the Sachdev-Ye-Kitaev model is presented as a solvable theoretical laboratory for non Fermi liquid behavior that captures Planckian dissipation, maximal quantum chaos, and characteristic transport signatures of strange metals. The chapter also highlights the connection between SYK physics and quantum gravity through shared features of phase transitions in charged Anti-de Sitter black holes. Finally, it outlines the advanced methodologies developed in later chapters, including large-N techniques, disorder averaging, Schwinger-Dyson equations, imaginary-time and Keldysh real-time formalisms, critical exponent analysis, quantum chaos diagnostics based on out of time ordered correlators, and contour deformations used for transport, thereby framing SYK as a pedagogical model for modern many body techniques.

Quantum many-body systems challenge our deepest intuitions about the physical world. When vast numbers of particles interact under quantum rules, they give rise to phenomena that transcend their microscopic constituents—superconductivity defying electrical resistance, fractional quantum hall states exhibiting exotic statistics, and quantum criticality where fluctuations span all length scales. These emergent behaviors reveal physics not encoded in fundamental equations but born from collective dynamics. Yet, our understanding remains fragmented. Conventional

R. Jha, *Theory of Strongly Correlated Systems*, Lecture Notes in Physics 1051,
https://doi.org/10.1007/978-3-032-20910-8_1

paradigms like Landau's Fermi liquid theory (FLT) succeed in describing metals where interactions merely renormalize electrons into "quasiparticles" but fail catastrophically in *strange metals*—ubiquitous in high-temperature superconductors and quantum critical materials—where *linear-in-temperature resistivity*, and anomalous transport laws signal the demise of particle-like excitations. This crisis demands frameworks that embrace strong correlations without quasiparticles.

The Sachdev-Ye-Kitaev (SYK) model serves as a paradigmatic framework for studying strongly correlated quantum phenomena, combining analytical tractability with rich non-integrable, chaotic, and ergodic dynamics. Originally conceived for quantum holography [15, 37], this effectively zero-dimensional model of N Majorana fermions with all-to-all random interactions exhibits maximal quantum chaos: a signature property of black holes in holographic duality [29, 39]. Its complex fermion generalization [9, 23] introduces a conserved $U(1)$ charge, enabling exploration of non-Fermi liquid transport and thermalization. While isolated large-q SYK systems (implying $q/2$-body interactions) thermalize instantaneously with respect to the Green's functions, the non-equilibrium dynamics and universality of coupled SYK lattices exhibit strange metallicity and rich thermalization dynamics. In order to appreciate the anomalous behavior of strange metals, we first give a brief overview of Landau's FLT and then proceed to comment on the experimental signatures that support FLT as well as those that violate it catastrophically.

1.1 Conventional Metals: Landau's Fermi Liquid Theory

Fermi liquid theory (FLT) constitutes the foundational framework for describing standard metallic systems. Pioneered by Landau [20, 21], this approach conceptualizes interacting fermions through the principle of *adiabatic continuity*. This principle asserts that turning on interactions gradually transforms non-interacting fermions into their interacting counterparts without abrupt changes in the system's essential properties. Such continuity enables direct comparisons between non-interacting and interacting regimes, as systematically outlined in Table 1.1. We briefly outline the conceptual underpinnings of FLT and we refer the reader to Ref. [3, 33] for a detailed deep dive into the topic.

Table 1.1 Contrasting excitation characteristics in non-interacting fermion systems versus interacting Fermi liquids. The ground state configuration exhibits full occupation of momentum states below p_F and complete vacancy above p_F. Elementary excitations are generated by relocating a fermion from an occupied state ($|\vec{p}| < p_F$) to an unoccupied state ($|\vec{p}| > p_F$), yielding complementary particle-like excitations above p_F and hole-like excitations below p_F

Non-interacting	Interacting
– electron (particle) $|\vec{p}| > p_F$	– quasiparticle $|\vec{p}| > p_F$
– hole ($|\vec{p}| < p_F$)	– quasihole $|\vec{p}| < p_F$
– charge $\pm e$, spin $\frac{1}{2}$	– charge $\pm e$, spin $\frac{1}{2}$
– always stable	–stable only at low energies ($\omega \to 0$)

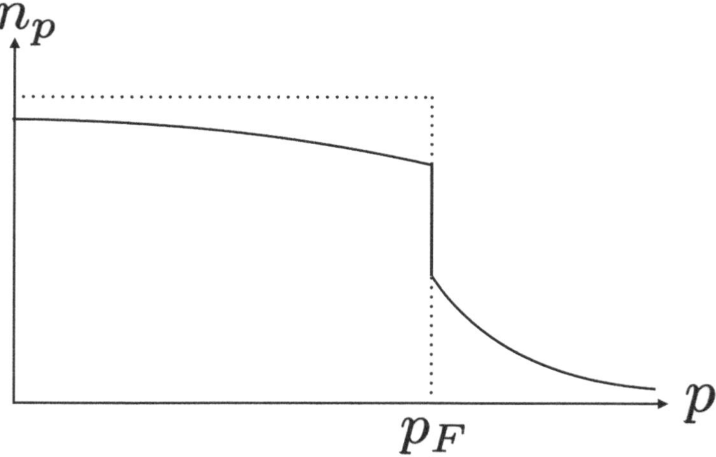

Fig. 1.1 Zero-temperature momentum distribution $n(\vec{p})$. Dashed line: non-interacting system; solid line: interacting case. Spherical symmetry is assumed for illustration

The FLT relies critically on two interconnected concepts:

- The Fermi Surface: A manifold in momentum space where the single-particle Green's function $\mathcal{G}(\vec{p}, \omega)$ exhibits singular behavior. For non-interacting systems at zero temperature, this coincides with the step discontinuity $n_0(\vec{p}) = \Theta\,(p_F - |\vec{p}|)$ in occupation number (see Fig. 1.1). Interactions modify the Fermi surface's shape while conserving its enclosed volume (Luttinger's theorem) [27, 28]:

$$\frac{1}{(2\pi)^d} \int_{|\vec{p}|<p_F} d^d p = \frac{N}{V}, \tag{1.1}$$

where N/V denotes particle density.
- Quasiparticles: These emergent excitations retain the charge and spin quantum numbers of bare electrons but feature renormalized kinetic properties (effective mass $m^\star$, velocity $v_F = p_F/m^\star$). Unlike bare particles, quasiparticles exhibit finite lifetimes near the Fermi surface at non-zero temperatures.

Quasiparticles dominate low-energy thermodynamics and transport. Their occupation distribution $n(\vec{p})$ deviates perturbatively from the non-interacting ground state $n_0(\vec{p})$ (Fig. 1.1). Defining $\delta n(\vec{p}) \equiv n(\vec{p}) - n_0(\vec{p})$, stability requires $\delta n(\vec{p})$ to be significant only near $|\vec{p}| \approx p_F$. The total energy $E = E_0 + \delta E$ then expands in powers of $\delta n(\vec{p})$

$$\delta E = \sum_{\vec{p}} \varepsilon(\vec{p})\delta n(\vec{p}) + \mathcal{O}(\delta n^2), \tag{1.2}$$

where spin indices are omitted for clarity.

The quasiparticle energy $\varepsilon(\vec{p})$ represents the energy shift from adding an excitation near the Fermi surface and satisfies

$$\varepsilon(\vec{p}) = \frac{\delta E}{\delta n(\vec{p})}. \tag{1.3}$$

At the Fermi momentum $|\vec{p}| = p_F$, $\varepsilon(\vec{p})$ equals the chemical potential μ, defined through

$$\mu = E(N+1) - E(N) = \frac{\partial E}{\partial N} = \varepsilon(p_F). \tag{1.4}$$

Though phenomenological, FLT applies to microscopically weakly interacting systems. A representative Hamiltonian includes (including spin labels σ and σ' that can take, for instance, $\uparrow$ and $\downarrow$ values for an electron)

$$\mathcal{H} = \underbrace{\sum_{\vec{p},\sigma} \varepsilon_0(\vec{p}) c^\dagger_{\vec{p}\sigma} c_{\vec{p}\sigma}}_{\text{Kinetic energy}} + \underbrace{\frac{1}{2} \sum \vec{p},\vec{p}',\vec{q}\ \sigma,\sigma' V(\vec{q}) c^\dagger_{\vec{p}+\vec{q}\sigma} c^\dagger_{\vec{p}'-\vec{q}\sigma'} c_{\vec{p}\sigma} c_{\vec{p}'\sigma'},}_{\text{Two-body interactions}} \tag{1.5}$$

where $\varepsilon_0(\vec{p})$ is the bare dispersion, $V(\vec{q})$ the interaction potential, and $c^\dagger_{\vec{p}\sigma}, c_{\vec{p}\sigma}$ are fermionic operators, that satisfy the standard anti-commutation relations. The energy expansion includes higher-order contributions representing interactions among quasiparticles, characterized by the symmetric Landau parameter function $f(\vec{p}, \vec{p}')$ (we again omit spin labels for clarity)

$$E(\delta n) = \sum_{\vec{p}} \varepsilon(\vec{p})\delta n(\vec{p}) + \frac{1}{2} \sum_{\vec{p},\vec{p}'} f(\vec{p}, \vec{p}')\delta n(\vec{p})\delta n(\vec{p}') + \mathcal{O}(\delta n^3). \tag{1.6}$$

This interaction term arises because the quasiparticle energy $\varepsilon(\vec{p})$ fundamentally depends on the configuration of other excitations in the system.

In the vicinity of the Fermi surface ($|\vec{p}| = p_F$ which sets the temperature scale T_F, also known as the Fermi temperature), the Fermi velocity $\vec{v}_F(\vec{p}) = \nabla_{\vec{p}}\varepsilon(\vec{p})$ determines the effective mass through

$$m^\star = \frac{p_F}{|\vec{v}_F(p_F)|}, \tag{1.7}$$

noting that this mass renormalization becomes direction-independent only for spherical Fermi surfaces. The quasiparticle energy itself contains many-body corrections expressed as

$$\varepsilon(\vec{p}) = \varepsilon_0(\vec{p}) + \sum_{\vec{p}'} f(\vec{p}, \vec{p}')\delta n(\vec{p}') + \cdots , \tag{1.8}$$

where $f\left(\vec{p}, \vec{p}'\right)$ quantifies the strength of interactions between excitations near the Fermi surface. This functional form embodies Landau theory's description of low-energy collective behavior in fermionic systems.

While our treatment has omitted spin degrees of freedom, Landau's theoretical framework readily accommodates their inclusion through systematic extension. Similarly, although spherical symmetry has been assumed throughout our analysis, generalizing to anisotropic configurations follows well-established formal procedures. These important extensions, while physically significant, exceed the boundaries of our present focus. Comprehensive treatments appear in Ref. [33]. Including spins, the quasiparticle density of states at the Fermi surface, denoted $N(0)$, is formally defined as

$$N(0) = \frac{1}{V}\sum_{\vec{p},\sigma} \delta\left(\varepsilon^0_{\vec{p}\sigma} - \mu\right) = -\frac{1}{V}\sum_{\vec{p},\sigma} \frac{\partial n^0_{\vec{p}\sigma}}{\partial \varepsilon_{\vec{p}\sigma}} \tag{1.9}$$

where $n^0_{\vec{p}\sigma}$ represents the zero-temperature Fermi distribution, and σ indexes the spin projection (typically $\uparrow$ or $\downarrow$ for electrons). Evaluation yields the closed-form expression

$$N(0) = \frac{m^\star p_F}{\pi^2 \hbar^3} \tag{1.10}$$

in terms of the effective mass $m^\star$. Crucially, $\varepsilon^0_{\vec{p}\sigma}$ signifies the quasiparticle energy at the Fermi surface, which differs fundamentally from the bare dispersion $\varepsilon_0(p)$.

The integrity of FLT critically depends on the asymptotic behavior of quasiparticle lifetimes τ. For the theory to remain valid, τ must diverge as excitation energies ω approach the chemical potential μ (with $\omega \to \mu \equiv E_F$ which also characterizes the Fermi temperature T_F) at $T = 0$. Analogously, at finite temperatures, τ must become infinite as $T \to 0$ when $\omega = \mu$. This divergence preserves the well-defined nature of quasiparticles near the Fermi surface. The decay rate $\Gamma(\omega - \mu, T) \equiv 1/\tau$, which depends on both energy deviation and temperature, appears in the quasiparticle Green's function $\mathcal{G}(p, \omega)$ through the Dyson equation:

$$\mathcal{G}(\vec{p}, \omega) = \left[\mathcal{G}_0^{-1}(\vec{p}, \omega) - \Sigma(\vec{p}, \omega)\right]^{-1} , \tag{1.11}$$

where $\mathcal{G}_0$ denotes the non-interacting propagator and Σ the self-energy. The decay rate is determined by the imaginary component of the self-energy

$$\Gamma(\vec{p}, \omega) = \text{Im}\Sigma(\vec{p}, \omega) = 1/\tau, \tag{1.12}$$

with the sign of ImΣ matching that of $\omega - \mu$ at the Fermi momentum $|\vec{p}| = p_F$.

Diagrammatically, finite lifetimes originate from scattering processes that convert a single quasiparticle into composite excitations (other quasiparticles and particle-hole pairs), with amplitudes governed by Landau parameters f. For $|\omega - \mu| \ll \mu$ and $T \ll \mu$, the momentum-specific decay rate $\Gamma_p \equiv 1/\tau_p$ at zero momentum transfer (forward scattering) follows

$$\Gamma_p \sim (\omega - \mu)^2 + (\pi k_B T)^2 + \cdots \tag{1.13}$$

Forward scattering events are characterized by negligible momentum transfer, defined as $\vec{q} = \vec{p}' - \vec{p} \approx 0$, where $\vec{p}'$ and $\vec{p}$ represent final and initial quasiparticle momenta near the Fermi surface, respectively. This condition implies conservation of momentum direction during scattering, with no net momentum exchanged between quasiparticles.

We get two fundamental regimes:

- Fermi Liquid Stability: Near the Fermi surface ($\omega \to \mu$) at low temperatures ($T \ll T_F$), Im $\Sigma \to 0$ ensures $\tau \to \infty$
- High-Energy Breakdown: For $|\omega - \mu| \gg \mu$, substantial ImΣ invalidates quasiparticles[1]

A key transport signature emerges in DC resistivity yielding the characteristic Fermi liquid scaling (see Chapter 6 of Ref. [3] for a heuristic derivation and discussion)

$$\rho \sim T^2. \tag{1.14}$$

The resistivity ρ is related to the scattering time z via the Drude formula $\rho = \frac{m^*}{ne^2 z}$ where m^* is the effective mass of quasiparticles, n is the carrier density, and e is the electron charge. Near the Fermi surface ($|\omega - \mu| \ll \mu$) at low temperatures, $\frac{1}{z} \propto T^2$. This is where we get the above scaling. Deviations from this quadratic temperature dependence indicate breakdown of quasiparticle dominance and potential non-Fermi liquid behavior.

A complementary hallmark is the level spacing Δ near the Fermi energy ($|\omega - \mu| \ll \mu$) at low temperatures for low-energy spectrum, given by

[1] If $T \gg T_F$, the system loses Fermi liquid behavior entirely (e.g., Fermi surface smears, quasiparticles undefined). However, this is a distinct, global breakdown—not specific to the high-energy regime. That's why we did not mention the condition of high temperatures here.

$$\Delta \sim \frac{1}{N}, \tag{1.15}$$

where N is the particle number. Anomalous N-dependence in Δ provides additional evidence of non-Fermi liquid physics.

1.1.1 Experimental Signatures

Landau's Fermi liquid theory has undergone extensive experimental verification across diverse systems. Key predictions—quadratic temperature dependence of resistivity ($\rho \propto T^2$), quasiparticle coherence peaks in spectroscopic measurements, and characteristic thermal/electrical transport ratios—have been confirmed in conventional metals, ultracold fermionic quantum gases, and liquid ^{3}He above its superfluid transition temperature [7, 8, 34, 36, 38].

A particularly stringent test comes from the Wiedemann-Franz law [6], which establishes a universal proportionality between the thermal-to-electrical conductivity ratio and temperature:

$$\frac{\kappa}{\sigma T} = L, \tag{1.16}$$

where L denotes the Lorenz number. Fermi liquid theory precisely predicts $L_{\mathrm{FL}} = \frac{\pi^2}{3}(k_B/e)^2 \approx 2.44 \times 10^{-8}\mathrm{V}^2\mathrm{K}^{-2}$ from fundamental constants. Remarkably, numerous metals obey this relation as demonstrated in Fig. 1.2, validating the theory's microscopic description of charge and heat transport [18].

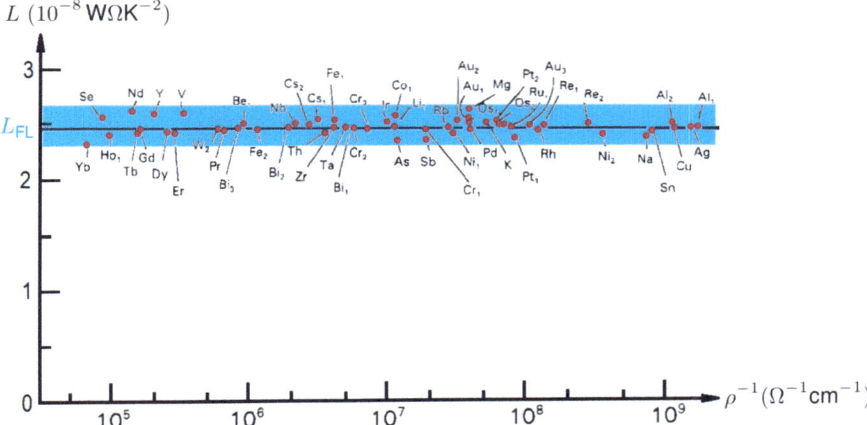

Fig. 1.2 Experimental validation of the Wiedemann-Franz law across multiple Fermi liquid systems. The measured ratio $\kappa/(\sigma T)$ exhibits linear temperature dependence with universal slope L, confirming theoretical prediction $L_{\mathrm{FL}} \approx 2.44 \times 10^{-8}\mathrm{V}^2\mathrm{K}^{-2}$. Data sourced from Ref. [18]

Despite these successes, mounting experimental evidence reveals systems where Fermi liquid theory fundamentally fails. Such breakdowns necessitate alternative frameworks collectively termed non-Fermi liquids. This work focuses specifically on the enigmatic class of materials exhibiting strange metal behavior theoretically modeled via the paradigmatic SYK model, whose properties starkly contradict Landau's paradigm. We refer the reader to a nice review [22] establishing strong experimental grounding for the SYK as a theory of Planckian dissipation in strange metals and how the SYK formalism bridges to cuprate phenomenology.

1.2 Strange Metals

Numerous correlated-electron superconductors with elevated transition temperatures display anomalous metallic phases above their superconducting critical temperature T_c. These systems violate fundamental Fermi liquid predictions and are consequently classified as strange metals. As exemplified in Fig. 1.3 (figure is taken from Ref. [40]) for copper-oxide superconductors, a continuous evolution connects conventional Fermi liquid and strange metal regimes across their phase diagrams.

The defining characteristic of strange metals is their linear temperature-dependent electrical resistivity ($\rho \propto T$), contrasting sharply with the quadratic

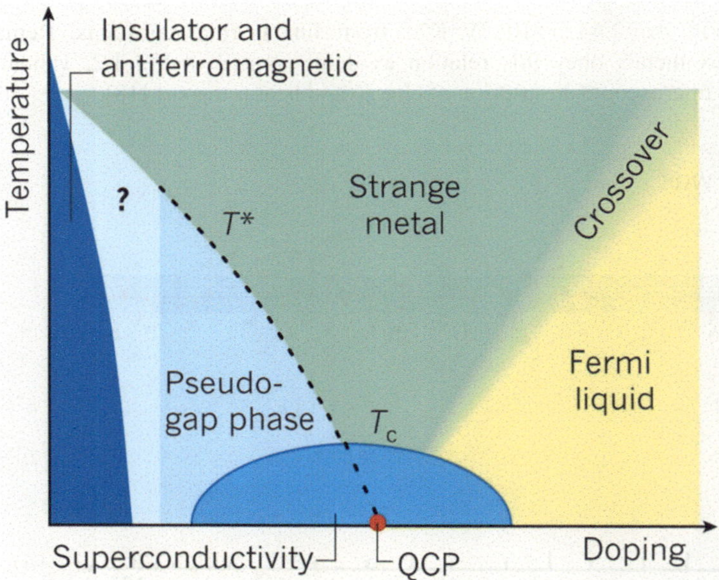

Fig. 1.3 Phase diagram of high-temperature copper-oxide superconductors (taken from Ref. [40]), highlighting the anomalous strange metallic regime characterized by linear-in-temperature DC resistivity—a signature departure from Fermi liquid behavior

dependence ($\rho \propto T^2$) predicted by Fermi liquid theory [35]. This anomalous scaling has been experimentally confirmed in multiple copper-oxide systems including

- Bismuth-strontium-calcium-copper-oxide [22]
- Bismuth-strontium-copper-oxide [31]
- Neodymium-doped lanthanum-strontium-copper-oxide [5]
- Lanthanum-strontium-copper-oxide [4]

Additional Fermi liquid violations include:

- Quantum critical transport anomalies in iron-arsenide systems [1]
- Crossover from non-Fermi- to Fermi-liquid resistivities phenomena in phosphorous-substituted iron arsenides [14] and monostrontium ruthenate [16]
- Anomalous heat capacity ($C_V \sim T \ln(1/T)$) near quantum criticality (in contrast with $C_V \sim T$ as predicted by the FLT)
- Violations of the Wiedemann-Franz law in cerium-praseodymium-copper-oxide [12]

Remarkably, these phenomena persist even in minimally disordered systems, as demonstrated in fluorinated barium-calcium-copper-oxide multilayers [19].

Table 1.2 contrasts fundamental properties of Fermi liquids with the Sachdev-Ye-Kitaev (SYK) model [15, 29, 37], which provides a theoretical framework for strange metals. Key distinctions include

- Spectral statistics (exponential vs. polynomial level spacing)
- Quasiparticle existence
- Characteristic scaling of equilibration rates
- Temperature dependence of resistivity

The SYK model specifically predicts Planckian dissipation dynamics with equilibration rate $\Gamma = 1 \cdot \frac{k_B T}{\hbar}$ (where prefactor 1 is intentionally shown as the SYK model predicts this proportionality constant). Note that equilibration rate Γ is also sometimes denoted by τ_{eq}^{-1} where τ_{eq} denotes the equilibration time for the system to reach thermal equilibrium. As shown in Table 1.3, experimental measurements across multiple strange metal compounds yield $\Gamma = \nu k_B T/\hbar$ with $\nu \approx 1$, consistent with SYK predictions within experimental uncertainty [10, 22].

Table 1.2 Comparative framework of Fermi liquid theory versus the Sachdev-Ye-Kitaev model, proposed as a microscopic description of strange metal phenomenology

	Fermi liquid	SYK model ($\sim$ Strange metal)
Energy level spacing	$\frac{1}{N}$	$e^{-\alpha N}$
Quasiparticles	Yes	No
Equilibration rate	$\sim T^2$	$\approx 1 \cdot \frac{k_B T}{\hbar}$
Electric resistivity	T^2	T

Table 1.3 Experimental verification of Planckian dissipation scaling $\Gamma = \nu k_B T/\hbar$ in strange metals. The SYK model prediction $\nu = 1$ agrees with measured values across multiple materials within experimental uncertainty. Data from Ref. [22]

	Bi2212	Bi2201	LSCO	Nd-LSCO	PCCO	LCCO	|TMTSF
ν	1.1 ± 0.3	1 ± 0.4	0.9 ± 0.3	0.7 ± 0.4	0.8 ± 0.2	1.2 ± 0.3	1 ± 0.3

1.2.1 Hints of Holography

The SYK model, originally conceived by Kitaev as a microscopic realization of gauge-gravity duality [15], has generated significant research demonstrating its critical connection to black hole physics. Specifically, the large-q SYK variant (detailed in later chapters) exhibits a continuous phase transition belonging to the same universality class as charged Anti-de Sitter (AdS) black holes. This analytical solvability enables exact thermodynamic solutions and identification of a precise gravitational dual [25]: a correspondence extending beyond thermodynamics to dynamical properties like maximal quantum chaos [30]. Within the grand canonical ensemble framework, the SYK phase transition manifests as a chemical-potential-driven discontinuity in charge density. This maps to charged AdS black holes through

- Chemical potential maps to black hole charge
- Charge density jump maps to surface charge discontinuity
- SYK's flavor-normalized charge jump reflects total charge change due to zero spatial extent
- Black hole transitions between horizon radii produce analogous discontinuities

A fundamental parallel emerges: both SYK models and charged AdS black holes exhibit first-order transitions between chaotic and non-chaotic phases. Their phase coexistence boundaries terminate at critical points with identical mean-field criticality (Landau-Ginzburg exponents $\alpha = 0,\ \beta = 1/2,\ \gamma = 1,\ \delta = 3$) [17]. A crucial divergence surfaces in the charge neutral limit: AdS black holes undergo complete evaporation into non-chaotic radiation at discrete Hawking-Page temperatures [11], while SYK systems sustain chaotic-nonchaotic transitions across a continuous low-temperature domain [24, 26]. This extended critical regime in SYK models effectively generalizes the Hawking-Page phenomenon beyond its gravitational instantiation. Our calculations—developed in subsequent chapters—demonstrate these results exclusively for the SYK framework, however we provide explicit commentary throughout this work on their gravitational implications and connections to black hole physics.

While studying the transport properties in SYK lattices in Chap. 5, we again encounter DC resistivity properties where an insulating phase emerges at low temperatures such that $\rho \sim T^{-2}$: scaling that aligns with holographic insulator predictions [2, 13, 32], potentially originating from low-temperature conformal invariance.

While this work adopts a condensed matter perspective, we acknowledge the profound connections between strange metal phenomenology (in particular, Planckian dissipation) and holographic duality [10, 35]. These intrinsic links permeate our analysis throughout this work, making explicit gravitational correspondence impossible to ignore. Rather than undertaking a comprehensive holographic treatment, we strategically embed commentary on these dualities throughout the text, providing specialized references for deeper exploration. Should resources allow and time permits, a dedicated sequel will examine the holographic implications of the low-temperature conformal symmetry inherent to SYK-like systems, building directly on the foundation laid down in this work.

1.3 Methodologies

The Sachdev-Ye-Kitaev model provides an unparalleled arena for deploying advanced analytical methodologies that transcend its immediate context and have wide ranging applicability from condensed matter to cosmology. This work leverages SYK's unique attributes—its non-integrability, maximal chaos, and exact solvability in the thermodynamic limit while being a quantum system—to offer a concrete laboratory for these techniques. Surprisingly solvable despite being ergodic and chaotic, the SYK paradigm affords explicit computations of phenomena ranging from thermalization dynamics to critical phase transitions.

Through this lens, we achieve dual objectives:

- Advanced Techniques & Methodologies: We implement powerful tools (as listed below) on a nontrivial quantum system
- Pedagogical Synthesis: SYK's analytical tractability demystifies abstract formalism, transforming generalized techniques into tangible procedures

This confluence positions SYK as an ideal pedagogical microcosm: a solvable yet chaotic quantum system where the reader can rigorously execute disorder averaging in interacting fermion systems, solve Kadanoff-Baym equations for far-from-equilibrium dynamics, compute critical exponents and quantum Lyapunov exponent, and so on and so forth. Accordingly, this review serves both as a technical exploration of SYK physics and as an introduction to modern quantum many-body methodologies, demonstrated through explicit calculations that retain full analytical tractability while illuminating conceptual foundations.

We develop the following methodologies within the SYK framework, explicitly demonstrating their general applicability across strongly correlated systems. This approach provides both technical exposition of the techniques themselves and concrete implementation protocols through worked examples in the realm of SYK physics. While the following list highlights essential aspects, full mathematical derivations and conceptual elaborations appear in subsequent chapters. Any specialized notation or formalism referenced here will be systematically unpacked in later sections in the context of SYK Hamiltonian.

- Large-N Techniques, Disorder Averaging & Effective Action: The core methodology includes starting with the partition function, for instance in real time, $\mathcal{Z} = \int \mathcal{D}\psi e^{\imath S[\psi, J]}$, where $S = \int dt \left[\frac{1}{2}\sum_i \psi_i \partial_t \psi_i - \mathcal{H}_J\right]$ is the action that depends on the fields ψ_i and disorder J and $\mathcal{H}_J$ is the disorder dependent Hamiltonian. We perform the disorder averaging and integrate out the fermionic fields, while introduce two bi-local functions: Green's function $\mathcal{G}(t, t')$ and self-energy $\Sigma(t, t')$. The final disorder-averaged partition function is $\langle \mathcal{Z} \rangle = \iint \mathcal{D}\mathcal{G}\mathcal{D}\Sigma e^{\imath S_{\text{eff}}[\mathcal{G}, \Sigma]}$ where $S_{\text{eff}}[\mathcal{G}, \Sigma]$ is the final effective action of the theory.
- Replica Trick: While performing the disorder-averaging, we make use of something called the replica trick where different replicas decouple for SYK-like systems.
- Schwinger-Dyson & Kadanoff-Baym Equations: By taking the large-N limit, the effective action $S_{\text{eff}}[\mathcal{G}, \Sigma]$ obtained is semi-classical. We extremize the action with respect to Σ and $\mathcal{G}$ to get the corresponding Euler-Lagrange equations, which are the Schwinger-Dyson equations of the theory. For SYK-like systems, we find that the Schwinger-Dyson equations are in closed form for large-N. This is what is generally implied when it's said that the SYK model is a solvable model. Furthermore, analogous to the large-N limit, there exists the large-$q/2$-body interacting SYK Hamiltonian for which these Schwinger-Dyson equations become analytically solvable across all temperatures in a closed form. The motivation comes from the observation that large-q results for physical entities are qualitatively similar to finite-q but offers the advantage of being analytically solvable which for the case of finite-q, one has to resort to numerics to solve the closed form Schwinger-Dyson equations. These Schwinger-Dyson equations in imaginary time can be analytically continue to real time that provides us with the Kadanoff-Baym equations—contour-ordered integro-differential equations governing the real-time evolution of interacting Green's functions. The Kadanoff-Baym equations allow us to study thermalization, transport, and quantum quenches in real time.
- Imaginary-Time (Matsubara) vs. Real-Time (Keldysh) Formalisms: Equilibrium solutions are analytically treated by going to the imaginary time $t \to -\imath\tau$ where all equilibrium properties such as thermodynamics or static correlations can be studied. In contrast, the non-equilibrium behaviors are studied via the analytic continuation to the real time where we have our Kadanoff-Baym equations. The real time formalism has a forward and backward time evolution that are explicitly kept track of. The bi-local Green's function is accordingly defined for different parts of the contour depending on their time arguments.
- Thermodynamics, KMS Relation & Grand Potential: The Kubo-Martin-Schwinger (KMS) condition is the cornerstone of quantum statistical mechanics, enforcing fundamental consistency between equilibrium dynamics and thermal states. For fermionic systems like the SYK model in imaginary time τ, the Green's function satisfies $\mathcal{G}(\tau) = -\mathcal{G}(\tau + \beta)$ anti-periodicity. The KMS relation directly gets related to the thermodynamics of the system where we can extract the equation of state from the KMS relation. In conjunction with the grand

potential, we extract exact thermodynamics from microscopic details and these provide necessary tools to probe whether or not the system undergoes any phase transitions (which are defined in the thermodynamic limit).

- Critical Exponents & Universality Classes: Near criticality (in a continuous phase transition), systems "forget" the microscopic details and their singular behavior are governed by universal critical exponents (α, β, γ). These exponents form the *Rosetta Stone* of phase transitions, classifying diverse systems (magnets, superfluids, quark-gluon plasmas) into universality classes. Remarkably, SYK reproduces Landau-Ginzburg exponents ($\alpha = 0, \beta = 1/2, \gamma = 1$), demonstrating shared criticality with van der Waals fluids and AdS black holes. This universality makes SYK an ideal laboratory for exploring mean-field (equivalently, Landau-Ginzburg) criticality beyond perturbation theory as the SYK is a model for strongly interacting fermions.
- Quantum Chaos, OTOCs & the MSS Bound: Quantum chaos transcends classical unpredictability and generalizes the classical chaos captured at the level of classical Lyapunov exponent to the quantum realm. Quantum chaos probes information scrambling and chaotic dynamics via out-of-time-ordered correlators (OTOCs). The exponential growth of OTOCs defines the quantum Lyapunov exponent λ_L, bounded by $\lambda_L \leq 2\pi k_B T/\hbar$ (MSS bound [30]). SYK achieves the maximal chaos $\lambda_L = 2\pi T$ at low temperatures, saturating the bound similar to black holes. That's why, the SYK model is sometimes dubbed as "maximally chaotic" in the literature.
- Keldysh Contour Deformations: As mentioned above, the Keldysh formalism provides a way to probe quantum systems driven far from equilibrium, where conventional Matsubara techniques fail. Additionally, the Keldysh formalism allows to modify or deform the Keldysh contours (forward and backward time evolution contours) that allows to evaluate expectation values and correlations (such as auto-correlations) of physical observables in real time. This has immense significance, for example, for transport properties of the system where current-current correlations can be evaluated using the Keldysh contour deformation techniques that leads to dynamical and DC conducting properties. In the SYK model, Keldysh contour deformations unlock exact non-equilibrium solutions in the thermodynamic limit—a rarity for chaotic quantum system.

1.4 Outline

We now explain the structure of this work. We introduce the prototypical SYK model of N Majorana fermions with all-to-all random $q/2$-body interactions in Chap. 2. After defining the Hamiltonian and symmetries, we derive the disorder-averaged effective action using large-N techniques and replica symmetry. The Schwinger-Dyson equations are solved in the infrared (IR) conformal limit, revealing emergent reparameterization invariance. We compute the effective Schwarzian action governing soft mode dynamics, discuss finite-temperature fluctuations, and solve the large-q limit analytically. The chapter concludes with real-time dynamics via the Keldysh formalism, including quench protocols.

We extend the formalism to complex fermions with conserved $U(1)$ charge in Chap. 3. We discuss the symmetry properly and, as for the Majorana variant, evaluate the effective action that gives us the Schwinger-Dyson equations. We analyze the IR conformal limit, derive the modified Liouville action, and solve for conformal Green's functions. The large-q limit is again leveraged for analytical solutions, setting the stage for equilibrium studies as well as non-equilibrium and transport studies.

Focusing on thermodynamic and chaotic behavior, we employ the Keldysh formalism to compute real-time Green's functions in Chap. 4. We solve the Kadanoff-Baym equations in equilibrium, extract scaling relations for entropy and energy, and derive the equation of state. We evaluate the grand potential, and together, we detect the presence of a first order phase transition that terminates in a critical point. The critical exponents are evaluated for the continuous phase transition and we find that the associated universality class is surprisingly a mean-field (Landau-Ginzburg) universality class which is satisfied by diverse systems such as classical van der Waals fluids and certain Anti-de Sitter (AdS) black holes. We compute the quantum Lyapunov exponent λ_L, saturating the Maldacena-Shenker-Stanford bound of quantum chaos $\lambda_L = 2\pi T$ at low temperatures [30] (explained in there). Holographic implications are highlighted. Finally we extend the setup to SYK chains which is needed to study transport properties later on. We solve for the chain in a general setup and specialize to uniform equilibrium situations, explaining the methodologies for obtaining analytic solution in the thermodynamic limit.

We explore far-from-equilibrium physics in Chap. 5, starting with quantum quenches in single dots that admit instantaneous thermalization with respect to the Green's function and chains that admit non-instantaneous thermalization with charge dynamics. Keldysh contour deformation techniques are developed to compute DC resistivity. For coupled SYK chains, we derive current-current correlations, solve continuity equations, and obtain the linear-in-temperature T resistivity, characteristic of strange metals. The low-temperature insulating phase ($\rho \sim T^{-2}$) for one of the considered chains is linked to holographic insulator models. The same chain admits a smooth crossover from insulating to Fermi liquid to strange metallic to bad metallic phases as the temperature is increased. We conclude and provide an outlook for the topics discussed in this work in Chap. 6.

Appendices provide self-contained technical supplements for derivations, identities, and computational frameworks referenced in the main text. To ensure reproducibility and pedagogical utility, we include commented Mathematica notebooks and Python codes implementing critical exponent calculations, Liouville equation verification, KMS relations, and DC resistivity solutions.

Finally, we explicitly restate critical conventions for Green's functions and SYK formalism. A comprehensive caution appears in the box below Eq. (3.10) which the readers is strongly encouraged to review before proceeding. Given the prevalence of conflicting sign conventions and variance choices for Gaussian ensembles in the literature, overlooking these details can lead to fundamental misunderstandings in later derivations. Critical conventions vary across SYK literature primarily in two key areas: (i) Disorder variance scaling, (ii) Green's function definitions (e.g., $-\imath$ vs. -1 prefactors). Crucially, physical results remain qualitatively convention-independent as they should (though numerical prefactors may differ depending on the convention chosen, for instance, in the definition of the disorder variance). However, the derivations fundamentally depend on the conventions chosen. Therefore, we have intentionally adopted the approach of using different conventions in different chapters. This provides the reader a hands-on experience with navigating the calculations and explicitly shows how and what change when the convention is changed. Every chapter starts by noting the conventions used, and they are applied consistently throughout. Notably, for the Majorana SYK, we define the Green's function as $\mathcal{G}(t_1, t_2) \equiv \frac{-\imath}{N} \sum_{j=1}^{N} \left\langle T_{\mathcal{C}} c_j(t_1)\, c_j^\dagger(t_2)\right\rangle$, while for the complex fermion generalization, we use $\mathcal{G}(t_1, t_2) \equiv \frac{-1}{N} \sum_{j=1}^{N} \left\langle T_{\mathcal{C}} c_j(t_1)\, c_j^\dagger(t_2)\right\rangle$.

With this structural roadmap established, we now commence our journey into the SYK framework and its advanced methodologies. Beginning with the Majorana SYK model in Chap. 2, we take a deep dive into its technical and mathematical formulations while elucidating physical insights and conceptual foundations. Throughout, a pedagogical approach is maintained and an attempt is made to keep the text self-contained, assuming prerequisites of advanced quantum mechanics and familiarity with Green's functions. Some sections are marked with a ⋆ symbol, indicating they require advanced background knowledge for full comprehension. Nevertheless, we provide conceptual foundations and technical roadmaps for these sections and include references to (preferably pedagogical, if possible) resources for deeper exploration and advanced discussions.

References

1. Analytis, J.G., Kuo, H.-H., McDonald, R.D., Wartenbe, M., Rourke, P.M.C., Hussey, N.E., Fisher, I.R.: Transport near a quantum critical point in BaFe2(As1-xPx)2. Nat. Phys. **10**, 194–197 (2014)
2. Andrade, T., Krikun, A., Schalm, K., Zaanen, J.: Doping the holographic Mott insulator. Nat. Phys. **14**, 1049–1055 (2018)
3. Coleman, P.: Introduction to Many-Body Physics. Cambridge University Press, Cambridge (2015)
4. Cooper, R.A., Wang, Y., Vignolle, B., Lipscombe, O.J., Hayden, S.M., Tanabe, Y., Adachi, T., Koike, Y., Nohara, M., Takagi, H., Proust, C., Hussey, N.E.: Anomalous criticality in the electrical resistivity of La2–xSrxCuO4. Science **323**(5914), 603–607 (2009)
5. Daou, R., Doiron-Leyraud, N., LeBoeuf, D., Li, S.Y., Laliberté, Cyr-Choinière, O., Jo, Y.J., Balicas, L., Yan, J.-Q., Zhou, J.-S., Goodenough, J.B., Taillefer, L.: Linear temperature dependence of resistivity and change in the Fermi surface at the pseudogap critical point of a high-Tc superconductor. Nat. Phys. **5**, 31–34 (2009)
6. Franz, R., Wiedemann, G.: Ueber die Wärme-Leitungsfähigkeit der Metalle. Ann. Phys. **165**(8), 497–531 (1853)
7. Greywall, D.S.: Specific heat of normal liquid ^{3}He. Phys. Rev. B **27**(5), 2747–2766 (1983)
8. Greywall, D.S.: Thermal conductivity of normal liquid ^{3}He. Phys. Rev. B **29**(9), 4933–4945 (1984)

9. Gu, Y., Kitaev, A., Sachdev, S., Tarnopolsky, G.: Notes on the complex Sachdev-Ye-Kitaev model. J. High Energy Phys. **2020**(2), 1–74 (2020). Publisher: Springer Berlin Heidelberg
10. Hartnoll, S.A., Mackenzie, A.P.: Colloquium: planckian dissipation in metals. Rev. Mod. Phys. **94**(4), 041002 (2022)
11. Hawking, S.W., Page, D.N.: Thermodynamics of black holes in anti-de Sitter space. Commun. Math. Phys. **87**(4), 577 (1982). Publisher: Springer
12. Hill, R.W., Proust, C., Taillefer, L., Fournier, P., Greene, R.L.: Breakdown of Fermi-liquid theory in a copper-oxide superconductor. Nature **414**, 711–715 (2001)
13. Horowitz, G.T., Santos, J.E., Tong, D.: Optical conductivity with holographic lattices. J. High Energy Phys. **2012**(7), 168 (2012)
14. Kasahara, S., Shibauchi, T., Hashimoto, K., Ikada, K., Tonegawa, S., Okazaki, R., Shishido, H., Ikeda, H., Takeya, H., Hirata, K., Terashima, T., Matsuda, Y.: Evolution from non-Fermi- to Fermi-liquid transport via isovalent doping in $BaFe_2(As_{1-x}P_x)_2$ superconductors. Phys. Rev. B **81**(18), 184519 (2010)
15. Kitaev, A.: A simple model of quantum holography. Talks given at "Entanglement in Strongly-Correlated Quantum Matter,". (Part1,. Part2), KITP (2015)
16. Kostic, P., Okada, Y., Collins, N.C., Schlesinger, Z., Reiner, J.W., Klein, L., Kapitulnik, A., Geballe, T.H., Beasley, M.R.: Non-Fermi-liquid behavior of $SrRuO_3$: evidence from infrared conductivity. Phys. Rev. Lett. **81**(12), 2498–2501 (1998)
17. Kubizňák, D., Mann, R.B.: P - V criticality of charged AdS black holes. J. High Energy Phys. **2012**(7), 033 (2012). Publisher: Springer-Verlag
18. Kumar, G.S., Prasad, G., Pohl, R.O.: Experimental determinations of the Lorenz number. J. Mater. Sci. **28**(16), 4261–4272 (1993)
19. Kurokawa, K., Isono, S., Kohama, Y., Kunisada, S., Sakai, S., Sekine, R., Okubo, M., Watson, M.D., Kim, T.K., Cacho, C., Shin, S., Tohyama, T., Tokiwa, K., Kondo, T.: Unveiling phase diagram of the lightly doped high-Tc cuprate superconductors with disorder removed. Nat. Commun. **14**(4064), 1–9 (2023)
20. Landau, L.D.: The theory of a fermi liquid. Soviet Phys. JETP-USSR **3**(6), 920–925 (1957)
21. Landau, L.D.: Oscillations in a fermi liquid. Soviet Phys. JETP-USSR **5**(1), 101–108 (1957)
22. Legros, A., Benhabib, S., Tabis, W., Laliberté, F., Dion, M., Lizaire, M., Vignolle, B., Vignolles, D., Raffy, H., Li, Z.Z., Auban-Senzier, P., Doiron-Leyraud, N., Fournier, P., Colson, D., Taillefer, L., Proust, C.: Universal T-linear resistivity and Planckian dissipation in overdoped cuprates. Nat. Phys. **15**, 142–147 (2019)
23. Louw, J.C., Kehrein, S.: Thermalization of many many-body interacting Sachdev-Ye-Kitaev models. Phys. Rev. B **105**(7), 075117 (2022)
24. Louw, J.C., Kehrein, S.: Shared universality of charged black holes and the complex large-q Sachdev-Ye-Kitaev model. Phys. Rev. B **107**(7), 075132 (2023)
25. Louw, J.C., Cao, S., Ge, X.-H.: Matching partition functions of deformed Jackiw-Teitelboim gravity and the complex SYK model. Phys. Rev. D **108**(8), 086014 (2023)
26. Louw, J.C., van Manen, L.M., Jha, R.: Thermodynamics and dynamics of coupled complex SYK models. J. Phys.: Condens. Matter **36**(49), 495601 (2024)
27. Luttinger, J.M.: Fermi surface and some simple equilibrium properties of a system of interacting fermions. Phys. Rev. **119**(4), 1153–1163 (1960)
28. Luttinger, J.M., Ward, J.C.: Ground-state energy of a many-fermion system. II. Phys. Rev. **118**(5), 1417–1427 (1960)
29. Maldacena, J., Stanford, D.: Remarks on the Sachdev-Ye-Kitaev model. Phys. Rev. D **94**(10), 106002 (2016)
30. Maldacena, J., Shenker, S.H., Stanford, D.: A bound on chaos. J. High Energy Phys. **2016**(8), 1–17 (2016)
31. Martin, S., Fiory, A.T., Fleming, R.M., Schneemeyer, L.F., Waszczak, J.V.: Normal-state transport properties of $Bi_{2+x}Sr_{2-y}CuO_{6+\delta}$ crystals. Phys. Rev. B **41**(1), 846–849 (1990)
32. Mefford, E., Horowitz, G.T.: Simple holographic insulator. Phys. Rev. D **90**(8), 084042 (2014)

33. Morandi, G., Sodano, P., Tagliacozzo, A., Tognetti, V. (Eds.): Field Theories for Low Dimensional Condensed Matter Systems: Spin Systems and Strongly Correlated Electrons. Springer, Berlin (2000)
34. Philippe, N., David, P.: Theory Of Quantum Liquids. CRC Press, Boca Raton (1999)
35. Phillips, P.W., Hussey, N.E., Abbamonte, P.: Stranger than metals. Science **377**, 6602 (2022)
36. Pickett, W.E., Singh, D.J., Krakauer, H., Cohen, R.E.: Fermi surfaces, fermi liquids, and high-temperature superconductors. Science **255**(5040), 46–54 (1992)
37. Sachdev, S., Ye, J.: Gapless spin-fluid ground state in a random quantum Heisenberg magnet. Phys. Rev. Lett. **70**(21), 3339 (1993)
38. Schulz, H.J.: Fermi liquids and non–fermi liquids (1995). https://arxiv.org/abs/cond-mat/9503150
39. Stanford, D.: Many-body chaos at weak coupling. J. High Energy Phys. **2016**(10), 1–18 (2016)
40. Varma, C.: Mind the pseudogap. Nature **468**, 184–185 (2010)

Majorana Variant

2

Abstract

This chapter develops the Majorana variant of the Sachdev-Ye-Kitaev (SYK) model as a minimal yet rich arena for non Fermi liquid physics, quantum chaos, and holographic duality. Starting from an SYK_q Hamiltonian of N disordered Majorana fermions with all to all $q/2$-body interactions drawn from a Gaussian ensemble, we construct the Euclidean path integral and perform explicit disorder averaging. Introducing bilocal fields for the Green's function and the self-energy, we integrate out the fermions to obtain a closed form effective action that becomes semiclassical in the large-N limit and yields exact Schwinger Dyson equations for general q. The free Majorana theory is solved via a generating functional, clarifying the structure of the inverse propagator and providing a benchmark for the interacting solution. Using the replica trick and an M flavored Majorana generalization, we show that replica symmetry leads to the same effective action and Schwinger-Dyson equations while giving direct access to the free energy. The chapter then analyzes the $O(N)$ symmetry of the effective action and show explicitly the application of Noether's theorem and its validity. Finally, we define the infrared or conformal limit, derive the corresponding reparameterization invariant effective action, and introduce the Matsubara imaginary-time and Keldysh real-time formalisms that will be used for thermodynamics and nonequilibrium dynamics in later chapters.

The Sachdev-Ye-Kitaev (SYK) model has emerged as a pivotal paradigm for studying non-Fermi liquids, quantum chaos, and holographic duality in strongly correlated systems [4]. This chapter focuses on the Majorana variant of the SYK model—a minimal yet rich quantum mechanical system of N disordered Majorana fermions with all-to-all $q/2$-body interactions.

R. Jha, *Theory of Strongly Correlated Systems*, Lecture Notes in Physics 1051,
https://doi.org/10.1007/978-3-032-20910-8_2

2.1 Model

The SYK model has been initially proposed for Majorana fermions by Kitaev [10]. We now present the model formally, following which we specialize to particular cases for further clarity. The *interacting* Hamiltonian[1] for the Majorana SYK model is given by

$$\mathrm{SYK}_q = \mathcal{H}_q = \imath^{q/2} \sum_{\{i_q\}_{\leq}} j_{q;\{i_q\}} \psi_{i_1} \ldots \psi_{i_q}, \tag{2.1}$$

where the subscript q (always even) denotes a label for $q/2$-body interaction (so, a total of $q/2$ creation and $q/2$ annihilation operators) and $\imath$ is the imaginary unit to keep the Hamiltonian Hermitian. Here, ψ_i denote Majorana fermions that satisfy the usual fermionic anti-commutation relation[2]

$$\{\psi_i, \psi_j\} = \delta_{ij}, \tag{2.2}$$

where they can in general be time-dependent. The interaction strength of $q/2$-body interaction is given by $j_{\{i_q\}}$. We have employed the notation

$$\{i_q\}_{\leq} \equiv 1 \leq i_1 < i_2 < \ldots < i_q \leq N, \tag{2.3a}$$

$$\{i_q\} \equiv \{i_1, i_2, \ldots, i_{q-1}, i_q\}. \tag{2.3b}$$

Since the SYK model is a model with disorder, accordingly the interaction strength $j_{q;\{i_q\}} = j_{q;i_1,i_2,\ldots,i_q}$ is taken to be a random variable. The random variable is derived from a Gaussian ensemble with following mean and variance:

$$\langle j_q \rangle = 0, \qquad \sigma_q^2 = \langle j_q^2 \rangle = \frac{J_q^2 (q-1)!}{N^{q-1}}, \tag{2.4}$$

where J_q is the strength capturing the variance. In other words, $j_{q;\{i_q\}} = j_{q;i_1,i_2,\ldots,i_q}$ is derived from the following normalized Gaussian probability distribution:

$$\mathcal{P}_q[j_{q;\{i_q\}}] = A \exp\left(-\frac{1}{2\sigma_q^2} \sum_{\{i_q\}_{\leq}} j_{q;\{i_q\}}^2 \right). \tag{2.5}$$

[1] There is also a kinetic term which we don't show here, but we will require it to calculate the action below. We will return to this later.

[2] There is another convention in the literature, high-energy physics in particular, where $\{\chi_i, \chi_j\} = 2\delta_{ij}$. Both conventions are physically equivalent and related by a rescaling $\psi_i = \frac{\chi_i}{\sqrt{2}} \Leftrightarrow \chi_i = \sqrt{2}\psi_i$. Our convention in Eq. (2.2) is the standard in the SYK literature, for example, see Ref. [6].

where $A = \sqrt{\frac{1}{2\pi\sigma_q^2}} = \sqrt{\frac{N^{q-1}q}{2\pi q! J_q^2}}$ is calculated via the normalization condition.[3]

In order to explicitly show the Hamiltonian, we consider the example for 2-body interaction where $q = 4$. Then the interacting Hamiltonian reduces to

$$\mathcal{H}_4 = - \sum_{1 \le i_1 < i_2 < i_3 < i_4 \le N}^{N} j_{4;i_1,i_2,i_3,i_4} \psi_{i_1} \psi_{i_2} \psi_{i_3} \psi_{i_4},$$

where we absorb the minus sign in the coupling constant $j_{4;i_1,i_2,i_3,i_4}$ to get

$$\mathrm{SYK}_4 = \mathcal{H}_4 = \sum_{1 \le i_1 < i_2 < i_3 < i_4 \le N}^{N} j_{4;i_1,i_2,i_3,i_4} \psi_{i_1} \psi_{i_2} \psi_{i_3} \psi_{i_4}. \tag{2.6}$$

The mean and variance of the Gaussian probability distribution from which the random variable $j_{4;i_1,i_2,i_3,i_4}$ is derived are zero and $\sigma_4^2 = \langle j_4^2 \rangle = \frac{J_4^2 3!}{N^3}$, respectively. We present the calculations for the SYK$_4$ model to keep things explicit, after which we will generalize all obtained results to an arbitrary $q/2$-body interacting model. That's why we suppress the subscript label 4 for brevity which denotes $q/2 = 2$-body interaction. Like any disorder model, we average over all possible realizations of the random variable, such that[4]

$$\langle j_{i_1,j_1,k_1,l_1} j_{i_2,j_2,k_2,l_2} \rangle = \sigma^2 \delta_{i_1 i_2} \delta_{j_1 j_2} \delta_{k_1 k_2} \delta_{l_1 l_2} = \frac{J^2 3!}{N^3} \delta_{i_1 i_2} \delta_{j_1 j_2} \delta_{k_1 k_2} \delta_{l_1 l_2}, \tag{2.7}$$

where the Kronecker delta function satisfies $\delta_{ij} = 1$, if $i = j$, otherwise it's zero.

2.2 The Schwinger-Dyson Equations

We wish to write down the partition function for the theory

$$\mathcal{Z}_r = \int \mathcal{D}\psi_i e^{\iota S_r[\psi_i]} \tag{2.8}$$

[3] Since probability must add to 1, we have the condition $\int \mathcal{D} j_{q;\{i_q\}} \mathcal{P}_q[j_{q;\{i_q\}}] = 1$ where $\mathcal{D} j_{q;\{i_q\}}$ is the measure over all possible instances/realizations of $j_{q;\{i_q\}}$, namely $\mathcal{D} j_{q;\{i_q\}} = \prod_{\{i_q\}_\le} d j_{q;\{i_q\}}$. Note the presence of $\{i_q\}_\le$: to avoid overcounting identical couplings, we restrict to ordered tuples. This is not required for the measure $\mathcal{D}\psi_i$ below.

[4] Any raising to odd powers vanish $\left\langle j_{i_1 i_2 i_3 i_4}^{2n+1} \right\rangle = 0$. The same generalization for arbitrary q holds.

where the measure is provided by $\mathcal{D}\psi_i \equiv \prod_{i=1}^{N} d\psi_i$ and $S[\psi_i]$ is the action given by

$$S_r[\psi_i] = \int dt \left[\frac{1}{2} \sum_i^N \psi_i \partial_t \psi_i - \mathcal{H}_4 \right]. \tag{2.9}$$

Here the subscript r denotes real time where integration is over real time $t \in \mathbb{R}$. Since we will not be dealing with non-equilibrium dynamics for the moment (later, we will), we prefer to use the imaginary time formalism (also known as the Euclidean time) where a *Wick rotation* is performed $t \rightarrow -\imath\tau$ (see Chapter 8 of Ref. [5] for a nice introduction to imaginary-time formalism). Here t denotes the real time while τ denotes the imaginary/Euclidean time. Accordingly the partition function in the Euclidean plane becomes

$$\mathcal{Z}_E = \int \mathcal{D}\psi_i e^{-S_E[\psi_i]} \tag{2.10}$$

where $S_E[\psi_i]$ is the Euclidean action[5]

$$S_E[\psi_i] = \int d\tau \left[\frac{1}{2} \sum_i^N \psi_i \partial_\tau \psi_i + \mathcal{H}_4 \right], \tag{2.11}$$

where integration is limited to the interval $\tau \in [0, \beta]$ (see Appendixes A and F to understand the properties of imaginary-time formalism). Now we perform a disorder-averaging which is defined as

$$\langle \mathcal{Z}_E \rangle_{J_4} = \int \mathcal{D} j_{4;i_1,i_2,i_3,i_4} \mathcal{P}[j_{4;i_1,i_2,i_3,i_4}] \mathcal{Z}_E \tag{2.12}$$

where $\mathcal{P}[j_{4;i_1,i_2,i_3,i_4}]$ is the Gaussian probability distribution from which the random couplings are drawn (see Eq. (2.5)). Plugging the probability distribution, we get

[5] Even using simple classical mechanics, we can see the interplay between real-time action and imaginary-time action. We start with real-time action $S = \int dt \left[\frac{1}{2} m \left(\frac{dx}{dt} \right)^2 - V(x) \right]$ where $V(x)$ is the interaction. Then WIck rotation simply implies a coordinate transformation, we perform the transformation, $\tau = \imath t$. Then $dt = \frac{d\tau}{\imath}$ and $\frac{dx}{dt} = \imath \frac{dx}{d\tau}$. Therefore, $S = \imath \int d\tau \left[\frac{1}{2} m \dot{x}^2 + V(x) \right] \equiv \imath S_E$. This is why the integrand in the path integral for the partition function behaves as $e^{\imath S} \rightarrow e^{-S_E}$.

$$\langle \mathcal{Z}_E \rangle_{J_4} = A \int \mathcal{D}\psi_i \exp\left\{-\frac{1}{2}\sum_{i}^{N}\int d\tau \psi_i \partial_\tau \psi_i\right\}$$
$$\times \int \mathcal{D}j_{4;i,j,k,l} \exp\left\{-\sum_{1\leq i<j<k<l\leq N}\left(\frac{j_{4;i,j,k,l}^2}{12\frac{J_4^2}{N^3}}\right.\right.$$
$$\left.\left. - j_{4;i,j,k,l}\int d\tau \psi_i(\tau)\psi_j(\tau)\psi_k(\tau)\psi_l(\tau)\right)\right\} \tag{2.13}$$

where we have allowed Majorana fields to be time-dependent. Now we can integrate out the coupling strengths by using the Gaussian integration $\int dx e^{-ax^2+bx} = \sqrt{\frac{\pi}{a}} e^{\frac{b^2}{4a}}$ where the pre-factor $\sqrt{\pi/a}$ exactly cancels the factor of A in front. We get

$$\langle \mathcal{Z}_E \rangle_{J_4} = \int \mathcal{D}\psi_i \exp\left\{-\frac{1}{2}\sum_{i}^{N}\int d\tau \psi_i \partial_\tau \psi_i\right\}$$
$$\times \exp\left\{+\frac{3J^2}{N^3}\left(\int d\tau \sum_{1\leq i<j<k<l\leq N}(\psi_i\psi_j\psi_k\psi_l)(\tau)\right)^2\right\}. \tag{2.14}$$

Now, the way we simplify the second exponential is

$$\exp\left\{+\frac{3J^2}{N^3}\left(\int d\tau \sum_{1\leq i<j<k<l\leq N}(\psi_i\psi_j\psi_k\psi_l)(\tau)\right)^2\right\}$$
$$= \exp\left\{+\frac{3J^2}{N^3}\iint d\tau' d\tau \sum_{1\leq i<j<k<l\leq N}(\psi_i\psi_j\psi_k\psi_l)(\tau)(\psi_i\psi_j\psi_k\psi_l)(\tau')\right\}. \tag{2.15}$$

We simplify the summation in the integral further where for the sake of generalization, we consider a general q term of the type

$$\sum_{\{i_q\}_\leq}\left(\psi_{i_1}\dots\psi_{i_q}\right)(\tau)\left(\psi_{i_1}\dots\psi_{i_q}\right)(\tau') = \frac{1}{q!}\sum_{\{i_q\}_{\neq}}\left(\psi_{i_1}\dots\psi_{i_q}\right)(\tau)\left(\psi_{i_1}\dots\psi_{i_q}\right)(\tau')$$
$$= \frac{(-1)^{q/2}}{q!}\left(\sum_{i=1}^{N}\psi_i(\tau)\psi_i(\tau')\right)^q \tag{2.16}$$

where we denote $i_1 \neq i_2 \cdots \neq i_q$ by $\{i_q\}_{\neq}$ for brevity. To illustrate the point, let's consider the two cases

- $q = 2$ simplifies to

$$\sum_{1\leq i_1<i_2\leq N}^{N} \left(\psi_{i_1}\psi_{i_2}\right)(\tau)\left(\psi_{i_1}\psi_{i_2}\right)(\tau') = \frac{1}{2!}\sum_{i_1\neq i_2}\left(\psi_{i_1}\psi_{i_2}\right)(\tau)\left(\psi_{i_1}\psi_{i_2}\right)(\tau')$$
$$= -\frac{1}{2!}\left(\sum_{i=1}^{N}\psi_i(\tau)\psi_i(\tau')\right)^2. \tag{2.17}$$

- $q = 4$ (the same as in Eq. (2.15) above) simplifies to

$$\sum_{1\leq i_1<i_2<i_3<i_4\leq N}^{N} \left(\psi_{i_1}\psi_{i_2}\psi_{i_3}\psi_{i_4}\right)(\tau)\left(\psi_{i_1}\psi_{i_2}\psi_{i_3}\psi_{i_4}\right)(\tau')$$
$$= \frac{1}{4!}\sum_{i_1\neq i_2\neq i_3\neq i_4}\left(\psi_{i_1}\psi_{i_2}\psi_{i_3}\psi_{i_4}\right)(\tau)\left(\psi_{i_1}\psi_{i_2}\psi_{i_3}\psi_{i_4}\right)(\tau')$$
$$= \frac{1}{4!}\left(\sum_{i=1}^{N}\psi_i(\tau)\psi_i(\tau')\right)^4. \tag{2.18}$$

Here comes the crucial step where we introduce the bi-temporal fields $\mathcal{G}(\tau, \tau')$[6]

$$\mathcal{G}(\tau, \tau') \equiv \frac{1}{N}\sum_{i=1}^{N}\psi_i(\tau)\psi_i(\tau'). \tag{2.19}$$

By definition of the Dirac delta function (in particular, $\int dx\delta(x) = 1$), this relation can be written as an integral

$$\int \mathcal{D}\mathcal{G}\delta\left(\mathcal{G}(\tau, \tau') - \frac{1}{N}\sum_{i}^{N}\psi_i(\tau)\psi_i(\tau')\right) = 1 \tag{2.20}$$

[6] This will be later identified as the Green's functions, but for now these can be treated as just bi-temporal fields. Also, there are multiple conventions for the definition of the Green's function in the literature where, for instance, there is a pre-factor of $\imath$ or a minus sign. We will always define the Green's function we are using and be consistent with it throughout. This will also be a good practice to show that how the convention chosen for the Green's function does not impact the physics, as it should not.

which enforces the definition of $\mathcal{G}$. Then we use the integral representation of the delta function, namely $\int \frac{dk}{2\pi} e^{\imath kx} dk = \delta(x)$, to re-write the above relation as

$$\int \mathcal{D}\mathcal{G}\mathcal{D}\Sigma \exp\left\{-\frac{N}{2}\iint d\tau d\tau' \Sigma(\tau,\tau')\left(\mathcal{G}(\tau,\tau') - \frac{1}{N}\sum_i^N \psi_i(\tau)\psi_i(\tau')\right)\right\} = 1 \tag{2.21}$$

where we have introduced another bi-temporal field $\Sigma(\tau, \tau')$.[7] Moreover, we have re-adjusted the measure by absorbing constants and the imaginary unit is also absorbed in Σ for convergence in the imaginary-time formalism. The measure is defined as $\mathcal{D}\mathcal{G} = \prod_{(\tau,\tau')\in[0,\beta]^2} d\mathcal{G}\left(\tau,\tau'\right)$ and $\mathcal{D}\Sigma = \prod_{(\tau,\tau')\in[0,\beta]^2} d\Sigma\left(\tau,\tau'\right)$. In practice, normalization factors (e.g., $N/(2\pi)$ per pair $\left(\tau,\tau'\right)$) are absorbed into the definition. Since $\mathcal{D}\mathcal{G}$ and $\mathcal{D}\Sigma$ integrate over all configurations (with τ, τ' independent), time-ordering generally matters for the product $\Sigma\left(\tau,\tau'\right) G\left(\tau',\tau\right)$, but not for the measure itself.

Proceeding, we can re-write the right-hand side of Eq. (2.16) as

$$\begin{aligned} \frac{(-1)^{q/2}}{q!}\left(\sum_{i=1}^N \psi_i(\tau)\psi_i(\tau')\right)^q &= \frac{(-1)^{q/2}N^q}{q!}\mathcal{G}(\tau,\tau')^q \\ &= \frac{N^4}{4!}\mathcal{G}(\tau,\tau')^4 \qquad (q=4). \end{aligned} \tag{2.22}$$

Now we are in a position to plug Eq. (2.22) in Eq. (2.15) which is plugged in Eq. (2.14), and we insert the identity in Eq. (2.21) to get for the averaged partition function (we have suppressed the subscript 4 for brevity but keep in mind that we solving for the case where $q = 4$)

$$\begin{aligned} \Rightarrow \langle \mathcal{Z}_E\rangle_J = \int \mathcal{D}\mathcal{G}\mathcal{D}\Sigma\mathcal{D}\psi_i \exp&\left\{-\frac{1}{2}\sum_i^N \int d\tau \psi_i(\tau)\partial_\tau\psi_i(\tau)\right\} \\ \times \exp&\left\{+\frac{3J^2}{N^3}\iint d\tau d\tau' \frac{N^4}{4!}\mathcal{G}(\tau,\tau')^4\right\} \\ \times \exp&\left\{-\frac{N}{2}\iint d\tau d\tau'\Sigma(\tau,\tau')\left(\mathcal{G}(\tau,\tau') - \frac{1}{N}\sum_i^N \psi_i(\tau)\psi_i(\tau')\right)\right\}. \end{aligned} \tag{2.23}$$

[7] This will later be identified as the self-energy, but for now this can be seen as a Lagrange multiplier.

Now we re-write the first exponential as

$$\exp\left\{-\frac{1}{2}\sum_i^N\int d\tau\psi_i(\tau)\partial_\tau\psi_i(\tau)\right\}$$
$$=\exp\left\{-\frac{1}{2}\sum_i^N\iint d\tau d\tau'\psi_i(\tau)\delta(\tau-\tau')\partial_\tau\psi_i(\tau')\right\}$$

where $\delta(\tau-\tau')$ collapses the $d\tau'$ integral. Then we re-arrange to get

$$\begin{aligned}\Rightarrow \langle\mathcal{Z}_E\rangle_J &= \int \mathcal{D}\mathcal{G}\mathcal{D}\Sigma\mathcal{D}\psi_i \\ &\times \exp\left\{-\frac{1}{2}\iint d\tau d\tau'\sum_i^N \psi_i(\tau)\left[\delta(\tau-\tau')\partial_\tau-\Sigma(\tau,\tau')\right]\psi_i(\tau')\right\} \\ &\times \exp\left\{-\frac{N}{2}\iint d\tau d\tau'\left(\Sigma(\tau,\tau')\mathcal{G}(\tau,\tau')-\frac{1}{4}J^2\mathcal{G}(\tau,\tau')^4\right)\right\}.\end{aligned} \tag{2.24}$$

The final step is to integrate out the fermionic field ψ_i using the identity

$$\begin{aligned}&\int \mathcal{D}\psi_i \exp\left\{-\frac{1}{2}\iint\int d\tau d\tau'\sum_i^N \psi_i(\tau)\left[\delta(\tau-\tau')\partial_\tau-\Sigma(\tau,\tau')\right]\psi_i(\tau')\right\} \\ &= e^{\frac{N}{2}\ln\det(\partial_\tau-\Sigma)}\end{aligned} \tag{2.25}$$

to finally get

$$\Rightarrow \langle\mathcal{Z}_E\rangle_J = \int \mathcal{D}\mathcal{G}\mathcal{D}\Sigma e^{-S_{E,\mathrm{eff}}[\mathcal{G},\Sigma]} \tag{2.26}$$

where $S_{E,\mathrm{eff}}$ is the effective (Euclidean) action given in closed form by

$$\boxed{\begin{aligned}&\frac{S_{E,\mathrm{eff}}[\mathcal{G},\Sigma]}{N} \\ &\equiv -\frac{1}{2}\ln\det[\partial_\tau-\Sigma]+\frac{1}{2}\iint d\tau d\tau'\left(\Sigma(\tau,\tau')\mathcal{G}(\tau,\tau')-\frac{1}{4}J^2\mathcal{G}(\tau,\tau')^4\right)\end{aligned}}, \tag{2.27}$$

where we can impose time-translation symmetry as we are in equilibrium (i.e., $\mathcal{G}(\tau,\tau')=\mathcal{G}(\tau-\tau')$ and $\Sigma(\tau,\tau')=\Sigma(\tau-\tau')$). As we can see, N plays the role of $1/\hbar$ in the averaged partition function, and therefore the action (accordingly the theory) becomes (semi-)classical in the large-N limit, allowing us to solve the

SYK model. Now we make use of the large-N limit where the saddle solutions dominate the theory which we calculate via the standard Euler-Lagrange equation, or equivalently by extremizing the action

$$\frac{\delta S_{E,\text{eff}}}{\delta \Sigma} \stackrel{!}{=} 0, \qquad \frac{\delta S_{E,\text{eff}}}{\delta \mathcal{G}} \stackrel{!}{=} 0. \tag{2.28}$$

Accordingly the (semi-)classical equations of motion are[8]

$$\mathcal{G} = [(\partial_\tau - \Sigma)]^{-1}, \qquad \Sigma = J^2 \mathcal{G}^3, \tag{2.29}$$

respectively. We can re-write the first equation by using the inverse Green's function for free fermion ($\mathcal{G}_0^{-1}$) to be equal to ∂_τ[9] in the Euclidean plane

$$\boxed{\mathcal{G}_0^{-1} = \mathcal{G}^{-1} + \Sigma, \qquad \Sigma = J^2 \mathcal{G}^3}, \tag{2.30}$$

where we recognize that the first equation is the famous *Dyson's equation*[10] [15] that connects the Green's function and the self-energy. That's why we mentioned above that we will later recognize $\mathcal{G}$ and Σ as the Green's function and the self-energy of the theory. Since the self-energy is in a closed form, we have effectively solved the theory in the large-N limit. Together, Eq. (2.30) are known as the *Schwinger-Dyson equations*.

We have tried to keep the steps as transparent as possible so that the immediate generalization to an arbitrary $q/2$-body interaction be straightforward. The averaged partition function is given by (we now restore the subscript q to denote that this is for $q/2$-body interaction)

$$\langle \mathcal{Z}_E \rangle_{J_q} = \int \mathcal{D}\mathcal{G}\mathcal{D}\Sigma e^{-S_{E,\text{eff}}[\mathcal{G},\Sigma]} \tag{2.31}$$

[8] We use the identities $\ln \det A = \operatorname{Tr} \ln A$ as well as $\frac{d \operatorname{Tr} f(A)}{dA} = f'(A^T)$ where superscript T denotes transpose and f is any arbitrary function of A.

[9] We have derived this in detail in the next Sect. 2.2.1.

[10] There is a nice diagrammatic expansion where one can deduce the Dyson's equation which can be found in Refs. [2, 15]. The basic idea is that if we define $A \cdot B(\tau, \tau') \equiv \int d\tau'' A(\tau, \tau'') B(\tau'', \tau')$, then the (infinite) Feynman diagrams can be summed as $\mathcal{G} = \mathcal{G}_0 + \mathcal{G}_0 \cdot \Sigma \cdot \mathcal{G}_0 + \mathcal{G}_0 \cdot \Sigma \cdot \mathcal{G}_0 \cdot \Sigma \cdot \mathcal{G}_0 + \cdots$ where $\mathcal{G}_0 \cdot \Sigma \cdot \mathcal{G}_0 = \mathcal{G}_0 \cdot (\Sigma \cdot \mathcal{G}_0)$. Then taking $\mathcal{G}_0$ common on the right-hand side, we get $\mathcal{G} = \mathcal{G}_0 \cdot \sum_{i=0}^{\infty} (\Sigma \cdot \mathcal{G}_0)^i$ which can be summed to $\mathcal{G} = \frac{\mathcal{G}_0}{1 - \Sigma \mathcal{G}_0}$. This can be re-arranged to get the Dyson's equation $\mathcal{G}_0^{-1} = \mathcal{G}^{-1} + \Sigma$.

where the effective action for a general $q/2$-body interaction is found to be extensive in N as it should (a useful benchmark) and given by

$$\boxed{\begin{aligned}\frac{S_{E,\text{eff}}[\mathcal{G},\Sigma]}{N} &\equiv -\frac{1}{2}\ln\det[\partial_\tau - \Sigma] \\ &+\frac{1}{2}\iint d\tau d\tau'\left(\Sigma(\tau-\tau')\mathcal{G}(\tau-\tau') - \frac{1}{q}J_q^2\mathcal{G}(\tau-\tau')^q\right)\end{aligned}} \tag{2.32}$$

and in the large-N limit, the (semi-)classical Schwinger-Dyson equations of the theory are

$$\boxed{\mathcal{G}_0^{-1} = \mathcal{G}^{-1} + \Sigma, \qquad \Sigma = J_q^2\mathcal{G}^{q-1}}. \tag{2.33}$$

2.2.1 Free Case

We now focus on the free case where the interacting Hamiltonian $\mathcal{H} = 0$. The partition function can be modified by adding a source term [16] to become a generating functional, which for free fermionic case is given in Euclidean plane by (superscript 0 denotes the free case)

$$\begin{aligned}\mathcal{Z}_E^0 &= \int \mathcal{D}\psi_i e^{-S_E^0[\psi_i] + \sum_{k=1}^{N}\int d\tau \psi_k(\tau)M_k(\tau)} \\ &= \int \mathcal{D}\psi_i e^{-\frac{1}{2}\sum_{i=1}^{N}\int d\tau\psi_i(\tau)\partial_\tau\psi_i(\tau) + \sum_{k=1}^{N}\int d\tau\psi_k(\tau)M_k(\tau)}\end{aligned} \tag{2.34}$$

where, again as above, $\mathcal{D}\psi_i \equiv \prod_{i=1}^{N} d\psi_i$ and M_k are the source terms. Now this is a quadratic integral in Majorana fermions (Grassmannian) ψ_i. We integrate out ψ_i by rewriting

$$\Rightarrow \mathcal{Z}_E^0 = \prod_{k=1}^{N}\left(\int d\psi_k e^{-\frac{1}{2}\int d\tau\psi_k(\tau)\partial_\tau\psi_k(\tau) + \int d\tau\psi_k(\tau)M_k(\tau)}\right) \tag{2.35}$$

where we can solve for each ψ_k and later take the product. In order to solve the integral, we introduce the inverse of ∂_τ operator given by $\Delta(\tau - \tau')$ defined via (exactly how Green's function/propagator in QFT is defined as the inverse of an operator [16])

$$\partial_\tau\Delta(\tau) = \delta(\tau). \tag{2.36}$$

To find an explicit expression of $\Delta(\tau)$, we take the Fourier transform of this equation to get[11]

$$\int \frac{d\omega}{2\pi}(-i\omega)\Delta(\omega)e^{-i\omega\tau} = \int \frac{d\omega}{2\pi}e^{-i\omega\tau}. \tag{2.37}$$

Equating the integrands gives

$$-i\omega\Delta(\omega) = 1 \Rightarrow \Delta(\omega) = \frac{+i}{\omega}. \tag{2.38}$$

The integral $\Delta(\omega) = -i/\omega$ is singular at $\omega = 0$. To resolve this, introduce a small $\epsilon > 0$ to shift the pole into the complex plane

$$\Delta(\omega) = \lim_{\epsilon \to 0} \frac{i}{\omega + i\epsilon}. \tag{2.39}$$

This enforces causality (retarded boundary conditions) [16]. Taking the inverse Fourier transform leads us to the following propagator for (Grassmannian) Majorana fermions

$$\Delta(\tau) = \lim_{\epsilon \to 0} \frac{\imath}{2\pi} \int_{-\infty}^{\infty} d\omega \frac{e^{-i\omega\tau}}{\omega + i\epsilon}. \tag{2.40}$$

Now we can solve this complex integral by consider two cases

- Case 1: $\tau > 0$

  1. Close the contour in the lower half-plane (where $e^{-i\omega\tau}$ decays).
  2. The pole at $\omega = -i\epsilon$ lies inside the contour.
  3. Residue:

  $$\text{Res}_{\omega=-i\epsilon}\left(\frac{e^{-i\omega\tau}}{\omega + i\epsilon}\right) = e^{-\epsilon\tau}.$$

  4. Integral evaluates to:

  $$\Delta(\tau) = \frac{i}{2\pi} \cdot 2\pi i \cdot e^{-\epsilon\tau} = -e^{-\epsilon\tau}.$$

  As $\epsilon \to 0$, this becomes $\Delta(\tau) = -1$.

[11] Since we are in imaginary-time formalism, ω denotes the Matsubara frequency. Conventionally, it is denoted by ω_n but for now we skip the subscript. Fourier transform in imaginary-time formalism is defined in Eq. (F.2). See Appendix F for further details.

- Case 2: $\tau < 0$

  1. Close the contour in the upper half-plane (no poles enclosed).
  2. Integral evaluates to zero: $\Delta(\tau) = 0$.

Therefore we find

$$\boxed{\Delta(\tau) = -\Theta(\tau)}, \tag{2.41}$$

where $\Theta(\tau)$ is the Heaviside step function given by 1 for $\tau > 0$, otherwise zero.

Now returning to Eq. (2.35), we complete the square (to solve for Gaussian integral) by shifting $\psi_k \to \psi_k + \int d\tau' \Delta\left(\tau - \tau'\right) M_k\left(\tau'\right)$ to get

$$\begin{aligned} &-\frac{1}{2}\int \psi_k \partial_\tau \psi_k d\tau + \int \psi_k M_k d\tau \\ &= -\frac{1}{2}\int \psi_k \partial_\tau \psi_k d\tau + \frac{1}{2}\int M_k(\tau)\Delta(\tau - \tau')M_k(\tau)d\tau d\tau' \end{aligned} \tag{2.42}$$

where we have used integration by parts and assumed that the boundary term vanishes. The Grassmann integral over ψ_k becomes $\int \mathcal{D}\psi_k e^{-\frac{1}{2}\int \psi_k \partial_\tau \psi_k d\tau} = \mathrm{Pf}\left(-\partial_\tau\right)$, where $\mathrm{Pf}\left(-\partial_\tau\right)$ is the Pfaffian of the operator, also written as $\sqrt{\det[-\partial_\tau]}$. This is absorbed into the measure. The remaining term gives:

$$\exp\left(\frac{1}{2}\int d\tau d\tau' M_k(\tau)\Delta\left(\tau - \tau'\right) M_k\left(\tau'\right)\right)$$

which when plugged into Eq. (2.35) to incorporate all N fields gives[12]

$$\Rightarrow \boxed{\mathcal{Z}_E^0 = \exp\left(\frac{1}{2}\sum_{k=1}^{N}\int d\tau d\tau' M_k(\tau)\Delta\left(\tau - \tau'\right) M_k\left(\tau'\right)\right)}. \tag{2.43}$$

We have the relation from Eq. (2.35) $\left.\frac{\delta}{\delta M_i(\tau)} \ln \mathcal{Z}_E^0\right|_{M=0} = \langle \psi_i(\tau)\rangle$, which vanishes for a free theory for $M = 0$. Then to get the free (connected[13]) Green's

[12] In general, for any operator A, we have $\int \mathcal{D}\psi e^{-\frac{1}{2}\int \psi A \psi + \int M\psi} \propto \exp\left(\frac{1}{2}\int MA^{-1}M\right)$.

[13] The connected Green's function $\mathcal{G}_{c,ij}$ encodes essential two-point correlations by subtracting disconnected parts and mean-field contributions from the full correlator. It is universally generated by $\delta^2 \ln \mathcal{Z}/\delta M_i \delta M_j = \left\langle \psi_i(\tau)\psi_j(0)\right\rangle - \langle \psi_i(\tau)\rangle \langle \psi_i(0)\rangle$, ensuring only physically linked processes contribute. We have removed the subscript c for brevity. In the Euclidean path integral formalism, functional derivatives automatically yield time-ordered correlation functions where $\mathcal{G}_{0,ij} =$

function $\mathcal{G}_{0,ij}$ for $\psi_i(\tau)$ and $\psi_j(0)$ (for a fixed i and j), we get

$$\mathcal{G}_{0,ij} = \frac{\delta}{\delta M_i(\tau)} \frac{\delta}{\delta M_j(0)} \ln \mathcal{Z}_E^0 \Big|_{M=0}. \tag{2.44}$$

where we set all source terms M_k to zero *after* taking the derivatives. So we get

$$\begin{aligned}
\mathcal{G}_{0,ij} =& \frac{\delta}{\delta M_i(\tau)} \Big[\Big(\frac{1}{2} \int d\tau'' \Delta(-\tau'') M_j(\tau'') - \frac{1}{2} \int d\tau' M_j(\tau') \Delta(\tau') \Big) \Big] \Big|_{M=0} \\
=& \Big(\frac{1}{2} \int d\tau'' \Delta(-\tau'') \delta_{ji} \delta(\tau'' - \tau) - \frac{1}{2} \int d\tau' \delta_{ji} \delta(\tau' - \tau) \Delta(\tau') \Big) \Big|_{M=0} \\
=& \frac{\delta_{ij}}{2} \Big[\Delta(-\tau) - \Delta(\tau) \Big],
\end{aligned} \tag{2.45}$$

where we used $\delta M_a(\tau)/\delta M_b(\tau') = \delta_{ab} \delta(\tau - \tau')$. As we have shown above in Eq. (2.41), we have $\Delta(-\tau) - \Delta(\tau) = -\Theta(-\tau) + \Theta(\tau) = \text{sgn}(\tau)$ where $\text{sgn}(\tau)$ is the sign function, given by $+1$ for $\tau > 0$ and -1 for $\tau < 0$. Therefore we get

$$\mathcal{G}_{0,ij}(\tau) = \frac{\delta_{ij}}{2} \text{sgn}(\tau) \quad \xrightarrow[\text{transform}]{\text{Fourier}} \quad \mathcal{G}_{0,ij}(\omega) = -\frac{\delta_{ij}}{\imath \omega} \tag{2.46}$$

where from the last relation, we see that $\mathcal{G}_{0,ij}(\omega)^{-1} = -\imath \omega \delta_{ij}$ which upon inverse Fourier transforming gives $\mathcal{G}_{0,ij}(\tau)^{-1} = \partial_\tau$. This is what we used above in Eqs. (2.30) and (2.33).

2.3 Replica Symmetry and Generalization to M Flavored Majorana SYK Model

We are interested in the thermodynamic properties of the SYK model for which we would like to evaluate the free energy given by the partition function through $\ln \mathcal{Z} = -\beta F$ where β is the inverse temperature. We find a problem that the disorder averaging lead to $\langle \ln \mathcal{Z} \rangle_J$, but we evaluated above $\langle \mathcal{Z} \rangle_J$. In general, we have

$$\langle (\ln \mathcal{Z}) \rangle_J \neq \ln \Big(\langle (\mathcal{Z}) \rangle_J \Big).$$

$\langle T\left(\psi_i(\tau)\psi_j(0)\right)\rangle_0 = \begin{cases} \psi_i(\tau)\psi_j(0) & \tau > 0 \\ -\psi_j(0)\psi_i(\tau) & \tau < 0 \end{cases} = \Theta(\tau)\psi_i(\tau)\psi_j(0) - \Theta(-\tau)\psi_j(0)\psi_i(\tau)$ where $\langle \cdot \rangle_0$ denotes the expectation value in the free theory, T denotes the time-ordering operation and $\Theta(x)$ denotes Heaviside step function which takes the value of 1 for $x > 0$, and vanishes otherwise.

Evaluating $\langle \ln \mathcal{Z} \rangle_J$ is significantly difficult. However, there exists something known as replica trick which implies

$$\langle \ln \mathcal{Z} \rangle_J = \lim_{M\to 0} \frac{\langle \mathcal{Z}^M \rangle - 1}{M} \tag{2.47}$$

which can be verified using L'Hospital rule. Equivalently, the replica trick can be written as[14]

$$\langle \ln \mathcal{Z} \rangle_J = \lim_{M\to 0} \frac{1}{M} \ln \left(\left\langle \mathcal{Z}^M \right\rangle_J \right). \tag{2.48}$$

This means that instead of calculating $\langle \ln \mathcal{Z} \rangle_J$, we can calculate $\langle \mathcal{Z}^M \rangle_J$. In the last section, we evaluated $\langle \mathcal{Z} \rangle_J$, which will not be sufficient because in general

$$\langle \mathcal{Z} \rangle_J^M \neq \left\langle \mathcal{Z}^M \right\rangle_J. \tag{2.49}$$

Since $\langle \mathcal{Z}^M \rangle_J$ is the object of interest, physically this means that we have to derive this in a more general setting where the Majorana fermions ψ_i has are flavored with M flavors ψ_i^α (Greek indices will denote flavors, running from 1 to M) for the SYK model. We consider SYK$_4$ model and start with the partition function and the action in Euclidean plane, same as in Eqs. (2.10) and (2.11), with the following generalization

$$\mathcal{Z}_E^M = \int \mathcal{D}\psi_i^\alpha e^{-S_E[\psi_i]} \tag{2.50}$$

where $\mathcal{D}\psi_i^\alpha = \prod_{\alpha=1}^{M} \prod_{i=1}^{N} d\psi_i^\alpha$ is the measure and effective action is given by

$$S_E[\psi_i] = \int d\tau \left[\frac{1}{2} \sum_{\alpha=1}^{M} \sum_{i=1}^{N} \psi_i^\alpha \partial_\tau \psi_i^\alpha + \mathcal{H}_4 \right]. \tag{2.51}$$

and the SYK interacting Hamiltonian is given by

$$\mathcal{H}_4 = \sum_{\alpha=1}^{M} \sum_{1\le i_1<i_2<i_3<i_4\le N}^{N} j_{4;i_1,i_2,i_3,i_4} \psi_{i_1}^\alpha \psi_{i_2}^\alpha \psi_{i_3}^\alpha \psi_{i_4}^\alpha, \tag{2.52}$$

[14] Proof: Using $A^n \simeq 1+n \ln A$ for small n, we have $\lim_{M\to 0} \frac{1}{M} \ln \left(\left\langle \mathcal{Z}^M \right\rangle_J \right) \simeq \lim_{M\to 0} \frac{1}{M} \ln(1 + M \langle \ln \mathcal{Z} \rangle_J) \simeq \langle \ln \mathcal{Z} \rangle_J$. This can also be seen using the identity $\ln(1+x) \simeq x$ for small x.

where $j_{4;i_1,i_2,i_3,i_4}$ are random variables derived from the same Gaussian ensemble as before (Eq. (2.5) for $q = 4$). Physically, this means that we are taking M copies of the system and evaluating the partition function for all M copies under the *assumption* that none of the copy are interacting among themselves (crucial to replica trick).

We now follow the same steps as in Sect. 2.2 where we show the intermediate steps for the reader to follow. Starting with the disorder-averaging, we get (ignoring the subscript 4 on J_4 for brevity)

$$\langle \mathcal{Z}_E^M \rangle_J = \int \mathcal{D} j_{4;i_1,i_2,i_3,i_4} \mathcal{P}[j_{4;i_1,i_2,i_3,i_4}] \mathcal{Z}_E^M . \tag{2.53}$$

We perform the integral as before to get

$$\Rightarrow \langle \mathcal{Z}_E^M \rangle_J = \int \mathcal{D}\psi_i^\alpha \exp\Bigg\{ -\frac{1}{2}\sum_{\alpha=1}^{M}\sum_{i=1}^{N} \int d\tau \psi_i^\alpha \partial_\tau \psi_i^\alpha + \frac{J^2 N}{8} \sum_{\alpha=1}^{M}\sum_{\beta=1}^{M} \iint d\tau d\tau' \left(\sum_{i=1}^{N} \psi_i^\alpha(\tau)\psi_i^\beta(\tau') \right)^4 \Bigg\}. \tag{2.54}$$

We again introduce the bi-local fields as in Eqs. (2.19) and (2.20) with following generalization

$$\mathcal{G}^{\alpha\beta}\left(\tau, \tau'\right) \equiv \frac{1}{N} \sum_{i=1}^{N} \psi_i^\alpha(\tau)\psi_i^\beta(\tau') \tag{2.55}$$

and self-energy as a Lagrange multiplier by using the integral representation of the δ-function

$$\int \mathcal{D}\mathcal{G}^{\alpha\beta} \delta \left(\mathcal{G}^{\alpha\beta} - \frac{1}{N} \sum_{i=1}^{N} \psi_i^\alpha(\tau)\psi_i^\beta(\tau') \right) = 1 \tag{2.56}$$

implying (absorbing the imaginary unit in the self-energy $\Sigma^{\alpha\beta}$ using the benefit of hindsight)

$$\int \mathcal{D}\mathcal{G}^{\alpha\beta} \mathcal{D}\Sigma^{\alpha\beta} \times \exp\Bigg\{ -\frac{N}{2} \iint d\tau d\tau' \Sigma^{\alpha\beta}(\tau, \tau') \left(\mathcal{G}^{\alpha\beta} - \frac{1}{N} \sum_{i=1}^{N} \psi_i^\alpha(\tau)\psi_i^\beta(\tau') \right) \Bigg\} = 1. \tag{2.57}$$

Plugging this in Eq. (2.54), we get

$$\Rightarrow \langle \mathcal{Z}_E^M \rangle_J = \int \mathcal{D}\psi_i^\alpha \mathcal{D}\mathcal{G}^{\alpha\beta} \mathcal{D}\Sigma^{\alpha\beta}$$
$$\exp\Big\{ - \sum_{\alpha,\beta=1}^{M} \sum_{i=1}^{N} \frac{1}{2} \iint d\tau d\tau' \Big[\psi_i^\alpha(\tau) \Big(\delta^{\alpha\beta} \delta(\tau - \tau') \partial_\tau - \Sigma^{\alpha\beta}(\tau, \tau') \Big) \psi_i^\beta(\tau') \Big]$$
$$- \frac{1}{2} \sum_{\alpha,\beta=1}^{M} \iint d\tau d\tau' \Big[N \Sigma^{\alpha\beta} \mathcal{G}^{\alpha\beta} - \frac{J^2 N}{4} \left(\mathcal{G}^{\alpha\beta}\right)^4 \Big] \Big\}. \tag{2.58}$$

Then we integrate out the fermions using the identity

$$\int \mathcal{D}\psi_i^\alpha \exp\Big\{ - \sum_{\alpha,\beta=1}^{M} \sum_{i=1}^{N} \frac{1}{2} \iint d\tau d\tau' \Big[\quad \psi_i^\alpha(\tau) \Big(\delta^{\alpha\beta} \delta(\tau - \tau') \partial_\tau$$
$$- \Sigma^{\alpha\beta}(\tau, \tau') \Big) \psi_i^\beta(\tau') \Big] \Big\} \tag{2.59}$$
$$= \exp\Big\{ \frac{N}{2} \sum_{\alpha,\beta=1}^{M} \ln \det[\delta^{\alpha\beta} \partial_\tau - \Sigma^{\alpha\beta}] \Big\}$$

to get

$$\Rightarrow \langle \mathcal{Z}_E^M \rangle_J = \int \mathcal{D}\mathcal{G}^{\alpha\beta} \mathcal{D}\Sigma^{\alpha\beta} \exp\Big\{ \frac{N}{2} \sum_{\alpha,\beta=1}^{M} \ln \det[\delta^{\alpha\beta} \partial_\tau - \Sigma^{\alpha\beta}] \Big\}$$
$$\times \exp\Big\{ - \frac{N}{2} \sum_{\alpha,\beta=1}^{M} \iint d\tau d\tau' \Big[\Sigma^{\alpha\beta} \mathcal{G}^{\alpha\beta} - \frac{J^2}{4} \left(\mathcal{G}^{\alpha\beta}\right)^4 \Big] \Big\}. \tag{2.60}$$

The last step is to *assume* a *replica symmetry saddle point* where the off-diagonal elements are sub-leading in N (thereby vanishing in large-N limit) in order to have $\mathcal{G}^{\alpha\beta} = \delta^{\alpha\beta}\mathcal{G}$ in the large-N limit (implying averaged diagonal Green's function $\mathcal{G} = \frac{1}{N}\sum_\alpha \mathcal{G}^{\alpha\alpha}$). This assumption is valid as long as replica symmetry is not broken and there are no stable spin glass solutions.[15] This allows us to get rid of the Greek indices and, therefore, we get

$$\Rightarrow \langle \mathcal{Z}_E^M \rangle_J = \int \mathcal{D}\mathcal{G}\mathcal{D}\Sigma e^{-M S_{E,\text{eff}}[\mathcal{G},\Sigma]}, \tag{2.61}$$

where a factor of M comes in the exponential denoting there are M (non-interacting) copies of the system for which we are evaluating the partition function. The effective

[15] We refer the reader to Ref. [3] for a nice introduction to spin-glass.

action is again found to be extensive in N and is given by

$$\boxed{\begin{aligned}&\frac{S_{E,\text{eff}}[\mathcal{G},\Sigma]}{N}\\&\equiv -\frac{1}{2}\ln\det[\partial_\tau-\Sigma]+\frac{1}{2}\iint d\tau d\tau'\left(\Sigma(\tau,\tau')\mathcal{G}(\tau,\tau')-\frac{1}{4}J^2\mathcal{G}(\tau,\tau')^4\right)\end{aligned}}, \tag{2.62}$$

which is exactly the same as in Eq. (2.27). In equilibrium, we can again take time-translational invariance. Accordingly the Schwinger-Dyson equations (2.30) remain unchanged. The generalization to $q/2$-body interacting case yields the same disorder-averaged partition function (Eq. (2.31)) and effective action (Eq. (2.33)), thereby yielding the exact Schwinger-Dyson equations as in Eq. (2.33).

Hence, the replica trick allows us to keep using the disorder-averaged partition function $\langle\mathcal{Z}\rangle_J$ instead of $\langle\mathcal{Z}^M\rangle_J$ as both yields the same Schwinger-Dyson equations. Therefore, the free energy is given by (using the replica trick Eq. (2.48))

$$\beta F = \langle\ln\mathcal{Z}\rangle_J = \lim_{M\to 0}\frac{1}{M}\ln\left(\left\langle\mathcal{Z}^M\right\rangle_J\right) = -S_{E,\text{eff}}[\mathcal{G},\Sigma]. \tag{2.63}$$

Hence knowing the effective action in a closed form allows us to study the thermodynamic properties (more on this later) of the SYK model exactly in the thermodynamic limit (large-N limit).

2.4 $O(N)$ Symmetry of the Effective Action

$O(N)$ symmetry refers to invariance under transformations by the orthogonal group $O(N)$, which consists of all $N\times N$ real matrices O satisfying $O^TO = I$ (i.e., rotations and reflections in N dimensional space). In this section, we investigate the symmetry of the SYK theory under $O(N)$ transformations.

2.4.1 Free Case

The (Euclidean) action is given by $S_E = \frac{1}{2}\int d\tau\psi_i(\tau)\dot{\psi}^i(\tau)$ where we take $O(N)$ transformation of the Majorana field $\psi_i \to O_i{}^j\psi_j$ where $O^i{}_j$ is a matrix representation of the group $O(N)$ and we have assumed Einstein summation convention where repeated indices are summed over[16] (therefore, the action is equivalent to $S_E = \frac{1}{2}\sum_i\int d\tau\psi_i(\tau)\dot{\psi}_i(\tau)$). We can perform an infinitesimal

[16] In this section, the position of indices will matter whether they are contravariant (superscript) or covariant (subscript).

expansion

$$\delta\psi_i = \xi_a (T^a)_i{}^j \psi_j + \mathcal{O}(\xi^2) \tag{2.64}$$

where $\delta\psi_i$ is the small change in the field component ψ_i under the symmetry transformation, and ξ_a are small, continuous parameters (e.g., angle increments) labeling the symmetry transformation. The index a runs over the generators of the symmetry group (e.g., $a = 1, \ldots, N$ for $O(N)$). Here, $(T^a)_i{}^j$ are the generators of the symmetry group in the representation of the field ψ_i. These are matrices encoding how the field components mix under the symmetry: index a labels the generator (e.g., rotation axes) while indices j and i act on the field components (summed over j). These are anti-symmetric in its indices, namely $(T^a)_i{}^j = -(T^a)^j{}_i$. Accordingly the transformation for $\delta\psi^i$ is given by

$$\delta\psi^i = \xi_a (T^a)^i{}_j \psi^j + \mathcal{O}(\xi^2) \tag{2.65}$$

Now we consider two cases where ξ_a is time-independent as well as time-dependent:

- ξ_a is a constant (time-independent): The change of action under $O(N)$ transformation (Eqs. (2.64) and (2.65)) becomes (dot represents time derivative with respect to τ)

$$\begin{aligned}\delta S_E =& \frac{1}{2}\int d\tau \left(\delta\psi_i(\tau)\dot{\psi}^i(\tau) + \psi_i(\tau)\delta\dot{\psi}^i(\tau)\right)\\ =& \frac{1}{2}\int d\tau \left(\xi_a (T^a)_i{}^j \psi_j \dot{\psi}^i + \psi_i \frac{d}{d\tau}(\underbrace{\xi_a (T^a)^i{}_j}_{\text{constant}} \psi^j)\right)\\ =& \frac{1}{2}\int d\tau \left(\xi_a (T^a)_i{}^j \psi_j \dot{\psi}^i + \psi_i (\xi_a \underbrace{(T^a)^i{}_j}_{=-(T^a)_j{}^i} \dot{\psi}^j)\right)\\ =& 0\end{aligned} \tag{2.66}$$

  where we used the property of dummy (repeated) indices i and j to switch them $i \leftrightarrow j$ in the first term, namely $(T^a)_i{}^j \psi_j \dot{\psi}^i = (T^a)_j{}^i \psi_i \dot{\psi}^j$. Note that ψ_i are fermionic variables and everytime they are interchanged, one gets a minus sign because $\{\psi_i, \psi_j\} = 0$ for $i \neq j$. However, we never made the fermionic fields pass through each other, so we did not have to encounter any additional minus sign. No integration by parts were required too (it will be required below) and we get $\delta S_E = 0$ identically. Hence the free theory possesses $O(N)$ symmetry.
- ξ_a is time-dependent ($\xi(\tau)$): Since there is a continuous symmetry of the (local) action, there must be an associated conserved Noether current. In order to

read-off the current associated with the $O(N)$ symmetry, we consider time-dependency for infinitesimal transformations:

$$\delta\psi_i = \xi_a(\tau)(T^a)_i{}^j\psi_j + \mathcal{O}(\xi^2), \qquad \delta\psi^i = \xi_a(\tau)(T^a)^i{}_j\psi^j + \mathcal{O}(\xi^2). \tag{2.67}$$

Then the variation of the action is

$$\delta S_E = \frac{1}{2}\int d\tau \left(\xi_a(T^a)_i{}^j\psi_j\dot{\psi}^i + \psi_i(\xi_a \underbrace{(T^a)^i{}_j}_{=-(T^a)_j{}^i} \dot{\psi}^j) + \dot{\xi}_a(\tau)(\psi_i(T^a)^i{}_j\psi^j) \right) \overset{!}{=} 0. \tag{2.68}$$

The first two terms cancel identically and we perform an integration by parts on the third term where we assume variations on the boundary to vanish. Then we read-off the associated conserved Noether current associated with $O(N)$ symmetry

$$\mathcal{I}^a = \frac{1}{2}\psi^i(\tau)T^a_{ij}\psi^j(\tau). \tag{2.69}$$

where we used the property of the dummy (repeated) indices $A^iB_i = A_iB^i$. The currents are conserved as $\dot{\mathcal{I}}^a = 0$.

This is consistent with the picture that the free case is still local and that's why Noether's theorem is applicable. As the SYK model is an all-to-all interacting model which makes the action non-local, we will see in the next subsection that Noether's theorem is not applicable there.

2.4.2 With Interactions

We begin by considering a naive calculation at the level of couplings where disorder-averaged partition function is given by Eq. (2.13) which we reproduce here for convenience

$$\langle \mathcal{Z}_E \rangle_{J_4} = A \int \mathcal{D}\psi_i \exp\left\{ -\frac{1}{2}\sum_i^N \int d\tau \psi_i \partial_\tau \psi_i \right\}$$
$$\times \int \mathcal{D}j_{4;i,j,k,l}$$
$$\times \exp\left\{ -\sum_{1\le i<j<k<l\le N} \left(\frac{j^2_{4;i,j,k,l}}{12\frac{J_4^2}{N^3}} - j_{4;i,j,k,l}\int d\tau \psi_i(\tau)\psi_j(\tau)\psi_k(\tau)\psi_l(\tau) \right) \right\}.$$

Then we might naively impose the $O(N)$ transformations as

$$\psi^i \longmapsto O^i{}_j \psi^j, \qquad j_{4;ijkl} \longmapsto O_i{}^a O_j{}^b O_k{}^c O_l{}^d j_{4;abcd}$$

but we see the problem immediately: $j_{4;ijkl}$ are not quantum fields and their variations on the boundary will not vanish ($\delta j_{4;ijkl}|_{\partial\mathcal{B}} \neq 0$) as they do for the case of fields ψ_i. Therefore we cannot directly deal at the level of bare action but require to integrate out the disorder $j_{4;ijkl}$. We start at the point where we arranged the fermions in Eq. (2.18) which we reproduce here for convenience

$$\Rightarrow \langle \mathcal{Z}_E \rangle_{J_4} = \int \mathcal{D}\psi_i \exp\Big\{ -\frac{1}{2}\sum_{i=1}^{N} \int d\tau \psi_i \partial_\tau \psi_i + \frac{J_4^2 N}{8} \iint d\tau d\tau' \Big(\sum_{i=1}^{N} \psi_i(\tau)\psi_i(\tau') \Big)^4 \Big\}.$$

from which we read off the action as

$$S_E = +\frac{1}{2}\sum_{i=1}^{N} \int d\tau \psi_i \partial_\tau \psi_i - \frac{J_4^2 N}{8} \iint d\tau d\tau' \Big(\sum_{i=1}^{N} \psi_i(\tau)\psi_i(\tau') \Big)^4 . \tag{2.70}$$

We again plug the $O(N)$ transformations in Eqs. (2.64) and (2.65) for the two cases where ξ_a is a constant and time-dependent:

- ξ_a is a constant (time-independent): We have the same transformations as in Eqs. (2.64) and (2.65) which we substitute in the variation of the action given by (using the Einstein summation convention for indices i, ℓ and j to suppress the summation sign)

$$\begin{aligned} \delta S =& \frac{1}{2} \int d\tau \underbrace{(\delta\psi_i \partial_\tau \psi^i + \psi_i \partial_\tau \delta\psi^i)}_{=0 \text{ because of Eq. (2.66)}} \\ & - \frac{J_4^2 N}{2} \iint d\tau d\tau' (\psi_\ell(\tau)\psi^\ell(\tau'))^3 \Big(\underbrace{\delta\psi_j(\tau)\psi^j(\tau') + \psi_j(\tau)\delta\psi^j(\tau')}_{=\xi_a (T^a)_j{}^m \psi_m \psi^j + \psi_j \xi_a \underbrace{(T^a)^j{}_m}_{=-(T^a)_m{}^j} \psi^m = 0} \Big) \\ =& 0 \end{aligned} \tag{2.71}$$

where we used the fact that m and j are dummy (repeated) indices therefore the second term in the under-brace could be written by swapping the indices $m \leftrightarrow j$ (without the need to swap fermionic fields through each other) $-\psi_j \xi_a (T^a)_m{}^j \psi^m = -\xi_a (T^a)_j{}^m \psi_m \psi^j$ which exactly cancels the first term in the under-brace. Again, we never needed to perform any integration by parts

(therefore, no assumption of variation of the fields vanishing on the boundary went in). We have $\delta S = 0$ identically. Therefore, the theory possesses $O(N)$ symmetry.
NOTE: Performing calculations for $q = 4$ immediately lead us to see that the same result holds true of vanishing variational action for arbitrary q where Eq. (2.70) would modify using Eq. (2.22) and all steps still goes through identically (without the need of integration by parts).

- ξ_a is time-dependent ($\xi(\tau)$): We have

$$
\begin{aligned}
\delta S &= \frac{1}{2}\int d\tau(\underbrace{\delta\psi_i\partial_\tau\psi^i + \psi_i\partial_\tau\delta\psi^i}_{\text{Same as free case (Eq. (2.66))}})\\
&\quad - \frac{J_4^2 N}{2}\iint d\tau d\tau'(\psi_\ell(\tau)\psi^\ell(\tau'))^3\Big(\underbrace{\delta\psi_j(\tau)\psi^j(\tau')+\psi_j(\tau)\delta\psi^j(\tau')}_{=\xi_a(\tau)(T^a)_j{}^m\psi_m\psi^j+\psi_j\xi_a(\tau')\underbrace{(T^a)^j{}_m}_{=-(T^a)_m{}^j}\psi^m}\Big)\\
&= \frac{1}{2}\int d\tau(\partial_\tau\xi_a(\tau))\psi_i(T^a)^i{}_j\psi^j\\
&\quad - \frac{J_4^2 N}{2}\iint d\tau d\tau'(\psi_\ell(\tau)\psi^\ell(\tau'))^3(\xi_a(\tau)-\xi_a(\tau'))(T^a)_j{}^m\psi_m\psi^j\\
&\neq 0 \qquad \text{(in general)}
\end{aligned}
\tag{2.72}
$$

We see the problem: we can never express δS in the form of $\dot{\xi}_a\mathcal{I}^a$ *due to the bi-locality of the action* (recall the bi-local fields $\mathcal{G}(\tau,\tau')$ and $\Sigma(\tau,\tau')$ we introduced in our derivation of the disorder-averaged partition function). Even if we attempt to expand $\xi_a(\tau')$ around τ as in $\xi_a(\tau') = \xi_a(\tau)+(\tau-\tau')\partial_\tau\xi_a(\tau)+\ldots$, we are not able to do this!
Conclusion: Even though there is a continuous symmetry of the action, there is no associated conserved current. Have we found a violation of the Noether's theorem? Absolutely not! Because Noether's theorem is formulated only for local actions!

2.5 The Infrared (~ Conformal) Limit

2.5.1 Definition

The infrared limit is defined when energy is low ($\omega \to 0$), thereby leading to[17] $|\omega| \ll |\Sigma|$, where Σ is the self-energy that appeared above in the Schwinger-Dyson equations (2.33). Physically, this implies that at low energies, the self-energy

[17] We work with natural units throughout, unless explicitly stated otherwise.

dominates over the bare kinetic term in the Schwinger-Dyson equations. This reflects the strongly interacting nature of the system in the IR.

Equivalently, the Schwinger-Dyson equations in the IR (infrared) limit implies a strong coupling limit satisfying $J|\tau| \to \infty$ (long time or strong coupling), where τ is the time separation and J is the coupling strength. This is same as when the dimensionless coupling is $J\beta \gg 1$ (low temperature or strong coupling), where $\beta = 1/T$ is the inverse temperature. As we shall see below, the IR limit makes the Schwinger-Dyson equations analytically solvable.

Let's start with the Schwinger-Dyson equations in Eq. (2.33) which we reproduce here for convenience

$$\mathcal{G}_0(\tau)^{-1} = \mathcal{G}(\tau)^{-1} + \Sigma(\tau), \qquad \Sigma(\tau) = J_q^2 \mathcal{G}(\tau)^{q-1},$$

where $\mathcal{G}_0(\tau)^{-1} = \partial_\tau$. Taking the Fourier transform, we get

$$\mathcal{G}(\omega)^{-1} = -\imath\omega - \Sigma(\omega), \qquad \Sigma(\omega) = J_q^2 \mathcal{G}(\omega)^{q-1}. \tag{2.73}$$

Therefore, based on the definition of the IR limit discussed above, we can ignore the first term on the right-hand side in the first equation in the IR limit. The Schwinger-Dyson equations simplify to

$$\mathcal{G}(\omega)^{-1} = -\Sigma(\omega), \qquad \Sigma(\omega) = J_q^2 \mathcal{G}(\omega)^{q-1} \quad \text{(IR limit)}. \tag{2.74}$$

Taking the inverse Fourier transform gives (written as bi-local)

$$\int d\tau'' \mathcal{G}(\tau, \tau'')\Sigma(\tau'', \tau') = -\delta(\tau - \tau'), \quad \Sigma(\tau, \tau') = J_q^2 \mathcal{G}(\tau, \tau')^{q-1}, \tag{2.75}$$

where Σ can be substituted using second equation into the first.

2.5.2 Effective Action in the IR Limit and Diff($\mathbb{R}$) Symmetry

A detailed discussion of re-parameterization is given in Ref. [11]. We provide a basic introduction here, but refer the reader to Ref. [11] for more advanced discussions. We have derived the effective action for general q in Eq. (2.32) where we can impose the IR limit, thereby ignoring the kinetic contributions and we get the effective action in the IR limit as (written in most general, non-time-translational invariant form)

$$\frac{S_{E,\text{eff}}[\mathcal{G}, \Sigma]}{N} \simeq -\frac{1}{2} \ln \det[-\Sigma]$$
$$+\frac{1}{2} \iint d\tau d\tau' \left(\Sigma(\tau, \tau')\mathcal{G}(\tau, \tau') - \frac{1}{q} J_q^2 \mathcal{G}(\tau, \tau')^q \right). \tag{2.76}$$

Now we would like to test for the (diffeomorphism group) Diff($\mathbb{R}$) symmetry of this action. It is defined via the group of all smooth, invertible coordinate transformations $f : \mathbb{R} \to \mathbb{R}$, i.e., reparameterizations. Since we are dealing with an effectively zero spatial dimensional (all-to-all) SYK dot, the total dimension becomes $0 + 1$D (only time serves as a dimension). Then in this 1D scenario, there is no intrinsic notion of angle or conformal structure (unlike in ≥ 2D). Thus, any diffeomorphism trivially preserves the "conformal structure" (since there is none to restrict transformations). That's why

$$\mathrm{Conf}(\mathbb{R}) \cong \mathrm{Diff}(\mathbb{R}) \quad \text{(in 1D, locally)}. \tag{2.77}$$

Accordingly testing for symmetries under diffeomorphism group effectively implies (in a trivial way) testing for conformal symmetry. So we start by considering the following reparameterizations

$$\tau \longmapsto f(\tau), \qquad \tau' \longmapsto f(\tau'). \tag{2.78}$$

Then we impose the transformation rules

$$\begin{aligned} \mathcal{G}\left(\tau, \tau'\right) &\longmapsto \left[f'(\tau) f'\left(\tau'\right)\right]^{\Delta} \mathcal{G}\left(f(\tau), f(\tau')\right) \\ \Sigma\left(\tau, \tau'\right) &\longmapsto \left[f'(\tau) f'\left(\tau'\right)\right]^{\Delta(q-1)} \Sigma\left(f(\tau), f(\tau')\right) \end{aligned} \tag{2.79}$$

where $'$ denotes derivative with respect to its argument. We can already see that the IR limit is important if the theory needs to have a diffeomorphism (or, equivalently conformal) symmetry because the kinetic contribution in the first term in Eq. (2.32) would have violated the aforementioned transformations. Since the kinetic contributions are negligible, the first term satisfies the time re-parameterization (see Ref. [11] for a detailed discussion). The second term becomes

$$\begin{aligned} &\frac{1}{2} \iint d\tau d\tau' \left(\Sigma(\tau, \tau')\mathcal{G}(\tau, \tau') - \frac{1}{q} J_q^2 \mathcal{G}(\tau, \tau')^q\right) \\ &\quad \longmapsto \frac{1}{2} \iint \underbrace{\frac{1}{\left|\frac{df}{d\tau}\right| \left|\frac{df}{d\tau'}\right|}}_{\text{Jacobian}} df(\tau) df(\tau') \left[\left(\left|\frac{df}{d\tau}\right| \left|\frac{df}{d\tau'}\right|\right)^{\Delta(q-1)} \left(\left|\frac{df}{d\tau}\right| \left|\frac{df}{d\tau'}\right|\right)^{\Delta}\right. \\ &\qquad \times \Sigma(f(\tau), f(\tau'))\mathcal{G}(f(\tau), f(\tau')) \\ &\qquad \left. - \frac{J^2}{q} \left(\left|\frac{df}{d\tau}\right| \left|\frac{df}{d\tau'}\right|\right)^{\Delta q} \mathcal{G}\left(f(\tau), f(\tau')\right)^q\right] \\ &\quad = \frac{1}{2} \iint df(\tau) df(\tau') \left[\Sigma(f(\tau), f(\tau'))\mathcal{G}(f(\tau), f(\tau')) - \frac{J^2}{q} \mathcal{G}\left(f(\tau), f(\tau')\right)^q\right] \\ &\qquad \times \left(\text{for } \Delta = \frac{1}{q}\right) \end{aligned} \tag{2.80}$$

where the integration variable are like dummy indices which we can replace by any other variable, so by switching back to $f(\tau) \to \tau$ and $f(\tau') \to \tau'$, we get the original effective action. Therefore, the action possesses a Diff($\mathbb{R}$) symmetry which for one spatio-temporal dimensional system implies a conformal symmetry (Eq. (2.79) with $\boxed{\Delta = 1/q}$).

Having verified that the effective action has the conformal symmetry, accordingly the associated Euler-Lagrange equations (the Schwinger-Dyson equations which can be explicitly verified too at the level of Eq. (2.75) where the second equation holds true $\forall\ \Delta$ and the first equation, namely the Dyson's equation imposes $\Delta = 1/q$) inherit the same symmetry[18] and we say that the theory in possesses the conformal symmetry in the IR limit.

2.5.3 General Form of Conformal Green's Function

Even though the Schwinger-Dyson equations have the conformal symmetry in the IR limit, we will see now that their solutions (i.e., the Green's functions) does not possess the full symmetry, but is invariant only under the subgroup of Conf($\mathbb{R}$) $\cong$ Diff($\mathbb{R}$) (in 1D), namely SL(2, $\mathbb{R}$)[19] (more on this below). This spontaneous symmetry breaking of the full symmetry to a smaller symmetric group is crucial to the SYK model and lead to Goldstone modes type situations.

With the benefit of hindsight, we know that the solution of the Schwinger-Dyson equations in the IR limit (Eq. (2.75)), namely the (two-point) Green's function does not possess the full symmetry of Schwinger-Dyson equations (Diff($\mathbb{R}$) which in 1D is locally congruent to Conf($\mathbb{R}$)). There is a spontaneous symmetry breaking to a subgroup of Diff($\mathbb{R}$), namely SL(2, $\mathbb{R}$) which generates global conformal transformations (translations, dilations, and special conformal transformations) on $\mathbb{R}$. These transformations are collectively called *Möbius transformations* which are smooth, invertible, and preserve the structure of $\mathbb{R}$. However, Diff($\mathbb{R}$) is infinite-dimensional (containing all smooth reparameterizations), while SL(2, $\mathbb{R}$) is a 3-dimensional subgroup corresponding to global conformal symmetry. They

[18] If the action $S[\phi]$ possesses a symmetry (i.e., invariance under a transformation of fields/coordinates), the Euler-Lagrange equations derived from it will necessarily inherit that symmetry. The reason is that the equations of motion are derived by extremizing the action. If the action is invariant under a transformation, the extremization process (variational principle) respects this invariance, leading to symmetric equations. However, the converse is not true (it is not a sufficient condition). The Euler-Lagrange equations may exhibit a symmetry even if the action does not. A trivial example is to consider the action $S = \int \left(\dot{x}^2 + x + f(t)\right) dt$, where $f(t)$ breaks time-translation symmetry (for time translation $t \to t+a$, $f(t+a) \neq f(t)$). However, the Euler-Lagrange equation $2\ddot{x} - 1 = 0$ is time-translational symmetric. Finally, even if the Euler-Lagrange equations having symmetry *does not guarantee* that their solutions will inherit that symmetry.

[19] Strictly, it's the Möbius group PSL(2, $\mathbb{R}$) $\cong$ SL(2, $\mathbb{R}$)/$\{\pm I\}$ where I is the 2×2 identity matrix. However, the distinction between SL(2, $\mathbb{R}$) and PSL(2, $\mathbb{R}$) is irrelevant for physical observables (e.g., Green's functions $G(\tau)$) since $-I$ acts trivially, therefore we will continue the convention found in physics literature of using SL(2, $\mathbb{R}$) instead of PSL(2, $\mathbb{R}$).

consists of real matrices of the form

$$\begin{pmatrix} a & b \\ c & d \end{pmatrix} \tag{2.81}$$

where $a, b, c, d \in \mathbb{R}$ and $ad - bc = 1$. The Green's function (solutions of Schwinger-Dyson equations) in the IR limit inherit this (smaller) symmetry. We will figure out the most general form of the conformal Green's function in this section.

We start with listing the transformations that constitute the SL$(2, \mathbb{R})$ group, namely the Möbius transformations, that, in general, preserve angles locally, making them conformal maps. In 1D, there are no angles, so this holds trivially. A general Möbius transformation is[20]

$$\tau \mapsto \frac{a\tau + b}{c\tau + d}, \qquad \text{where } a, b, c, d \in \mathbb{R} \text{ and } ad - bc = 1. \tag{2.82}$$

This transformation can be decomposed into the following translations, dilations, and special conformal transformations, which generate SL$(2, \mathbb{R})$:

- Translations: $\tau \mapsto \tau + a$, with matrix: $\begin{pmatrix} 1 & a \\ 0 & 1 \end{pmatrix} \in$ SL$(2, \mathbb{R})$. The physical meaning is to shift τ by a constant a.
- Dilations:[21] $\tau \mapsto \lambda\tau$, with matrix: $\begin{pmatrix} \sqrt{\lambda} & 0 \\ 0 & 1/\sqrt{\lambda} \end{pmatrix} \in$ SL$(2, \mathbb{R})$. This means rescaling τ by λ, preserving $\det = 1$.
- Special Conformal Transformations: $\tau \mapsto \frac{\tau}{1+b\tau}$, with matrix: $\begin{pmatrix} 1 & 0 \\ b & 1 \end{pmatrix} \in$ SL$(2, \mathbb{R})$. Physically, this implies inverting and shifting τ, akin to a "boost" in 1D.

where a, λ and b are real constants.

Accordingly, any element of SL$(2, \mathbb{R})$ can be expressed as a product of these three transformations. For example:

$$\begin{pmatrix} a & b \\ c & d \end{pmatrix} = \begin{pmatrix} 1 & 0 \\ c/a & 1 \end{pmatrix} \begin{pmatrix} \sqrt{a} & 0 \\ 0 & 1/\sqrt{a} \end{pmatrix} \begin{pmatrix} 1 & b/a \\ 0 & 1 \end{pmatrix},$$

[20] It's easy to see the justification behind footnote 19 as $\{a, b, c, d\} \longmapsto \{-a, -b, -c, -d\}$, neither Eqs. (2.81) nor (2.82) change, that's why strictly speaking, there is a PSL$(2, \mathbb{R}) \cong$ SL$(2, \mathbb{R})/\{\pm I\}$ symmetry instead of SL$(2, \mathbb{R})/\{\pm I\}$. However, as stated in footnote 19, the difference is irrelevant for physical observables, that's why we will continue to use the terminology SL$(2, \mathbb{R})/\{\pm I\}$.

[21] A dilation is a function $f : N \to N$ where N is a metric space such that $d(f(x), f(y)) = r d(x, y)$ $\forall x, y \in N$ where $d(x, y)$ is the distance between x & y and r is some positive real number.

assuming $a \neq 0$. This decomposition shows that

- Translations $\tau \longmapsto \tau + b/a$ (upper-triangular matrices);
- Dilations $\tau \longmapsto a\tau$ (diagonal matrices);
- Special conformal transformations $\tau \longmapsto \frac{\tau}{1+\frac{c}{a}\tau}$ (lower-triangular matrices),

generate the full group SL(2, $\mathbb{R}$).

Let's consider a generic two-point function $\langle \psi_1(\tau_1)\psi_2(\tau_2)\rangle$. We will impose the aforementioned three fundamental transformations that form the basis for SL(2, $\mathbb{R}$ and find the most general form of the two-point functions that respects SL(2, $\mathbb{R}$ symmetry.

- Translations: This immediately constraints the form to a time-translational invariant function

$$\langle \psi_1(\tau_1)\psi_2(\tau_2)\rangle = f(|\tau_1 - \tau_2|), \tag{2.83}$$

where f is some (unknown) function.
- Dilations: They imply $\psi(\lambda\tau_i) = \lambda^{-\Delta_i}\psi(\tau)$ for some Δ_i. This implies

$$\begin{aligned} &\langle \psi_1(\lambda\tau_1)\psi_2(\lambda\tau_2)\rangle = \lambda^{-\Delta_1-\Delta_2}\langle \psi_1(\tau_1)\psi_2(\tau_2)\rangle \\ &\Rightarrow \langle \psi_1(\tau_1)\psi_2(\tau_2)\rangle = \lambda^{+\Delta_1+\Delta_2}\langle \psi_1(\lambda\tau_1)\psi_2(\lambda\tau_2)\rangle. \end{aligned} \tag{2.84}$$

But the left-hand side = $f(|\tau_1 - \tau_2|)$, so in order to satisfy this condition, we need

$$\langle \psi_1(\tau_1)\psi_2(\tau_2)\rangle = f(|\tau_1 - \tau_2|) = \frac{B}{|\tau_1 - \tau_2|^{\Delta_1+\Delta_2}} \tag{2.85}$$

for some scalar B.
- Special Conformal Transformations: $\tau \longmapsto \tau' = \frac{\tau}{1+b\tau}$ implies

$$|\tau_1' - \tau_2'| = \left|\frac{\tau_1}{1+b\tau_1} - \frac{\tau_2}{1+b\tau_2}\right| = \left|\frac{\tau_1 - \tau_2}{(1+b\tau_1)(1+b\tau_2)}\right|. \tag{2.86}$$

The Jacobian for this transformation is

$$\left|\frac{\partial \tau_i'}{\partial \tau_i}\right| = \frac{1}{(1+b\tau_i)^2} \text{ for } i = 1, 2. \tag{2.87}$$

Therefore, we have the form for the Green's function

$$
\begin{aligned}
f\left(|\tau_1-\tau_2|\right) &= \left|\frac{\partial\tau_1'}{\partial\tau_1}\right|^{\Delta_1}\left|\frac{\partial\tau_2'}{\partial\tau_2}\right|^{\Delta_2} f\left(\left|\tau_1'-\tau_2'\right|\right) \\
&= \left|\frac{1}{1+b\tau_1}\right|^{2\Delta_1}\left|\frac{1}{1+b\tau_2}\right|^{2\Delta_2}\left|\frac{(1+b\tau_1)(1+b\tau_2)}{\tau_1-\tau_2}\right|^{\Delta_1+\Delta_2} B \\
&\overset{!}{=} \frac{B}{|\tau_1-\tau_2|^{2\Delta}} \quad \text{(if and only if } \Delta_1=\Delta_2=\Delta \text{ for some } \Delta\text{)}.
\end{aligned}
\tag{2.88}
$$

So we have found a constraint on the exponents Δ_i for the two-point function to be SL(2, $\mathbb{R}$) symmetric.

Hence, we conclude the most general form of (conformal) two-point function with SL(2, $\mathbb{R}$) symmetry is given by

$$
\boxed{\langle\psi_1(\tau_1)\psi_2(\tau_2)\rangle = f\left(|\tau_1-\tau_2|\right) = \frac{B}{|\tau_1-\tau_2|^{2\Delta}}}
\tag{2.89}
$$

for some scalar B and exponent Δ.

2.5.4 Solving the Schwinger-Dyson Equations in the IR Limit

We solve the Schwinger-Dyson equations in the IR limit, given by Eq. (2.75) which we reproduce here for convenience, for the corresponding Green's function $\mathcal{G}_c(\tau-\tau')$.

$$
\int d\tau''\mathcal{G}(\tau,\tau'')\Sigma(\tau'',\tau') = -\delta(\tau-\tau'), \quad \Sigma(\tau,\tau') = J_q^2\mathcal{G}(\tau,\tau')^{q-1}.
\tag{2.90}
$$

We have put the subscript c to denote the conformal (IR) limit and already imposed the time-translational invariance emerging from the condition for translations above. We impose the ansatz of the most general conformal Green's function that we derived above. The ansatz will lead to a set of consistency relations and if they resolve smoothly, we have found our solution and explicitly shown that the solutions break the symmetry of Conf($\mathbb{R}$) of the Schwinger-Dyson equations to the subgroup SL(2, $\mathbb{R}$) (global conformal transformations). The ansatz is (we have used τ for $\tau-\tau'$)

$$
\boxed{\mathcal{G}_c(\tau) = \frac{b}{|\tau|^{2\Delta}}\mathrm{sgn}(\tau)}
\tag{2.91}
$$

where sgn(τ) enters our ansatz due to the form of the free Green's function that we derived in Eq. (2.46). As we can see, sgn(τ) makes the ansatz almost conformal symmetric, but not exact.

For $\omega > 0$, we have the Fourier transform[22]

$$\mathcal{G}_c(\omega) = b\int_{-\infty}^{+\infty} d\tau \frac{\text{sgn}(\tau)}{|\tau|^{2\Delta}} e^{\imath\omega\tau} = b\int_0^{\infty} d\tau \frac{e^{\imath\omega\tau}}{\tau^{2\Delta}} - \underbrace{\int_{-\infty}^{0} d\tau \frac{e^{\imath\omega\tau}}{|\tau|^{2\Delta}}}_{\tau\to-\tau' \text{ where } \tau'>0}$$

$$= b\int_0^{\infty} d\tau \frac{e^{\imath\omega\tau}}{\tau^{2\Delta}} - \int_0^{\infty} d\tau' \frac{e^{-\imath\omega\tau'}}{\tau'^{2\Delta}}$$

$$= b\int_0^{\infty} d\tau \frac{e^{\imath\omega\tau} - e^{-\imath\omega\tau}}{\tau^{2\Delta}} \qquad (\tau' \text{ is a dummy variable, so we can } \tau' \to \tau)$$

$$= 2\imath b\int_0^{\infty} d\tau \frac{\sin(\omega\tau)}{\tau^{2\Delta}} = 2\imath b \,\text{Im}\left\{\int_0^{\infty} d\tau \frac{e^{\imath\omega\tau}}{\tau^{2\Delta}}\right\} \tag{2.92}$$

where Im[...] denotes the imaginary part of the argument. Now we substitute $\tau = \imath t/\omega$ and $\Delta = (1-x)/2$ to get for the measure $d\tau = \imath dt/\omega$ and $\frac{1}{\tau^{2\Delta}} = (\frac{\imath}{\omega})^{-2\Delta} t^{-2\Delta}$. We have

$$\mathcal{G}_c(\omega) = 2\imath b\text{Im}\left\{\left(\frac{\imath}{\omega}\right)^{1-2\Delta}\int_0^{\infty} t^{-2\Delta}e^{-t}dt\right\} = 2\imath b\text{Im}\left\{\left(\frac{\imath}{\omega}\right)^{1-2\Delta}\underbrace{\int_0^{\infty} t^{x-1}e^{-t}dt}_{\equiv\Gamma(x)=\Gamma(1-2\Delta)}\right\}$$

$$= 2\imath b\text{Im}\left\{\left(\frac{\imath}{\omega}\right)^{1-2\Delta}\Gamma(1-2\Delta)\right\} = 2\imath b\frac{\Gamma(1-2\Delta)}{\omega^{1-2\Delta}}\text{Im}(\imath^{1-2\Delta}), \tag{2.93}$$

where we identified the definition of Γ-function which is a real function for real inputs ($1 - 2\Delta$ is real) and that's why we took it out from the Im[...] alongside the factor of $1/\omega^{1-2\Delta}$. Next

$$\text{Im}\left(\imath^{1-2\Delta}\right) = \text{Im}\left\{e^{\imath\frac{\pi}{2}(1-2\Delta)}\right\} = \text{Im}\left\{\imath e^{-\imath\pi\Delta}\right\} = \cos(-\pi\Delta) = \cos(\pi\Delta).$$

Finally generalizing to $\omega < 0$ by repeating the same steps as above, finally yields the form of conformal Green's function up to two unknown constants b and Δ, which we will figure out later:

$$\boxed{\mathcal{G}_c(\omega) = 2\imath b\cos(\pi\Delta)\frac{\Gamma(1-2\Delta)}{|\omega|^{1-2\Delta}}\text{sgn}(\omega)}. \tag{2.94}$$

We use this to substitute in the second Schwinger-Dyson equations in Eq. (2.75) to get (using the ansatz in Eq. (2.91))

[22] Recall the footnote 11.

$$\boxed{\Sigma_c(\tau) = J_q^2 \mathcal{G}_c^{q-1} = J_q^2 \frac{b^{q-1}}{|\tau|^{2\Delta(q-1)}} \operatorname{sgn}(\tau)}, \tag{2.95}$$

where subscript c denotes conformal limit and we used $\operatorname{sgn}(\tau)^{q-1} = \operatorname{sgn}(\tau)$ because q is always taken to be even. Repeating the steps for $\mathcal{G}_c$ for Σ for both $\omega > 0$ and $\omega < 0$, we find the conformal self-energy up to two unknown constants b and Δ as follows

$$\boxed{\Sigma_c(\omega) = 2\imath J_q^2 b^{q-1} \cos(\pi\Delta(q-1)) \frac{\Gamma(1-2\Delta(q-1))}{\omega^{1-2\Delta(q-1)}}}. \tag{2.96}$$

Finally, we solve for b using the IR limit of Dyson's equation in Fourier space, namely $\mathcal{G}_c(\omega)\Sigma_c(\omega) = -1$ (see Eq. (2.74)) where we plug the form of $\mathcal{G}_c(\omega)$ and $\Sigma_c(\omega)$ from above to get

$$-4J_q^2 b^q \cos\left(\pi - \frac{\pi}{q}\right)\cos\left(\frac{\pi}{q}\right)\Gamma\left(-\left(1-\frac{2}{q}\right)\right)\Gamma\left(1-\frac{2}{q}\right) = -1. \tag{2.97}$$

However, we have the identity of Γ-function: $\Gamma(z)\Gamma(-z) = -\frac{\pi}{z\sin(\pi z)}$ which can simplify the left-hand side by identifying $z = 1 - 2/q$ to get

$$\frac{4\pi J_q^2 b^q \cos\left(\pi - \frac{\pi}{q}\right)\cos\left(\frac{\pi}{q}\right)}{\left(1-\frac{2}{q}\right)\sin\left(\pi - \frac{2\pi}{q}\right)} = -1 \tag{2.98}$$

which we further simplify using trigonometric identities $\cos\left(\frac{\pi}{q}\right) = -\cos(\pi/q)$, $\sin\left(\pi - \frac{2\pi}{q}\right) = \sin\left(\frac{2\pi}{q}\right)$ and $\sin\left(\frac{2\pi}{q}\right) = 2\cos(\pi/q)\sin(\pi/q)$ to get

$$\begin{gathered} \frac{-\pi J^2 b^q}{\left(\frac{1}{2}-\frac{1}{q}\right)\tan\left(\frac{\pi}{q}\right)} = -1 \\ \Rightarrow \boxed{b = \left(\frac{1}{\pi J^2}\left(\frac{1}{2}-\frac{1}{q}\right)\tan\left(\frac{\pi}{q}\right)\right)^{1/q}}. \end{gathered} \tag{2.99}$$

Finally, we find Δ. We use dimensional analysis to figure this out. In terms of energy dimension, $[\tau] = -1$ while any unit of energy, say $[\omega] = +1$. Accordingly from Eqs. (2.94) and (2.96), we get[23]

[23] Note that in imaginary-time domain (Eqs. (2.91) and (2.95)), $\mathcal{G}_c(\tau) \sim \tau^{-2\Delta} \Rightarrow [\mathcal{G}_c(\tau)] = 2\Delta$. Similarly $\Sigma_c(\tau) \sim \tau^{-2\Delta(q-1)} \Rightarrow [\Sigma_c(\tau)] = 2\Delta(q-1)$.

$$\begin{aligned} \mathcal{G}_c(\omega) &\sim \omega^{2\Delta-1} &&\Rightarrow [\mathcal{G}_c(\omega)] = 2\Delta - 1, \\ \Sigma_c(\omega) &\sim \omega^{2\Delta(q-1)-1} &&\Rightarrow [\Sigma_c(\omega)] = 2\Delta(q-1) - 1. \end{aligned} \tag{2.100}$$

Then using the Dyson's equation in frequency space in the IR limit (Eq. (2.74)), we have $\mathcal{G}_c(\omega)\Sigma_c(\omega) = -1$. The right-hand side has zero energy dimension, so the left-hand side must also have $[\mathcal{G}_c(\omega)\Sigma_c(\omega)] = 0$. This gives

$$2\Delta - 1 + 2\Delta(q-1) - 1 = 0 \quad \Rightarrow \boxed{\Delta = \frac{1}{q}}. \tag{2.101}$$

This justifies using the symbol Δ in the ansatz Eq. (2.91) because we found the same result while imposing Diff($\mathbb{R}$) ($\cong$ Conf($\mathbb{R}$)) symmetry in Eq. (2.80). Since $q = 4$ model is heavily studied, we present the result for the conformal Green's function as

$$\mathcal{G}_c(\tau) = -\left(\frac{1}{4\pi J_4^2}\right)^{1/4} \frac{1}{\sqrt{\tau}}\,\mathrm{sgn}(\tau). \tag{2.102}$$

Summary: We have found exact solutions for arbitrary q for Green's function and the self-energy in the IR limit where the action has a conformal symmetry and the solutions have a spontaneously broken emergent (global) conformal symmetry, namely SL(2, $\mathbb{R}$).[24] The exact solutions (both in imaginary-time and Matsubara/-Fourier plane) in the thermodynamic limit (since N is taken to infinity) are provided in Eqs. (2.91), (2.94), (2.95) and (2.96) with constants b and Δ found in Eqs. (2.99) and (2.101), respectively.

2.5.5 A Brief Note on Interpolation to Finite Temperature*

All results that we derived above are for temperature $T \to 0$. Generalizing to finite temperature is particularly straightforward for conformal systems. Partition function of d-dimensional quantum system at finite temperature is same as the Euclidean quantum field theory at $T = 0$ in $d+1$-dimensions with Euclidean time τ being periodic or identified as $\tau \sim \tau + \beta$. As we have seen, the $0+1$-dimensional SYK model is conformal at $T \to 0$ limit where $-\infty < \tau < +\infty$ and this can be related to finite temperatures via time reparameterization as considered in Eqs. (2.78) and (2.79). The particular choice of function f is given by (β is the

[24] If we consider leading order corrections to $\mathcal{G}_c$ by solving the full Schwinger-Dyson equations (without taking the IR limit) and expand perturbatively in the IR (low temperatures $\beta J_q \gg 1$) to take into account the (UV at high temperatures $\beta J_q \ll 1$) leading order correction to the IR's emergent conformal symmetry, we find that the Green's function at leading order does not remain conformal anymore and violates SL(2, $\mathbb{R}$).

inverse temperature) [11]

$$\tau \longmapsto f(\tau) = \frac{\beta}{\pi} \sin\left(\frac{\pi \tau}{\beta}\right) \tag{2.103}$$

to get (using the ansatz Eq. (2.91) and $\Delta = 1/q$ from Eq. (2.101))

$$\mathcal{G}_{c,\beta}(\tau) = b \left[\frac{\pi}{\beta \sin\left(\frac{\pi \tau}{\beta}\right)} \right]^{\frac{2}{q}} \operatorname{sgn}(\tau), \tag{2.104}$$

where $0 \leq \tau < \beta$ and $\mathcal{G}$ satisfies the anti-commutation relations (see Appendix A). Here b is given by Eq. (2.99). Therefore the scaling transformation in Eq. (2.103) maps $\mathbb{R} \to S^1$.

NOTE: This technique is strictly valid if conformal symmetry is exact. However, as we have seen, the SYK model has emergent conformal symmetry in its Green's function which is not exact and only becomes exact as $T \to 0$ and N is taken to be large. Accordingly, the result in Eq. (2.104) is not exact but an approximate one that works well for very low temperatures only. We will consider exact finite temperature results later when we will take large-q limit. That's why we only briefly mentioned about this technique without going in further details.

2.6 Effective Schwarzian Action

We showed above that the solutions of the Schwinger-Dyson equations in the IR limit, i.e. the 2-point functions $\mathcal{G}_c(\tau)$, breaks the symmetry (but not completely) of Diff($\mathbb{R}$) ($\cong$ Conf($\mathbb{R}$) in 1D) to SL(2, $\mathbb{R}$) $\subset$ Diff($\mathbb{R}$). This spontaneously symmetry breaking leads us with (infinite) degrees of freedom (one for each broken generator) that are massless (gapless in the sense of having no energy gap and power-law decay instead of exponential decay) Goldstone modes in the IR. Accordingly, the Goldstone modes are soft temporal (no spatial dimension to propagate) fluctuations whose governing action (the so-called *Schwarzian action*) we intend to derive in this section.[25] Note that the Goldstone modes are not bosonic particles but collective reparameterizations of time $f(\tau)$ (see Eq. (2.78)). They are "massless" in the sense that their effective action (Schwarzian) has no energy gap (gapless), but they do not propagate like particles.

In the IR regime of the SYK model, we have seen that an approximate conformal symmetry emerges for the Green's function (Eq. (2.91)), marked by invariance

[25] We mention in passing that there have been constructions of coupled SYK models with two independent Majorana fields that do not admit Schwarzian action dominance, see [13]. However these are beyond the scope of this review as we focus on the same species of the Majorana fields to build our SYK-like models throughout this work, as is also the case throughout this section.

under smooth time reparameterizations $\tau \to f(\tau)$. However, this symmetry is not perfect—it is subtly disrupted by quantum and thermal fluctuations. These disruptions lead to a low-energy description dominated by the *Schwarzian derivative*, a geometric tool that measures how far a transformation deviates from "simple" Möbius transformations. As seen above in Eqs. (2.83), (2.84) and (2.86), Möbius transformations include three basic time transformations: translations ($\tau \to \tau + a$), dilations ($\tau \to \lambda\tau$), and special conformal transformations ($\tau \to \frac{t}{1+bt}$), all of which leave the Schwarzian derivative unchanged (it vanishes for these three cases). While the conformal fixed point remains formally invariant under arbitrary $f(\tau)$, fluctuations around this configuration break the symmetry. The resulting dynamics of the soft mode $f(\tau)$, which encodes these deviations, are governed by the Schwarzian action, capturing the residual low-energy behavior of the system. These Goldstone modes dominate the low-energy physics of the SYK model. They encode the soft, nearly conformal fluctuations around the saddle-point solution (the conformal Green's function). As an example, the Schwarzian action leads to the quantum Lyapunov exponent (introduced later) to have the value $2\pi T$ which saturates the Maldacena-Shenker-Stanford (MSS) bound of quantum chaos [12], thereby being referred to as "maximally chaotic".

2.6.1 Properties of the Effective Schwarzian Action

We list down the properties of the effective Schwarzian action, following which we will derive an explicit form. The collective reparameterizations of time $f(\tau)$ represent the Goldstone modes, fluctuating in time whose behavior is governed by the Schwarzian action $S_{\text{Sch}} = S_{\text{Sch}}[f]$.

We know that if $f \in \text{SL}(2, \mathbb{R})$, then the most general form of $f(\tau)$ is given in Eq. (2.82), namely $f(\tau) = \frac{a\tau+b}{c\tau+d}$ such that $ad - bc = 1$ for $a, b, c, d \in \mathbb{R}$. Then the Schwarzian action is identically zero

$$f(\tau) \in \text{SL}(2, \mathbb{R}) \quad \Rightarrow S_{\text{Sch}}[f] = 0. \tag{2.105}$$

If $f \notin \text{SL}(2, \mathbb{R})$, then the transformed field f should have the same dynamics as the original, leading to

$$f(\tau) \notin \text{SL}(2, \mathbb{R}) \quad \Rightarrow \delta S_{\text{Sch}}[f] = 0 \quad \text{for } f(\tau) \longmapsto \mathcal{F}[f] = \frac{af(\tau)+b}{cf(\tau)+d} \tag{2.106}$$

where $ad - bc = 1$ for $a, b, c, d \in \mathbb{R}$.

Therefore we have to find $S_{\text{Sch}}[f]$ such that the conditions in Eqs. (2.105) and (2.106) are satisfied. Accordingly, we need a combination of derivatives of $\mathcal{F}[f]$, defined in Eq. (2.106), that yields the same functional form for $S_{\text{Sch}}[f]$ if we rename $\mathcal{F} \to f$.

2.6.2 Finding the Effective Schwarzian Action

We start with the definition of $\mathcal{F}[f] = \frac{af(\tau)+b}{cf(\tau)+d}$ (where $f(\tau) \notin \mathrm{SL}(2,\mathbb{R})$, $a, b, c, d \in \mathbb{R}$ and $ad - bc = 1$) and take its derivative to get ($'$ denotes derivative with respect to τ, namely $' = \frac{d}{d\tau}$)

$$\begin{aligned}
\mathcal{F}' &= \frac{(cf+d)af' - (af+b)cf'}{(cf+d)^2} = \frac{f'}{(cf+d)^2} \quad (\because ad - bc = 1) \\
\Rightarrow \mathcal{F}'' &= \frac{f''}{(cf+d)^2} - \frac{2c\left(f'\right)^2}{(cf+d)^3} \\
\Rightarrow \mathcal{F}''' &= \frac{f'''}{(cf+d)^2} - \frac{6cf'f''}{(cf+d)^3} + \frac{6c^2f'^3}{(cf+d)^4}.
\end{aligned} \tag{2.107}$$

Then the only combination of derivatives of $\mathcal{F}$ such that $\delta S_{\mathrm{Sch}}[f]$ for $f \to \mathcal{F}$ where $f \notin \mathrm{SL}(2,\mathbb{R})$ is

$$\boxed{f \notin \mathrm{SL}(2,\mathbb{R}) \quad \Rightarrow \frac{\mathcal{F}'''}{\mathcal{F}'} - \frac{3}{2}\left(\frac{\mathcal{F}''}{\mathcal{F}'}\right)^2 = \frac{f'''}{f'} - \frac{3}{2}\left(\frac{f''}{f'}\right)^2}. \tag{2.108}$$

Thus condition (2.106) is satisfied. With this, we can also immediately see that the condition (2.105) is satisfied: if $f \in \mathrm{SL}(2,\mathbb{R})$, then $f(\tau) = \frac{m\tau+n}{o\tau+p}$ (using Eq. (2.82)) where $m, n, o, p \in \mathbb{R}$ and $mp - on = 1$, leads to the Schwarzian derivative to vanish (by explicitly plugging the form of $f(\tau)$):

$$\boxed{f \in \mathrm{SL}(2,\mathbb{R}) \quad \Rightarrow \frac{f'''}{f'} - \frac{3}{2}\left(\frac{f''}{f'}\right)^2 = 0}. \tag{2.109}$$

This combination of derivatives satisfying the two requirements has a special name and it goes by *Schwarzian derivative*, which is denoted for any function $g(\tau)$ as $\{g, \tau\}$: ($'$ denotes derivative with respect to the argument, in this case τ)

$$\boxed{\{g,\tau\} \equiv \frac{g'''(\tau)}{g'(\tau)} - \frac{3}{2}\left(\frac{g''(\tau)}{g'(\tau)}\right)^2} \quad \text{(Schwarzian derivative)} \tag{2.110}$$

and the action governing the behavior of $g(\tau)$, the Schwarzian action $S_{\mathrm{Sch}}[g]$ depends on $\{g, \tau\}$:

$$\boxed{S_{\mathrm{Sch}}[g] = \gamma \int d\tau \{g, \tau\}} \quad \text{(Schwarzian action)} \tag{2.111}$$

where γ is the proportionality (normalizing) factor, given by (without proof, see Ref. [11])

$$\gamma = \frac{-\alpha_q N}{J_q}\sqrt{\frac{2^{q-1}}{q}} = \frac{-\alpha_q N}{\mathcal{J}_q} \quad (\mathcal{J}_q \equiv J_q\sqrt{\frac{q}{2^{q-1}}}). \tag{2.112}$$

Here N and J_q are the same model parameters as used in the SYK Hamiltonian (e.g., Eq. (2.1)). Physically, the Schwarzian action describes the fluctuations around $T = 0$ ground state. Since these calculations are based for temperature $T \to 0$ limit, if we wish to deal with the free energy, we must generalize this to finite T.

2.6.3 The Effective Schwarzian Action at Finite Temperature*

We again make use of diffeomorphism symmetry and the emergent conformal symmetry SL(2, $\mathbb{R}$) in the IR limit to extrapolate to finite temperature via (see Sect. 2.5.5)

$$\tau \longmapsto \mathcal{F}[f] = e^{2\pi\imath f(\tau)/\beta}, \tag{2.113}$$

where we employ the following Schwarzian identity whose proof is provided in Appendix C

$$\boxed{\left\{f(g(\tau)), \tau\right\} = \left(g'(\tau)\right)^2 \{f, g\} + \{g, \tau\}}. \tag{2.114}$$

Then the Schwarzian action gives (subscript β denotes finite temperature)

$$S_{\text{Sch}}[\mathcal{F}]_\beta = -\frac{\alpha_q N}{\mathcal{J}_q}\int d\tau\{\mathcal{F}, \tau\} = -\frac{\alpha_q N}{\mathcal{J}_q}\int d\tau\left\{e^{2\pi\imath f(\tau)/\beta}, \tau\right\}. \tag{2.115}$$

Applying the Schwarzian identity gives

$$\begin{aligned}
\left\{e^{2\pi\imath f(\tau)/\beta}, \tau\right\} &= (f'(\tau))^2\{e^{2\pi\imath f/\beta}, f\} + \{f, \tau\} \\
&= (f'(\tau))^2\left[\frac{\frac{d^3 e^{2\pi\imath f/\beta}}{df^3}}{\frac{de^{2\pi\imath f/\beta}}{df}} - \frac{3}{2}\left(\frac{\frac{d^2 e^{2\pi\imath f/\beta}}{df^2}}{\frac{de^{2\pi\imath f/\beta}}{df}}\right)^2\right] + \{f, \tau\} \\
&= (f'(\tau))^2\left[\frac{(2\pi\imath/\beta)^3}{(2\pi\imath/\beta)} - \frac{3}{2}\left(\frac{(2\pi\imath/\beta)^2}{(2\pi\imath/\beta)}\right)^2\right] + \{f, \tau\} \\
&= -(f'(\tau))^2\imath^2\frac{1}{2}\left(\frac{2\pi}{\beta}\right)^2 + \{f, \tau\} = (f'(\tau))^2\frac{1}{2}\left(\frac{2\pi}{\beta}\right)^2 + \{f, \tau\}
\end{aligned} \tag{2.116}$$

to finally get the Schwarzian action at finite temperature $1/\beta$

$$\boxed{S_{\text{Sch}}[f]_\beta = -\frac{\alpha_q N}{\mathcal{J}_q}\int d\tau\left[\frac{1}{2}\left(\frac{2\pi}{\beta}\right)^2\left(f'(\tau)\right)^2 + \frac{f'''}{f'} - \frac{3}{2}\left(\frac{f''}{f'}\right)^2\right].} \tag{2.117}$$

This clearly boils down to Eq. (2.111) when $T \to 0$ (equivalently $\beta \to \infty$).

As an aside, we can perform integration by parts on the second term in the right-hand side while ignoring the boundary terms to get

$$\int d\tau \frac{f^{(''')}}{f'} = +\int d\tau \left(\frac{-1}{f'^2}\right) f'' \times f'' = -\int d\tau \left(\frac{f''}{f'}\right)^2, \tag{2.118}$$

which simplifies Eq. (2.117) to[26]

$$S_{\text{Sch}}[f]_\beta = +\frac{\alpha_q N}{2\mathcal{J}_q}\int d\tau\left[\left(\frac{f''}{f'}\right)^2 - \left(\frac{2\pi}{\beta}\right)^2 (f'(\tau))^2\right]. \tag{2.119}$$

Then we obtain the expression for free energy for the Schwarzian effective action F_{Sch} using Eq. (2.63) to get

$$\frac{\beta F_{\text{Sch}}}{N} = \frac{S_{\text{Sch}}[f]_\beta}{N}, \tag{2.120}$$

which can be used to study the thermodynamic properties of the Goldstone modes (collective time reparameterization $f(\tau)$) at finite temperatures.

2.6.4 Fluctuations at Finite Temperature and Hint of Holography*

We can vary the Schwarzian action at finite temperature in Eq. (2.119) for small reparameterizations of the form

$$f(\tau) = \tau + \epsilon(\tau) \tag{2.121}$$

where ϵ is small. Then we get for the Schwarzian action

$$S_\epsilon = \frac{\alpha_q N}{2\mathcal{J}_q}\int d\tau\left[\left(\frac{\epsilon''(\tau)}{1+\epsilon'(\tau)}\right)^2 - \left(\frac{2\pi}{\beta}\right)^2 (1+\epsilon'(\tau))^2\right]. \tag{2.122}$$

[26] There is another systematic way to derive the effective Schwarzian action using the 4-point correlation function. We refer the reader to Ref. [11] for a detailed exposition.

Keeping to the second order in ϵ, we get

$$\boxed{S_\epsilon = \frac{\alpha_q N}{2\mathcal{J}_q}\int d\tau \left[\left(\epsilon''(\tau)\right)^2 - \left(\frac{2\pi}{\beta}\right)^2\left(\epsilon'(\tau)\right)^2\right] + \mathcal{O}(\epsilon^3)}. \tag{2.123}$$

The near-horizon gravitational dynamics of extremal black holes in AdS (anti-de Sitter) spacetime is given by the same Schwarzian action when we map the dual theory of the strongly coupled SYK model via gauge-gravity duality to the IR limit [11, 14]. Strongly coupled systems described by holography flow to the same IR fixed point as the SYK model and that IR fixed point action is this Schwarzian.

2.7 The Large-q Limit

The Hamiltonian for $q/2$-body Hamiltonian is given in Eq. (2.1) where we solved the theory in large-N limit and obtained the Schwinger-Dyson equations in closed form in Eq. (2.33). We solved the Schwinger-Dyson equations in the IR limit in Sect. 2.5.4. Now, we will take another limit, namely the large-q limit in the $q/2$-body interaction which will allow us to solve for the Green's function across all temperatures (not just the IR limit). Note that the ordering of taking the limits matter (they don't commute): first the large-N limit needs to be taken, followed by the large-q limit. We reproduce the Schwinger-Dyson equations for arbitrary q for convenience

$$\mathcal{G}_0^{-1} = \mathcal{G}^{-1} + \Sigma, \qquad \Sigma = J_q^2\mathcal{G}^{q-1}$$

where we assume time-translations invariance (since we are considering equilibrium) such that $\mathcal{G}$ and Σ are functions of τ. We have solved for the equilibrium free Green's function in Eq. (2.46) where $\mathcal{G}_0(\tau) = \frac{1}{2}\mathrm{sgn}(\tau)$. We make the following ansatz for the Green's function *across all temperatures* in the large-q limit (not just the ansatz in Eq. (2.91) which holds in the IR limit/low temperature only)

$$\boxed{\mathcal{G}(\tau) = \frac{1}{2}\,\mathrm{sgn}(\tau)e^{g(\tau)/q} = \mathcal{G}_0(\tau)e^{g(\tau)/q}}, \tag{2.124}$$

where the exponential captures deviations from the free-fermion Green's function $\mathcal{G}_0(\tau) = \frac{1}{2}\,\mathrm{sgn}(\tau)$ and $g(\tau)$ is an even function: $g(-\tau) = g(\tau)$ (originating from $\mathcal{G}(-\tau) = -\mathcal{G}(\tau)$) and real: $g(\tau)^\star = g(\tau)$. Since g completely determines the Green's function $\mathcal{G}$, sometimes g is also referred to as the little Green's function, or simply the Green's function for brevity. Accordingly, using the Schwinger-Dyson equations, the self-energy becomes (recall that q is even, therefore $\mathrm{sgn}(\tau)^{q-1} =$

$\mathrm{sgn}(\tau))$

$$\Sigma(\tau) = J_q^2 \mathcal{G}^{q-1} = 2^{1-q} J_q^2 \,\mathrm{sgn}(\tau) e^{g(\tau)}. \tag{2.125}$$

By construction, we have $g = \mathcal{O}(q^0)$. Then we take the large-q limit and keep until orders of $1/q$ and ignore anything $\mathcal{O}(1/q^2)$ and above:

$$\mathcal{G}(\tau) = \frac{1}{2}\,\mathrm{sgn}(\tau)\Big[1 + \frac{g(\tau)}{q} + \mathcal{O}\left(\frac{1}{q^2}\right)\Big]. \tag{2.126}$$

The boundary condition for $g(\tau)$ comes the Majorana fermion anticommutation relations $\{\psi_i(\tau), \psi_j(\tau)\} = \delta_{ij}$. which implies for $i = j$:

$$\{\psi_i(0), \psi_i(0)\} = 2\psi_i(0)\psi_i(0) = 1 \quad \Rightarrow \quad \psi_i(0)\psi_i(0) = \psi_i^2(0) = \frac{1}{2}. \tag{2.127}$$

Taking average on both sides gives $\langle \psi_i(0)\psi_i(0)\rangle = 1/2$. This is an exact algebraic constraint from the fermionic operator algebra. It holds at any temperature and for any Hamiltonian. The Euclidean Green's function is defined as the time-ordered correlator

$$\mathcal{G}(\tau) = \frac{1}{N}\sum_{i=1}^{N} \langle T\psi_i(\tau)\psi_i(0)\rangle\,, = \begin{cases} \frac{1}{N}\sum_{i=1}^{N} \langle \psi_i(\tau)\psi_i(0)\rangle\,, & \tau > 0, \\ -\frac{1}{N}\sum_{i=1}^{N} \langle \psi_i(0)\psi_i(\tau)\rangle\,, & \tau < 0. \end{cases} \tag{2.128}$$

As $\tau \to 0^+$, the time ordering enforces

$$\mathcal{G}\left(0^+\right) = \frac{1}{N}\sum_{i=1}^{N} \langle \psi_i(0)\psi_i(0)\rangle = \frac{1}{N}\sum_{i=1}^{N} \frac{1}{2} = \frac{1}{N}\cdot N \cdot \frac{1}{2} = \frac{1}{2}. \tag{2.129}$$

Accordingly, from Eq. (2.124), we find the boundary condition for $g(\tau)$ (since $\mathrm{sgn}(\tau \to 0^+) = +1$):

$$\boxed{g(\tau = 0) = 0}. \tag{2.130}$$

The purpose is to find $g(\tau)$ that satisfies this boundary condition. Once found, we have solved the theory across all temperatures by calculating exactly the Green's function (accordingly, the self-energy) via Eqs. (2.124) and (2.125).

Moreover, we know that the Green's function in Eq. (2.124) is anti-periodic in τ (see Appendix A), namely $\mathcal{G}(\tau + \beta) = -\mathcal{G}(\tau)$. This ensures that $g(\tau)$ remains symmetric under imaginary-time translations[27]—a fact we are going to use later.

[27] We are dealing with Majorana fermions, things are going to generalize once we go to complex ($U(1)$ charged) fermions in Chap. 3.

2.7.1 Differential Equation for the Green's Function

We take the Fourier transform of Eq. (2.126) to get (recall footnote 11)

$$\begin{aligned} \mathcal{G}(\omega) &= \int d\tau \frac{1}{2}\,\text{sgn}(\tau)\left[1+\frac{g(\tau)}{q}\right] e^{\imath\omega\tau} \\ &= \frac{-1}{\imath\omega}\left[1-\frac{\imath\omega}{2q}F[\text{sgn}(\tau)g(\tau)]\right] = \text{Free} + \text{1st order} \end{aligned} \tag{2.131}$$

where $F[x]$ denotes the Fourier transform of the x and $F[\text{sgn}(\tau)] = -\frac{2}{\imath\omega}$.

Proof We have the definition of $\text{sgn}(\tau)$ as

$$\text{sgn}(\tau) = \begin{cases} 1 & \tau > 0, \\ -1 & \tau < 0. \end{cases}$$

Then the Fourier transform is given by

$$\mathcal{G}(\omega) = \int_{-\infty}^{\infty} d\tau e^{\imath\omega\tau}\,\text{sgn}(\tau) = \int_{0}^{\infty} d\tau e^{\imath\omega\tau} - \underbrace{\int_{-\infty}^{0} d\tau e^{\imath\omega\tau}}_{\tau\to-\tau'}.$$

This gives

$$\begin{aligned} \mathcal{G}(\omega) &= \int_0^{\infty} d\tau e^{\imath\omega\tau} - \int_0^{\infty} d\tau' e^{-\imath\omega\tau'} \\ &= \int_0^{\infty} \left(e^{\imath\omega\tau} - e^{-\imath\omega\tau}\right) d\tau = 2\imath \int_0^{\infty} \sin(\omega\tau) d\tau. \end{aligned}$$

We regulate with convergence factor of $e^{-\epsilon|\tau|}$ ($\epsilon > 0$) tp get

$$\mathcal{G}_{\epsilon}(\omega) = 2\imath \int_0^{\infty} \sin(\omega\tau) e^{-\epsilon\tau} d\tau$$

where the integral is given by

$$\int_0^{\infty} e^{-\epsilon\tau}\sin(\omega\tau)d\tau = \text{Im}\left[\int_0^{\infty} e^{-(\epsilon-\imath\omega)\tau}d\tau\right] = \text{Im}\left[\frac{1}{\epsilon-\imath\omega}\right] = \frac{\omega}{\epsilon^2+\omega^2}.$$

Accordingly, we finally get

$$\mathcal{G}(\omega) = \lim_{\epsilon\to 0^+} \mathcal{G}_{\epsilon}(\omega) = \lim_{\epsilon\to 0^+} \frac{2i\omega}{\epsilon^2+\omega^2} = \frac{2\imath}{\omega} = -\frac{2}{\imath\omega} \text{ for } \omega \neq 0.$$

□

Resuming from Eq. (2.131), we get the inverse

$$\begin{aligned}\mathcal{G}(\omega)^{-1} &= -\imath\omega\left[1 - \frac{\imath\omega}{2q}F[\mathrm{sgn}(\tau)g(\tau)]\right]^{-1} \\ &= -\imath\omega + \frac{\omega^2}{2q}F[\mathrm{sgn}(\tau)g(\tau)] + \mathcal{O}\left(\frac{1}{q^2}\right).\end{aligned} \tag{2.132}$$

We match against the exact second Dyson equation (not just the IR/low-temperature limit but exact across all temperatures), namely $\mathcal{G}(\omega)^{-1} = -\imath\omega - \Sigma(\omega)$ (see Eq. (2.73)) to get

$$\Sigma(\omega) = \frac{-\omega^2}{2q}F[\mathrm{sgn}(\tau)g(\tau)]. \tag{2.133}$$

whose inverse Fourier transform gives (recall $\partial_\tau \leftrightarrow -\imath\omega$)

$$\Sigma(\tau) = \partial_\tau^2\left[\frac{1}{2q}\,\mathrm{sgn}(\tau)g(\tau)\right]. \tag{2.134}$$

However from Eq. (2.125), we deduce that

$$\partial_\tau^2\left[\frac{1}{2q}\,\mathrm{sgn}(\tau)g(\tau)\right] = 2^{1-q}J_q^2\,\mathrm{sgn}(\tau)e^{g(\tau)}. \tag{2.135}$$

The presence of $\mathrm{sgn}(\tau)$ suggests caution about the sign for positive and negative τ, however by explicit calculations for both cases verify that the sign remains consistent for both $\tau < 0$ and $\tau > 0$, with a discontinuity at $\tau = 0$. Therefore, we can cancel $\mathrm{sgn}(\tau)$ on both sides and this leads us to the differential equation for $g(\tau)$

$$\boxed{\partial_\tau^2 g(\tau) = 2\mathcal{J}_q^2 e^{g(\tau)}} \quad (\mathcal{J}_q \equiv J_q\sqrt{\frac{q}{2^{q-1}}}), \tag{2.136}$$

where $\mathcal{J}_q$ first showed up in Eq. (2.112) while evaluating the Schwarzian action. This also brings us to the definition of the large-q limit: the $q \to \infty$ limit is taken such that the quantity $J_q^2\frac{q}{2^{q-1}}$ remains finitely constant. Recall that large-q limit only makes sense after large-N limit has been employed; both limits don't commute.

2.7.2 Solving for the Green's Function

We multiply both sides of Eq. (2.136) by $\frac{dg(\tau)}{d\tau}$ and integrate with respect to τ to get

$$\int \frac{dg(\tau)}{d\tau^2}\frac{d^2g(\tau)}{d\tau^2}d\tau = 2\mathcal{J}_q^2\int e^{g(\tau)}\frac{dg}{d\tau}d\tau, \tag{2.137}$$

where the left-hand side can be simplified by integration by parts

$$\text{Left-hand side} = \int \frac{dg}{d\tau}\frac{d^2g}{d\tau^2}d\tau = \left(\frac{dg}{d\tau}\right)\left(\frac{dg}{d\tau}\right) - \int \frac{d^2g}{d\tau^2}\frac{dg}{d\tau}d\tau$$
$$\Rightarrow \text{Left-hand side} = \int \frac{dg}{d\tau}\frac{d^2g}{d\tau^2}d\tau = \frac{1}{2}\left(\frac{dg}{d\tau}\right)^2. \tag{2.138}$$

The right-hand side is

$$\text{Right-hand side} = 2\mathcal{J}_q^2 \int e^{g(\tau)}\, \mathrm{d}g(\tau) = 2\mathcal{J}_q^2 \left(e^{g(\tau)} + c_1\right) \tag{2.139}$$

where c_1 is a constant of integration. Accordingly, Eq. (2.136) reduces to a first-order differential equation

$$\frac{1}{2}\left(\frac{dg}{d\tau}\right)^2 = 2\mathcal{J}_q^2\left(e^{g(\tau)} + c_1\right) \quad \Rightarrow \frac{dg}{d\tau} = 2\mathcal{J}_q\sqrt{e^{g(\tau)} + c_1}, \tag{2.140}$$

which can be integrated to

$$\int \frac{dg(\tau)}{\sqrt{e^{g(\tau)} + c_1}} = 2\mathcal{J}_q \int d\tau. \tag{2.141}$$

We consider two cases

- $c_1 > 0$: Eq. (2.141) integrates to[28] (c_2 is another constant of integration)

$$\frac{-1}{\sqrt{c_1}}\tanh^{-1}\left(\sqrt{\frac{e^{g(\tau)} + c_1}{c_1}}\right) = \mathcal{J}_q\left(\tau + c_2\right)$$
$$\Rightarrow e^{g(\tau)} = c_1\left[\tanh^2\left(\mathcal{J}_q\sqrt{c_1}\left(\tau + c_2\right)\right) - 1\right]. \tag{2.142}$$

  Plugging this in Eq. (2.124) gives us the full Green's function $\mathcal{G}(\tau)$. However, we immediately see the problem: the Green's function is not anti-periodic in τ when $\tau \to \tau + \beta$ (see Appendix A). In other words, $g(\tau)$ hereby obtained is not symmetric and also tanh is not periodic, thereby disallowing any periodicity. Therefore, $c_1 > 0$ cannot work.

[28] A software program such as Mathematica can help verify this, or just take the result and plug in Eq. (2.141) to check.

- $c_1 < 0$: We again integrate Eq. (2.141) to get

$$\begin{aligned}
&\frac{2}{\sqrt{-c_1}}\cot^{-1}\left(\frac{+\sqrt{-c_1}}{\sqrt{c_1+e^{g(\tau)}}}\right) = 2\mathcal{J}_q\left(\tau + c_2\right) \\
\Rightarrow &\frac{-c_1}{c_1+e^{g(\tau)}} = \cot^2\left(\sqrt{-c_1}\mathcal{J}_q\left(\tau+c_2\right)\right) \\
\Rightarrow &e^{g(\tau)} = -c_1\left[\tan^2\left(\sqrt{-c_1}\mathcal{J}_q\left(\tau+c_2\right)\right)+1\right].
\end{aligned} \tag{2.143}$$

We use the identity $\tan^2(x) + 1 = \frac{1}{\sin^2(x+\frac{\pi}{2})}$ where $x = \sqrt{-c_1}\mathcal{J}_q\left(\tau + c_2\right)$ and substitute $(-c_1) = \left(\frac{c}{\mathcal{J}_q}\right)^2 > 0$ and $c_2 = \tau_0 - \frac{\pi}{2c}$ to get

$$e^{g(\tau)} = \frac{-c_1}{\sin^2\left(\sqrt{-c_1}\mathcal{J}_q\left(\tau+c_2\right)+\frac{\pi}{2}\right)} = \frac{c^2}{\mathcal{J}_q^2}\frac{1}{\sin^2\left(c\left(\tau+\tau_0\right)\right)} \tag{2.144}$$

where new constants of integration are $\{c, \tau_0\}$. Recall that in Eq. (2.135), we showed that $\mathrm{sgn}(\tau)$ does not lead to a sign problem for both $\tau > 0$ and $\tau < 0$. Accordingly, here we have to make $g(\tau)$ symmetric, so we can do so by $\tau \to |\tau|$ and the differential equation is still satisfied. Accordingly we get

$$e^{g(\tau)} = \frac{c^2}{\mathcal{J}_q^2}\frac{1}{\sin^2\left[c\left(|\tau|+\tau_0\right)\right]} \tag{2.145}$$

Now we impose the boundary condition (Eq. (2.130)) for $g(\tau)$ as well as the periodicity in τ to find constraints between the parameters $\{c, \tau_0\}$

1. $g(\tau = 0) = 0$: This leads to $\left(\frac{c}{\mathcal{J}_q}\right)^2 = \sin^2\left(c\tau_0\right)$.
2. Imposing periodicity in τ leads to $\sin^2\left[c\tau_0\right] = \sin^2\left[c\left(\beta+\tau_0\right)\right]$ which leads to $g(\beta) = 0$.

Combining both constraints, we get

$$\left(\frac{c}{\mathcal{J}_q}\right)^2 = \sin^2\left(c\tau_0\right) = \sin^2\left[c\left(\beta+\tau_0\right)\right]. \tag{2.146}$$

Therefore $\{c, \tau_0\}$ are not independent constraints (in fact, they are connected via temperature) and we can reduce by one. Without loss of generality, we can impose (and boil down to one parameter ν)

$$c = \frac{\pi\nu}{\beta}, \quad \tau_0 = \frac{\beta(1-\nu)}{2\nu}, \tag{2.147}$$

where the first and last terms in Eq. (2.146) gives

$$\left(\frac{\pi\nu}{\beta\mathcal{J}_q}\right)^2 = \sin^2\left(\frac{\pi\nu}{\beta}\left(\beta+\frac{\beta(1-\nu)}{2\nu}\right)\right) = \sin^2\left(\pi\nu+\frac{\pi}{2}-\frac{\pi\nu}{2}\right). \tag{2.148}$$

Using $\sin^2\left(x+\frac{\pi}{2}\right)=\cos^2(x)$, we get

$$\boxed{\beta\mathcal{J}_q = \frac{\pi\nu}{\cos\left(\frac{\pi\nu}{2}\right)}} \tag{2.149}$$

which determines the value of ν at a given temperature. Recall $\mathcal{J}_q \equiv J_q\sqrt{\frac{q}{2^{q-1}}}$. Therefore, we solve for $g(\tau)$ using Eq. (2.145) (where $\cos(-x)=\cos(x)$ and taking natural logarithm on both sides)

$$\boxed{g(\tau) = 2\ln\left\{\frac{\cos\left(\frac{\pi\nu}{2}\right)}{\cos\left[\pi\nu\left(\frac{1}{2}-\frac{|\tau|}{\beta}\right)\right]}\right\}} \tag{2.150}$$

where ν is calculated via Eq. (2.149). This satisfies all conditions and constraints we have on the Green's function and therefore, this is our solution. Plugging this in Eqs. (2.124) and (2.125) gives us the exact Green's function and the self-energy for all temperatures at leading order in $1/q$ (recall that the derivation is based at $\mathcal{O}(1/q)$, starting Eq. (2.132)).

We briefly comment on the physical content of the parameter ν that forms an integral part of the solution. From Eq. (2.149), we deduce: (1) ν runs from 0 and 1 as dimensionless coupling runs from 0 to ∞; (2) $\nu=0$ denotes a free theory; (3) $\nu=1$ implies a zero temperature limit with infinitely strong coupling.

2.8 The Free Energy

We already saw in Eq. (2.63) how to relate the free energy F to the effective action which we reproduce here for convenience

$$\beta F = -S_{E,\text{eff}}[\mathcal{G},\Sigma],$$

where for $q/2$-body interaction, the effective action is given in Eq. (2.32). Combining, we get

$$\begin{aligned}-\frac{\beta F}{N} &= +\frac{1}{2}\ln\det[\partial_\tau-\Sigma]\\ &\quad-\frac{1}{2}\iint d\tau d\tau'\left(\Sigma(\tau-\tau')\mathcal{G}(\tau-\tau')-\frac{1}{q}J_q^2\mathcal{G}(\tau-\tau')^q\right).\end{aligned} \tag{2.151}$$

We are in equilibrium, that's why we are making use of time-translational invariance as well as $\int d\tau' = \beta$. To get rid of the logarithmic term, we make $J_q \frac{\partial}{\partial J_q}$ on both sides

$$J_q \partial_{J_q} \left(-\frac{\beta F}{N}\right) = \frac{J_q^2 \beta}{q} \int_0^\beta d\tau G(\tau)^q. \tag{2.152}$$

We use the full Schwinger-Dyson equation in Eq. (2.33), namely $\Sigma = J_q^2 \mathcal{G}^{q-1}$ to get

$$J_q \partial_{J_q} \left(-\frac{\beta F}{N}\right) = \frac{\beta}{q} \int_0^\beta d\tau \Sigma(\tau) G(\tau). \tag{2.153}$$

The purpose of this section is to simplify this differential equation and find an exact expression for the free energy.

We start by simplifying the right-hand side where the full Schwinger-Dyson equations in Fourier space reads (Eq. (2.73)) $\mathcal{G}(\omega)^{-1} = -\imath\omega - \Sigma(\omega)$ whose Fourier transform gives

$$\delta(\tau) = \partial_\tau \mathcal{G}(\tau) - \int_0^\beta d\tau' \Sigma(\tau, \tau') \mathcal{G}(\tau'), \tag{2.154}$$

where we take the limit $\tau \to 0^+$ to get

$$\int_0^\beta d\tau' \Sigma(\tau') G(\tau') = \lim_{\tau \to 0^+} \partial_\tau \mathcal{G}(\tau) \tag{2.155}$$

where $\Sigma(\tau') = \Sigma(\tau \to 0^+, \tau')$. Accordingly, Eq. (2.153) becomes

$$J_q \partial_{J_q} \left(-\frac{\beta F}{N}\right) = \frac{\beta}{q} \lim_{\tau \to 0^+} \partial_\tau \mathcal{G}(\tau). \tag{2.156}$$

We now use the large-q limit where $\mathcal{G}$ is given by Eq. (2.124) and solution for $g(\tau)$ is given in Eq. (2.150) where ν is calculated in Eq. (2.149). The right-hand side becomes

$$\begin{aligned} \text{Right-hand side} =& \frac{\beta}{q} \lim_{\tau \to 0^+} \partial_\tau \left[\frac{1}{2} \operatorname{sgn}(\tau) e^{\frac{g(\tau)}{q}}\right] \\ =& \frac{\beta}{2q} \lim_{\tau \to 0^+} \left[2\delta(\tau) e^{\frac{g(\tau)}{q}} + \operatorname{sgn}(\tau) \frac{\partial_\tau g(\tau)}{q} e^{\frac{g(\tau)}{q}}\right] \\ =& \frac{\beta}{2q^2} \lim_{\tau \to 0^+} \operatorname{sgn}(\tau) \partial_\tau g(\tau) e^{\frac{g(\tau)}{q}} \end{aligned} \tag{2.157}$$

where $\partial_\tau \text{sgn}(\tau) = 2\delta(\tau)$. We calculate $\partial_\tau g(\tau)$ via Eq. (2.150) to get

$$\partial_\tau g(\tau) = -\frac{2\pi\nu}{\beta}\,\text{sgn}(\tau)\tan\left[\pi\nu\left(\frac{1}{2} - \frac{|\tau|}{\beta}\right)\right] \tag{2.158}$$

which gives $(\text{sgn}(\tau)^2 = +1)$

$$\text{Right-hand side} = -\frac{\pi\nu}{q^2}\tan\left(\frac{\pi\nu}{2}\right) \tag{2.159}$$

to get to the differential equation for the free energy F:

$$J_q \partial_{J_q}\left(-\frac{\beta F}{N}\right) = -\frac{\pi\nu}{q^2}\tan\left(\frac{\pi\nu}{2}\right). \tag{2.160}$$

We now re-write the left-hand side using the chain rule

$$J_q \partial_{J_q} = J_q \frac{\partial(\mathcal{J}_q\beta)}{\partial J}\frac{\partial\nu}{\partial(\mathcal{J}_q\beta)}\partial_\nu \tag{2.161}$$

where $\mathcal{J}_q \equiv J_q\sqrt{\frac{q}{2^{q-1}}}$ showed up in two different contexts above, namely Eqs. (2.112) and (2.136). However from Eq. (2.149), we have $\beta\mathcal{J}_q = \frac{\pi\nu}{\cos(\frac{\pi\nu}{2})}$, therefore we get

$$\begin{aligned}
J_q \partial_{J_q} &= J_q \frac{\partial}{\partial J_q}\left(J_q\sqrt{\frac{q}{2^{q-1}}}\beta\right)\frac{1}{\frac{\partial}{\partial\nu}\left(\frac{\pi\nu}{\cos(\frac{\pi\nu}{2})}\right)}\partial_\nu \\
&= \underbrace{\underbrace{J_q\sqrt{\frac{q}{2^{q-1}}}}_{\equiv\mathcal{J}_q}\beta}_{\mathcal{J}_q\beta = \frac{\pi\nu}{\cos(\frac{\pi\nu}{2})}}\frac{1}{\frac{\partial}{\partial\nu}\left(\frac{\pi\nu}{\cos(\frac{\pi\nu}{2})}\right)}\partial_\nu \\
&= \frac{\nu}{1 + \frac{\pi\nu}{2}\tan\left(\frac{\pi\nu}{2}\right)}\partial_\nu.
\end{aligned} \tag{2.162}$$

Therefore the differential equation in Eq. (2.160) becomes

$$\frac{\nu}{1 + \frac{\pi\nu}{2}\tan\left(\frac{\pi\nu}{2}\right)}\frac{\partial}{\partial\nu}\left(-\frac{\beta F}{N}\right) = -\frac{\pi\nu}{q^2}\tan\left(\frac{\pi\nu}{2}\right). \tag{2.163}$$

This can be integrated to get the expression for the free energy

$$\frac{\beta F}{N} = \frac{\pi\nu}{q^2}\left[\tan\left(\frac{\pi\nu}{2}\right) - \frac{\pi\nu}{4}\right] + C \tag{2.164}$$

where C is a constant of integration which is evaluated in the limit $\nu \to 0$ when the theory becomes free and $F \to F_0$ (F_0 is the free energy for the non-interacting case):

$$\frac{\beta F_0}{N} = \frac{1}{N}\ln(\mathcal{Z}_0) = \frac{1}{N}\ln\left(2^{N/2}\right) = \frac{1}{2}\ln(2) = C. \tag{2.165}$$

where $\mathcal{Z}_0$ is the partition function for the free case, which is equal to the dimension of the Hilbert space $\mathcal{Z}_0 = \dim(H) = 2^{N/2}$. This arises because N Majorana fermions can be grouped into $N/2$ Dirac fermions, each with a 2-dimensional Hilbert space (occupied/unoccupied). Therefore, the expression for the free energy at large-q for all temperatures is

$$\boxed{\frac{\beta F}{N} = \frac{\pi\nu}{q^2}\left[\tan\left(\frac{\pi\nu}{2}\right) - \frac{\pi\nu}{4}\right] + \frac{1}{2}\ln(2)} \tag{2.166}$$

where ν is evaluated at a given temperature $\frac{1}{\beta} = T$ and the coupling strength $\mathcal{J}_q = J_q\sqrt{\frac{q}{2^{q-1}}}$ via the relation (2.149): $\beta\mathcal{J}_q = \frac{\pi\nu}{\cos(\frac{\pi\nu}{2})}$. All thermodynamics flows from here.

2.8.1 Connections to Energy

We start with the definition of the Green's function for the Majorana fermions ψ_i (where T denotes time-ordering)

$$\mathcal{G}(\tau) = \frac{1}{N}\sum_{i=1}^{N}\langle T\psi_i(\tau)\psi_i(0)\rangle \tag{2.167}$$

whose derivative gives

$$\begin{aligned}\lim_{\tau\to 0^+}\partial_\tau\mathcal{G}(\tau) &= \frac{1}{N}\sum_{i=1}^{N}\lim_{\tau\to 0^+}\partial_\tau\left[\Theta(\tau)\langle\psi_i(\tau)\psi_i(0)\rangle - \Theta(-\tau)\langle\psi_i(0)\psi_i(\tau)\rangle\right]\\ &= \frac{1}{N}\sum_{i=1}^{N}\lim_{\tau\to 0^+}\left[\delta(\tau)\langle\{\psi_i(\tau)\psi_i(0)\}\rangle + \Theta(\tau)\langle\partial_\tau\psi_i(\tau)\psi_i(0)\rangle\right.\\ &\quad\left. - \Theta(-\tau)\langle\psi_i(0)\partial_\tau\psi_i(\tau)\rangle\right].\end{aligned} \tag{2.168}$$

But the derivative of the field is given by

$$\partial_\tau\psi_i(\tau) = [\mathcal{H}_q, \psi_i](\tau) = \mathcal{H}_q\psi_i - \psi_i\mathcal{H}_q \tag{2.169}$$

where $\mathcal{H}_q$ is the Majorana SYK Hamiltonian for $q/2$-body interactions, given in Eq. (2.1), namely

$$\mathcal{H}_q = \imath^{q/2} \sum_{\{i_q\}_\leq} j_{q;\{i_q\}} \psi_{i_1} \dots \psi_{i_q} = \frac{\imath^{q/2}}{q!} \sum_{i_1, i_2, \dots, i_q = 1}^{N} j_{q;\{i_q\}} \psi_{i_1} \dots \psi_{i_q}.$$

Therefore we have

$$\begin{aligned}
\lim_{\tau \to 0^+} \partial_\tau \mathcal{G}(\tau) =& \frac{1}{N} \sum_{i=1}^{N} \lim_{\tau \to 0^+} \left[\delta(\tau) \left\langle \{\psi_i(\tau)\psi_i(0)\} \right\rangle + \left\langle T[\mathcal{H}_q, \psi_i](\tau)\psi_i(0) \right\rangle \right] \\
=& \frac{1}{N} \sum_{i=1}^{N} \left\langle [\mathcal{H}_q, \psi_i](0^+)\psi_i(0) \right\rangle .
\end{aligned} \tag{2.170}$$

We now have to evaluate the commutator $\left\langle [\mathcal{H}_q, \psi_i](0)\psi_i(0) \right\rangle$ where we show the calculations for $q = 2$:

$$\begin{aligned}
\sum_{i=1}^{N} \left\langle [\mathcal{H}_{q=2}, \psi_i](0^+)\psi_i(0) \right\rangle =& \sum_{i=1}^{N} \left\langle \frac{\imath}{2!} \left[\sum_{i_1, i_2 = 1}^{N} j_{2;i_1 i_2} \psi_{i_1} \psi_{i_2}, \psi_i \right] \psi_i \right\rangle \\
=& \left\langle \frac{\imath}{2!} \left[\sum_{i_1, i_2 = 1}^{N} j_{2;i_1 i_2} \psi_{i_1} \psi_{i_2}, \psi_{i_1} \right] \psi_{i_1} \right\rangle \\
&+ \left\langle \frac{\imath}{2!} \left[\sum_{i_1, i_2 = 1}^{N} j_{2;i_1 i_2} \psi_{i_1} \psi_{i_2}, \psi_{i_2} \right] \psi_{i_2} \right\rangle ,
\end{aligned} \tag{2.171}$$

where we use the property $\{\psi_i, \psi_j\} = \delta_{ij}$ at equal time (leading to $\psi_i^2 = 1/2$) from which commutation relation can be deduced at equal time to get

$$\left[\psi_i, \psi_j \right] = 2\psi_i \psi_j - \delta_{ij} \mathbb{1} \quad \Rightarrow [\psi_i, \psi_i] = 0 \tag{2.172}$$

that leads to (recall $[AB, C] = A[B, C] + [A, C]B$)

$$\begin{aligned}
&\left\langle \frac{\imath}{2!} \left[\sum_{i_1, i_2 = 1}^{N} j_{2;i_1 i_2} \psi_{i_1} \psi_{i_2}, \psi_{i_1} \right] \psi_{i_1} \right\rangle \\
&= \left\langle \frac{\imath}{2!} \sum_{i_1, i_2 = 1}^{N} j_{2;i_1 i_2} \psi_{i_1} \left(2\psi_{i_2} \psi_{i_1} \right) \psi_{i_1} \right\rangle \quad \left(\psi_{i_1}^2 = \frac{1}{2} \right)
\end{aligned}$$

$$= \left\langle \frac{\imath}{2!} \sum_{i_1,i_2=1}^{N} j_{2;i_1i_2} \psi_{i_1} \psi_{i_2} \right\rangle = \langle \mathcal{H}_2 \rangle$$

and

$$\left\langle \frac{\imath}{2!} \left[\sum_{i_1,i_2=1}^{N} j_{2;i_1i_2} \psi_{i_1} \psi_{i_2}, \psi_{i_2} \right] \psi_{i_2} \right\rangle = \left\langle \frac{\imath}{2!} \sum_{i_1,i_2=1}^{N} j_{2;i_1i_2} \psi_{i_1} \psi_{i_2} \right\rangle = \langle \mathcal{H}_2 \rangle$$

to finally get

$$\sum_{i=1}^{N} \big\langle [\mathcal{H}_{q=2}, \psi_i](0^+) \psi_i(0) \big\rangle = 2 \langle \mathcal{H}_2 \rangle. \tag{2.173}$$

This can be generalized to arbitrary q as

$$\boxed{\sum_{i=1}^{N} \big\langle [\mathcal{H}_q, \psi_i](0^+) \psi_i(0) \big\rangle = q \langle \mathcal{H}_q \rangle}. \tag{2.174}$$

Therefore we get by plugging this in Eq. (2.170)

$$\boxed{\lim_{\tau \to 0^+} \partial_\tau \mathcal{G}(\tau) = \frac{q}{N} \langle \mathcal{H}_q \rangle = \frac{q}{N} U}, \tag{2.175}$$

where U is the energy of the system. This is one of the simple version of a much powerful result known by the name of (generalized) *Galitskii-Migdal sum rule*. We will come to this later in Chap. 4 in the box below Eq. (4.32). Accordingly, we can relate this to the free energy using Eq. (2.156) to get

$$J_q \partial_{J_q} \left(-\frac{\beta F}{N} \right) = \frac{\beta}{q} \lim_{\tau \to 0^+} \partial_\tau \mathcal{G}(\tau) = \frac{\beta}{N} \langle \mathcal{H}_q \rangle = \frac{\beta}{N} U. \tag{2.176}$$

2.9 Real-Time Formalism

We have focused on systems in equilibrium for which the imaginary-(Euclidean-)time formalism is suited. This is what we have employed until now. The theoretical description of equilibrium many-body systems relies on the *adiabatic hypothesis*: interacting states evolve continuously from non-interacting states in the distant past and future. This principle underlies the imaginary-time formalism (also known as the Matsubara formalism), the established framework for equilibrium quantum statistical mechanics. For systems driven away from equilibrium, however, the adiabatic connection fails. In such cases, the real-time formalism (also known as the Keldysh formalism) becomes essential to capture general non-equilibrium dynamics. The purpose of this section is to not derive the two formalism from first principles as that will take a book in itself, but to provide a brief motivation and conceptual understanding of topics relevant for us, before we start applying the formalism. For a detailed deep dive, the reader is referred to Refs. [9, 15].

2.9.1 A Brief Overview of Matsubara vs. Keldysh Formalism

We begin by outlining the Matsubara formulation (imaginary-time formalism), then highlight its limitations for non-equilibrium systems to motivate the Keldysh formulation (real-time formalism). As mentioned above, we restrict ourselves to providing motivation and conceptual flow; comprehensive details appear in Ref. [9].

The Matsubara formalism originates from the von Neumann equation of motion

$$\partial_t \hat{\rho}(t) = -i\left[\mathcal{H}(t), \hat{\rho}(t)\right], \tag{2.177}$$

where $\mathcal{H}$ is the time-dependent Hamiltonian of the system. This equation is solved formally by

$$\hat{\rho}(t) = \hat{U}_{t,-\infty}\hat{\rho}_{-\infty}\left[\hat{U}_{t,-\infty}\right]^{\dagger} = \hat{U}_{t,-\infty}\hat{\rho}_{-\infty}\hat{U}_{-\infty,t}, \tag{2.178}$$

Here $\hat{U}_{t',t}$ is the unitary time evolution operator, and $\hat{\rho}_{-\infty}$ denotes the initial density matrix prepared in equilibrium at $t \to -\infty$. The evolution operator satisfies

$$\partial_t' \hat{U}_{t',t} = -\imath\mathcal{H}(t')\hat{U}_{t',t}, \quad \partial_t \hat{U}_{t',t} = +\imath\hat{U}_{t',t}\mathcal{H}(t). \tag{2.179}$$

Unitarity implies $\hat{U}_{t',t}\left[\hat{U}_{t',t}\right]^{\dagger} = 1$, from which one can deduce the identities $\left[\hat{U}_{t',t}\right]^{\dagger} = \left[\hat{U}_{t',t}\right]^{-1} = \hat{U}_{t,t'}$, applied in the last equality in Eq. (2.178). The expectation value of observable $\hat{\mathcal{O}}$ is

$$\langle\hat{\mathcal{O}}(t)\rangle \equiv \frac{\mathrm{Tr}[\hat{\mathcal{O}}\hat{\rho}(t)]}{\mathrm{Tr}[\hat{\rho}(t)]} = \frac{\mathrm{Tr}\left[\hat{U}_{-\infty,t}\hat{\mathcal{O}}\hat{U}_{t,-\infty}\hat{\rho}_{-\infty}\right]}{\mathrm{Tr}[\hat{\rho}_{-\infty}]} \tag{2.180}$$

where the equality uses: (a) Eq. (2.178), (b) trace cyclicity $\mathrm{Tr}[ABC] = \mathrm{Tr}[BCA] = \mathrm{Tr}[CAB]$, and (c) trace conservation under unitary evolution (per Eq. (2.177)). Rewriting $\hat{U}_{-\infty,t} = \hat{U}_{-\infty,+\infty}\hat{U}_{+\infty,t}$ in the numerator yields:

$$\text{Numerator} = \mathrm{Tr}[\overbrace{\hat{U}_{-\infty,+\infty}}^{\text{backward evolution}} \underbrace{\hat{U}_{+\infty,t}\hat{\mathcal{O}}\hat{U}_{t,-\infty}\hat{\rho}_{-\infty}]}_{\text{forward evolution}} \tag{2.181}$$

Equilibrium systems are based on a *central dogma*, namely *adiabatic continuity*: interacting states connect adiabatically to non-interacting states at $t \to \pm\infty$. For non-interacting state $|0\rangle$, this implies:

$$\hat{U}_{+\infty,-\infty}|0\rangle = e^{\imath\phi}|0\rangle \text{ (forward evolution)} \Rightarrow \hat{U}_{-\infty,+\infty}|0\rangle$$
$$= e^{-\imath\phi}|0\rangle \text{ (backward evolution)} \tag{2.182}$$

where ϕ is a phase. As shown in Ref. [9], this trivializes backward evolution in Eq. (2.181), reducing the expectation value to:

$$\langle\hat{O}(t)\rangle^{\text{eq}} = \frac{\text{Tr}\left[\hat{U}_{+\infty,t}\hat{O}\hat{U}_{t,-\infty}\hat{\rho}_{-\infty}\right]}{\text{Tr}[\hat{\rho}_{-\infty}]} \tag{2.183}$$

We have, explicitly, $\rho_{-\infty} = e^{-\beta\mathcal{H}}$ and $\hat{U}_{t,t'} = e^{-i(t-t')\mathcal{H}}$. The Matsubara formulation follows from Wick rotation $t \to -i\tau$ $(0 \le \tau \le \beta)$, combining $\hat{\rho}_{-\infty}$ and $\hat{U}_{t,t'}$ into a single exponential for diagrammatics (see Ref. [2] for a great exposition to diagrammatics), hence the name "imaginary-time formalism".

Having established the Matsubara formalism, we now address its failure for non-equilibrium systems. The key pillar enabling the Matsubara formulation to avoid backward evolution—the assumption of *adiabatic continuity*—breaks down in non-equilibrium settings. This implies that if interactions are switched off at a future time, the final state at $t \to +\infty$ does not return to the original non-interacting state at $t \to -\infty$. Consequently, Eq. (2.182) fails for non-equilibrium systems.

The backward evolution in Eq. (2.181) must therefore be retained, and the combined forward-backward time path constitutes a closed time contour. This framework, formulated in real time (hence "real-time formalism"), is known as the Martin-Schwinger-Kadanoff-Baym-Keldysh formalism. For brevity, we refer to it as the Keldysh formalism (or real-time formalism).

The Keldysh contour (Fig. 2.1) comprises of three paths

1. A forward real-time branch $\mathcal{C}_+$ $(-\infty \to +\infty)$,
2. A backward real-time branch $\mathcal{C}_-$ $(+\infty \to -\infty)$,
3. An imaginary-time branch $\mathcal{C}_{\text{imag}}$ from t_0 to $t_0 - i\beta$ (with $t_0 \to -\infty$).

The Keldysh time ordering has the following chronological ordering: first, the events happening on $\mathcal{C}_+$ come the earliest, followed by events on $\mathcal{C}_-$ and then at last with events on the imaginary contour. The notation to denote such Keldysh time ordering between $t_1 \in \mathcal{C}_+$ and $t_2 \in \mathcal{C}_-$ is $t_1 <_{\mathcal{C}} t_2$.

The imaginary branch represents the initial equilibrium state. For general non-equilibrium dynamics prepared at $t_0 \to -\infty$, this branch is typically omitted due to Bogoliubov's principle of weakening correlations. Bogoliubov's principle of weakening correlations states that for systems prepared in equilibrium in the distant past $(t_0 \to -\infty)$, correlations between observables at finite times and initial-

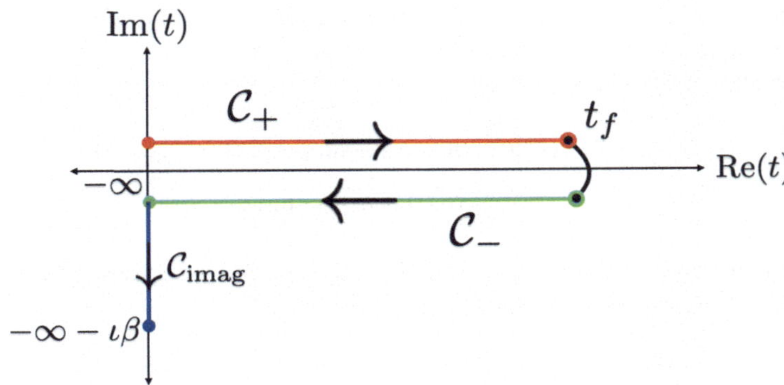

Fig. 2.1 The Schwinger-Keldysh contour $\mathcal{C} = \mathcal{C}_+ + \mathcal{C}_- + \mathcal{C}_{\text{imag}}$ consists of three segments: (1) Forward real-time branch ($\mathcal{C}_+$): extends from $-\infty$ to $+\infty$; (2) Backward real-time branch ($\mathcal{C}_-$): extends from $+\infty$ to $-\infty$; (3) Imaginary-time branch ($\mathcal{C}_{\text{imag}}$): represents the equilibrium state. The horizontal branches ($\mathcal{C}_+$ and $\mathcal{C}_-$) are strictly real-valued; their vertical offset in diagrams is purely a visualization aid. In contour ordering, points on the vertical branch $\mathcal{C}_{\text{imag}}$ are later than all points on both horizontal branches ($\mathcal{C}_+$ and $\mathcal{C}_-$). If one has to arrange the Keldysh contour chronologically then events on $\mathcal{C}_+$ happen earliest, followed by events on $\mathcal{C}_-$ and lastly on $\mathcal{C}_{\text{imag}}$. In other words, $t_+ < t_- < t_{\text{imag}}$ where t_{imag} is obtained by Wick's rotation $t \to -\imath\tau$. The time ordering notation in Keldysh plane, for instance for $t_1 \in \mathcal{C}_+$ and $t_2 \in \mathcal{C}_-$, is $t_1 <_{\mathcal{C}} t_2$

state details decay as $|t - t_0| \to \infty$. This justifies neglecting explicit initial-state terms (e.g., the imaginary-time branch in Keldysh contour) when studying long-time non-equilibrium dynamics, as their influence becomes negligible and eliminates any redundant initial-state dependencies. As is common in the literature [1], the system is always prepared in equilibrium in this work too at $t_0 \to -\infty$.

The central object in Keldysh formalism is the partition function:

$$\mathcal{Z} \equiv \frac{\text{Tr}[\hat{U}_{\mathcal{C}}\hat{\rho}_{-\infty}]}{\text{Tr}\hat{\rho}_{-\infty}} \tag{2.184}$$

where $\hat{U}_{\mathcal{C}} \equiv \hat{U}_{-\infty,+\infty}\hat{U}_{+\infty,-\infty}$ (subscript $\mathcal{C}$ denotes the closed Keldysh contour as in Fig. 2.1 excluding the imaginary branch which, as justified above, has been deprecated due to Bogoliubov's principle of weakening correlations). When the Hamiltonian is identical on the forward ($-\infty \to +\infty$) and backward ($+\infty \to -\infty$) branches, this implies $\hat{U}_{\mathcal{C}} = \mathbb{1}$. This serves as a useful benchmark in calculations to check for consistency, as this means that the Keldysh partition function is normalized to unity $\mathcal{Z} = 1$.

Expressed in field variables ψ:

$$\mathcal{Z} = \int D[\psi, \overline{\psi}]e^{\imath S[\psi,\overline{\psi}]}, \tag{2.185}$$

where $\overline{\psi}$ is the conjugate field, and the Keldysh action is:[29]

$$S[\psi, \overline{\psi}] = \int_C dt \left(\frac{\imath}{2} \psi(t) \partial_t \overline{\psi}(t) - \mathcal{H}(t) \right). \tag{2.186}$$

The integral traverses the Schwinger-Keldysh contour $\mathcal{C}$ (Fig. 2.1), reduced to $\mathcal{C} = \mathcal{C}_+ + \mathcal{C}_-$ (forward/backward branches) by omitting the imaginary branch per Bogoliubov's principle.

To compute an observable $\hat{\mathcal{O}}$ expectation, we insert it into the contour via:

$$\mathcal{H}_\eta^\pm = \mathcal{H} \pm \hat{\mathcal{O}}\eta(t), \tag{2.187}$$

where $\eta(t)$ is a source field. The $+(-)$ sign corresponds to the forward (backward) branch. This branch-dependent Hamiltonian implies $\hat{U}_\mathcal{C} \neq 1$ and therefore a nontrivial partition function $\mathcal{Z}[\eta] \neq 1$, which serves as a generating functional:

$$\mathcal{Z}[\eta] \equiv \frac{\mathrm{Tr}\left[\hat{U}_\mathcal{C}[\eta]\hat{\rho}_{-\infty}\right]}{\mathrm{Tr}[\hat{\rho}_{-\infty}]}. \tag{2.188}$$

For an operator on the forward branch(modifying the Hamiltonian as $\mathcal{H} \to \mathcal{H}_\eta^- = \mathcal{H} - \hat{\mathcal{O}}\eta(t)$, while keeping it the unchanged along the backward contour), its expectation is

$$\langle \hat{\mathcal{O}}(t) \rangle = \imath \left. \frac{\delta \mathcal{Z}[\eta]}{\delta \eta(t)} \right|_{\eta=0}. \tag{2.189}$$

The process of introducing a source field $\eta(t)$ and making the partition function a generating functional have close resemblance to the imaginary-time formalism, where we performed similar manipulations to calculate the free Green's function in Sect. 2.2.1. However there is one crucial difference that makes the Keldysh formalism more straightforward: unlike equilibrium Matsubara formalism, Keldysh directly uses the generating functional *without logarithms* (contrast Eq. (2.189) with Eq. (2.44)), significantly simplifying calculations.

[29] Contrast with Eqs. (2.8) and (2.9) where the partition function and the action are provided in real time for real Grassmann variables while here, these are the Keldysh partition function and the Keldysh action, respectively. Accordingly, these are different conceptually. Instead of integrating over real time $t \in \mathbb{R}$ for real Grassmann variables, we integrate here over the Keldysh contour $\mathcal{C}$ for independent complex Grassmann variables ψ and $\overline{\psi}$. Similarly the measure becomes $D[\psi, \overline{\psi}] = \prod_t d\psi(t) d\overline{\psi}(t)$. That's why we put a subscript r in Eqs. (2.8) and (2.9) to denote real time integration over $\mathbb{R}$ while we could use a subscript K to denote Keldysh, we chose to avoid such labeling to reduce clutter. It's understood that the Keldysh formalism is the real-time formalism where integration is over the Keldysh contour $\mathcal{C}$. The Keldysh formalism is the powerful methodology that treats forward and backward evolution separately, as can be the case in non-equilibrium dynamics where adiabatic continuity fails.

2.9.2 Green's Functions in Real-Time

We evaluated the equations of motion in the imaginary-time formalism, namely the Schwinger-Dyson equations which are the Euler-Lagrange equation for the effective action derived at large-N. We are now interested in formulating the dynamical equations in the real-time formalism in the Keldysh plane,[30] known as the Kadanoff-Baym equations. The building blocks for real-time dynamics are the Green's functions. On the Keldysh contour, the two time arguments of the bi-local Green's function may lie on the forward branch ($\mathcal{C}_+$) or backward branch ($\mathcal{C}_-$). This branching generates four distinct components within the Keldysh-contour-ordered Green's function $\mathcal{G}\left(t, t'\right)$-a structure applicable to all contour functions

$$\mathcal{G}\left(t, t'\right) = \begin{cases} \mathcal{G}^{>}\left(t, t'\right), \; t \in \mathcal{C}_-, t' \in \mathcal{C}_+ & (\because t > t') \\ \mathcal{G}^{<}\left(t, t'\right), \; t \in \mathcal{C}_+, t' \in \mathcal{C}_- & (\because t < t') \\ \mathcal{G}_{\mathrm{t}}\left(t, t'\right), \quad t, t' \in \mathcal{C}_+ & \text{(time-ordered)} \\ \mathcal{G}_{\tilde{\mathrm{t}}}\left(t, t'\right), \quad t, t' \in \mathcal{C}_- & \text{(anti-time-ordered)} \end{cases} . \tag{2.190}$$

The time-ordered ($\mathcal{G}_{\mathrm{t}}$) and anti-time-ordered $\left(\mathcal{G}_{\tilde{\mathrm{t}}}\right)$ components are defined via the Heaviside function $\Theta(t)$ as

$$\begin{aligned} \mathcal{G}_{\mathrm{t}}\left(t, t'\right) &\equiv +\Theta\left(t - t'\right) G^{>}\left(t, t'\right) + \Theta\left(t' - t\right) G^{<}\left(t, t'\right) \\ \mathcal{G}_{\tilde{\mathrm{t}}}\left(t, t'\right) &\equiv +\Theta\left(t' - t\right) G^{>}\left(t, t'\right) + \Theta\left(t - t'\right) G^{<}\left(t, t'\right) \end{aligned} \tag{2.191}$$

where $\Theta(t)$ is the Heaviside function. Therefore, we have the following identity

$$\mathcal{G}_{\mathrm{t}}\left(t, t'\right) + \mathcal{G}_{\tilde{\mathrm{t}}}\left(t, t'\right) = \mathcal{G}^{>}\left(t, t'\right) + \mathcal{G}^{<}\left(t, t'\right), \tag{2.192}$$

establishing that only three of the four Green's functions are linearly independent in the Keldysh plane. While conventions vary, we work with the most suitable set $\left\{\mathcal{G}, \mathcal{G}^{<}, \mathcal{G}^{R}, \mathcal{G}^{A}\right\}$, where the retarded $\left(\mathcal{G}^{R}\right)$ and advanced $\left(\mathcal{G}^{A}\right)$ Green's functions are

$$\begin{aligned} \mathcal{G}^{R}\left(t, t'\right) &\equiv +\Theta\left(t - t'\right)\left[\mathcal{G}^{>}\left(t, t'\right) - \mathcal{G}^{<}\left(t, t'\right)\right] \\ \mathcal{G}^{A}\left(t, t'\right) &\equiv +\Theta\left(t' - t\right)\left[\mathcal{G}^{<}\left(t, t'\right) - \mathcal{G}^{>}\left(t, t'\right)\right], \end{aligned} \tag{2.193}$$

where again there are three linearly independent Green's functions due to the following constraint/identity (flowing from their respective definitions)

$$\mathcal{G}^{R}(t, t') - \mathcal{G}^{A}(t, t') = \mathcal{G}^{>}(t, t') - \mathcal{G}^{<}(t, t'). \tag{2.194}$$

[30] We will ignore the imaginary contour $\mathcal{C}_{\mathrm{fig}}$ in Fig. 2.1 always, unless explicitly stated otherwise.

A useful property of the lesser/greater Green's function (Majorana or otherwise, both in and out-of-equilibrium) is

$$\left[\mathcal{G}^{\gtrless}(t_1, t_2)\right]^{\star} = -\mathcal{G}^{\gtrless}(t_2, t_1), \tag{2.195}$$

while there is an additional constraint for Majorana fermions (both in and out-of-equilibrium), namely

$$\mathcal{G}^{>}(t_1, t_2) = -\mathcal{G}^{<}(t_2, t_1) \quad \text{(Majorana condition)}. \tag{2.196}$$

NOTE: These are true for the convention adoped for the Green's function here. Namely, we work in the convention $\mathcal{G}(t_1, t_2) \equiv \frac{-\iota}{N} \sum_{j=1}^{N} \left\langle T_{\mathcal{C}} c_j(t_1) c_j^{\dagger}(t_2) \right\rangle$. We stick to a convention consistently throughout the chapter where the convention is stated at the start. In later chapters, we will practice with other conventions which we will state at the beginning of each chapter. This will allow us to get a hands-on experience in dealing with different conventions prevalent in the literature.

This simplifies the life while dealing with Majorana fermions because we only need to determine either the greater or the lesser Green's function and we have effectively solved the theory because we have successfully evaluated the lesser, the greater as well as the retarded and the advanced Green's functions using Eq. (2.193).

Note that the identities for lesser and greater Green's functions are convention dependent (convention comes in while choosing the prefactor in the definition of the Green's function). Throughout this chapter, we adopt the consistent convention. Later we will try other conventions too to get a hands-on experience of how to deal with different scenarios. In general, the reader is encouraged to first check the definition to clarify the convention before proceeding. We will always explicitly state the convention of the Green's function used in each chapter.

These real-time Green's functions form the basis for writing down the equations of motion, namely the *Kadanoff-Baym equations*, for non-equilibrium dynamics.

2.9.3 The Kadanoff-Baym Equations

In order to write the real-time dynamical equations, we start with the partition function in Keldysh plane. The partition function as well as the action in Keldysh plane are provided in Eqs. (2.185) and (2.186) which we reproduce here for convenience (we recommend to revisit the footnote 29)

$$\mathcal{Z} = \int D[\psi, \overline{\psi}] e^{\imath S[\psi, \overline{\psi}]},$$

where $\overline{\psi}$ is the conjugate field, and the Keldysh action is given by

$$S[\psi, \overline{\psi}] = \int_C dt \left(\frac{\imath}{2} \psi(t) \partial_t \overline{\psi}(t) - \mathcal{H}(t) \right).$$

Here $\mathcal{H}_q$ is the Hamiltonian for $q/2$-body interacting SYK model given in Eq. (2.1) which we also reproduce for convenience

$$\text{SYK}_q = \mathcal{H}_q = \imath^{q/2} \sum_{\{i_q\}_\leq} j_{q;\{i_q\}} \psi_{i_1} \dots \psi_{i_q}, \tag{2.197}$$

where the notation in the summation is explained in Eq. (2.3) while the random variables $j_{q;\{i_q\}}$ are derived from the Gaussian ensemble whose mean and variance are given in Eq. (2.4).

We now proceed in the same fashion as we did in Sect. 2.2 but in Keldysh plane. Accordingly, we can use the same identity as in Eq. (2.21) but in real time (see Eq. (2.212) later on the form of the Green's function, also refer Appendix F of Ref. [8] for a detailed discussion of analytic continuation of the imaginary-time ansatz in Eq. (2.124) to the real-time Green's function ansatz in Eq. (2.212))

$$\iint D\mathcal{G} D\Sigma \exp \left\{ \frac{-N}{2} \int_C dt_1 dt_2 \Sigma(t_1, t_2) \left(\mathcal{G}(t_1, t_2) + \frac{\imath}{N} \sum_j \psi_j(t_1) \psi_j(t_2) \right) \right\} = 1, \tag{2.198}$$

where integration is over the Keldysh contour C and we have used the real-time continuation of the Green's function, following the standard convention in the literature, namely $\mathcal{G}(t_1, t_2) = -\frac{\imath}{N} \sum_j \psi_j(t_1) \psi_j(t_2)$.[31] We refer to the discussion about the measures of integration below Eq. (2.21). Then the partition function takes the form

$$\overline{\mathcal{Z}} = \int D\mathcal{G} D\Sigma e^{\imath S_{\text{eff}}[\mathcal{G}, \Sigma]} \tag{2.199}$$

[31] Alternative conventions in the literature define the Green's function with different prefactors. Using these conventions implies adjusting the definition of the self-energy in Eq. (2.198) to incorporate the relevant factors. For this chapter, we maintain the convention with $-\imath$ in the Green's function. In later chapters, we will adopt different conventions to gain practical experience in handling them while providing practical insight into managing these differences.

where the action is

$$\frac{S_{\text{eff}}[\mathcal{G},\Sigma]}{N} \equiv \frac{-\imath}{2}\ln\det(\partial_t + \imath\Sigma) + \frac{\imath(-1)^{q/2}J_q^2}{2q}\int dtdt'\mathcal{G}(t,t')^q$$
$$+\frac{\imath}{2}\int dtdt'\Sigma(t,t')\mathcal{G}(t,t'). \tag{2.200}$$

where time variables are real and integration is performed over the Keldysh contour. Then extremizing the action with respect to Σ and $\mathcal{G}$, namely $\frac{\delta S_{\text{eff}}[\mathcal{G},\Sigma]}{\delta\Sigma}\Big|_{\mathcal{G}} \stackrel{!}{=} 0$ and $\frac{\delta S_{\text{eff}}[\mathcal{G},\Sigma]}{\delta\mathcal{G}}\Big|_{\Sigma} \stackrel{!}{=} 0$ gives the Euler-Lagrange equation in the large-N limit where the saddle point solutions dominate. These are the Schwinger-Dyson equations

$$\boxed{\mathcal{G}_0(t,t')^{-1} = \mathcal{G}(t,t')^{-1} + \Sigma(t,t') \quad \Sigma(t,t') = -(-1)^{q/2}J_q^2\mathcal{G}(t,t')^{q-1}}, \tag{2.201}$$

where $\mathcal{G}_0^{-1}(t,t') = \imath\delta(t-t')\partial_t$ is the free Majorana Green's function where the origin of $\imath$ is because there is an imaginary unit in the free Majorana real-time action $S_0 = \frac{\imath}{2}\int dt\psi_i(t)\partial_t\psi_i(t)$ (see Eq. (2.9)).

We can now derive the Kadanoff-Baym equations starting from these Schwinger-Dyson equations. We take the Dyson equation, namely $\mathcal{G}_0^{-1}(t_1,t_3) = \mathcal{G}^{-1}(t_1,t_3) + \Sigma(t_1,t_3)$ and take a convolution product from the right with respect to $\mathcal{G}(t_3,t_2)$ to get

$$\int_{\mathcal{C}} dt_3\,\mathcal{G}_0^{-1}(t_1,t_3)\mathcal{G}(t_3,t_2) = \int_{\mathcal{C}} dt_3\,\mathcal{G}^{-1}(t_1,t_3)\mathcal{G}(t_3,t_2) + \int_{\mathcal{C}} dt_3\,\Sigma(t_1,t_3)\mathcal{G}(t_3,t_2)$$
$$= \delta_{\mathcal{C}}(t_1,t_2) + \int_{\mathcal{C}} dt_3\,\Sigma(t_1,t_3)\mathcal{G}(t_3,t_2), \tag{2.202}$$

where $\delta_{\mathcal{C}}(t_1,t_2)$ is the delta-function on the Keldysh contour where it's non-vanishing if and only if both t_1 and t_2 belong to the same Keldysh contour, namely either on $\mathcal{C}_+$ or on $\mathcal{C}_-$.

Mathematical Functions in Keldysh Plane

We define some relevant mathematical functions in the Keldysh plane, starting with the Heaviside step function $\Theta(x)$ whose derivative gives the δ-function. The generalization to the Keldysh plane is as follows:

$$\Theta_{\mathcal{C}}(t_1-t_2) = \begin{cases} 1 & \text{if } t_1 >_{\mathcal{C}} t_2 \\ 0 & \text{if } t_1 <_{\mathcal{C}} t_2 \end{cases} \tag{2.203}$$

(continued)

where the notation $<_{\mathcal{C}}$ and $>_{\mathcal{C}}$ show Keldysh time ordering, as explained in the caption of Fig. 2.1. In other words, if $t_1, t_2 \in \mathcal{C}_+$ (forward contour in Fig. 2.1), then this is just the ordinary step function. If $t_1, t_2 \in \mathcal{C}_-$ (backward contour), then we have $\Theta_{\mathcal{C}}(t_1 - t_2) = \Theta(t_2 - t_1)$ where Θ without the subscript $\mathcal{C}$ denotes the standard Heaviside function. The derivative in the Keldysh plane acts as the distributional derivative with respect to contour parameter s: $\frac{d}{ds}\Theta_{\mathcal{C}}(t(s) - t_0) = \delta(s - s_0) = \left|\frac{dt}{ds}\right| \delta_{\mathcal{C}}(t - t_0)$, where s_0 is the parameter value at the "reference" t_0, and $\delta(s - s_0)$ is the Dirac delta in s. Also $\left|\frac{dt}{ds}\right|$ is the speed of the contour parameterization. For a unit-speed parameterization $\left(\left|\frac{dt}{ds}\right| = 1\right)$: $\frac{d}{ds}\Theta_{\mathcal{C}}(t(s) - t_0) = \delta_{\mathcal{C}}(t - t_0)$.[a]

A related concept is the sign function $\text{sgn}(x) = 2\Theta(x) - 1 = \Theta(x) - \Theta(-x)$ (which enforces the value $\Theta(x = 0) = \frac{1}{2}$) whose Keldysh generalization is given by

$$\text{sgn}_{\mathcal{C}}(t_1 - t_2) = \begin{cases} \text{sgn}(t_1 - t_2) & \text{if } t_1, t_2 \in \mathcal{C}_+ \\ \text{sgn}(t_2 - t_1) & \text{if } t_1, t_2 \in \mathcal{C}_- \\ -1 & \text{if } t_1 \in \mathcal{C}_+, t_2 \in \mathcal{C}_- \\ +1 & \text{if } t_1 \in \mathcal{C}_-, t_2 \in \mathcal{C}_+ \end{cases} \tag{2.204}$$

where the identities carry forward directly: $\text{sgn}_{\mathcal{C}}(t_1 - t_2) = 2\Theta_{\mathcal{C}}(t_1 - t_2) - 1 = \Theta_{\mathcal{C}}(t_1 - t_2) - \Theta_{\mathcal{C}}(t_2 - t_1)$. Again, this enforces $\Theta_{\mathcal{C}}(t_1 - t_2 = 0) = \frac{1}{2}$ when $t_1 = t_2$ and they belong on the same branch of the Keldysh contour.

[a]Note that $\Theta_{\mathcal{C}}(t)$ is meaningless and ill-defined because $\Theta_{\mathcal{C}}$ is fundamentally a function of two contour times (e.g., $\Theta_{\mathcal{C}}(t_1 - t_2)$), not a single variable t. A single time argument t lacks a reference point for "order" on $\mathcal{C}$.

We continue from Eq. (2.202) where without loss of generality, we take $t_1 \in \mathcal{C}_-$ and $t_2 \in \mathcal{C}_+$. Accordingly, the δ-function vanishes. The left-hand side simplifies using the real-space representation of the free Majorana Green's function, namely $\mathcal{G}_0^{-1}(t_1, t_3) = \imath\delta(t_1 - t_3)\partial_{t_1}$:

$$\text{Left-hand side} = \int dt_3\, \mathcal{G}_0^{-1}(t_1, t_3)\mathcal{G}(t_3, t_2) = \imath\partial_{t_1}\mathcal{G}(t_1^{\mathcal{C}_-}, t_2^{\mathcal{C}_+}) = \imath\partial_{t_1}\mathcal{G}^>(t_1, t_2). \tag{2.205}$$

Next, we apply the Langreth rule (derived in Appendix D, Eq. (D.6)) to the right-hand side, deriving the first Kadanoff-Baym equation:

$$\imath\partial_{t_1}\mathcal{G}^>(t_1, t_2) = \int_{-\infty}^{\infty} dt_3 \left(\Sigma^R(t_1, t_3)\mathcal{G}^>(t_3, t_2) + \Sigma^>(t_1, t_3)\mathcal{G}^A(t_3, t_2)\right), \tag{2.206}$$

where Eq. (2.193) defines the retarded and advanced Green's functions. The same definition holds for other functions in Keldysh plane, such as Σ (simply replace $\mathcal{G} \to \Sigma$ in Eq. (2.193)). Similarly, convolving the Dyson equation with from the left (with $t_1 \in \mathcal{C}_-$, $t_2 \in \mathcal{C}_+$) gives the second Kadanoff-Baym equation:

$$- \imath \partial_{t_2} \mathcal{G}^>(t_1, t_2) = \int_{-\infty}^{\infty} dt_3 \left(\mathcal{G}^R(t_1, t_3) \Sigma^>(t_3, t_2) + \mathcal{G}^>(t_1, t_3) \Sigma^A(t_3, t_2) \right). \tag{2.207}$$

Note that for the Majorana fermions, we have Eq. (2.196) which implies that Eq. (2.207) is redundant and solving Eq. (2.206) is enough. However, this only holds for Majorana fermions.

Similarly, there are dynamical equations of motion in real-time for $\mathcal{G}^<$ with respect to both time arguments which we present in a unified manner, the full Kadanoff-Baym equations (depreciating the imaginary contour in Fig. 2.1)

$$\begin{aligned} \imath \partial_{t_1} \mathcal{G}^{\gtrless}(t_1, t_2) &= \int_{-\infty}^{\infty} dt_3 \left(\Sigma^R(t_1, t_3) \mathcal{G}^{\gtrless}(t_3, t_2) + \Sigma^{\gtrless}(t_1, t_3) \mathcal{G}^A(t_3, t_2) \right), \\ -\imath \partial_{t_2} \mathcal{G}^{\gtrless}(t_1, t_2) &= \int_{-\infty}^{\infty} dt_3 \left(\mathcal{G}^R(t_1, t_3) \Sigma^{\gtrless}(t_3, t_2) + \mathcal{G}^{\gtrless}(t_1, t_3) \Sigma^A(t_3, t_2) \right). \end{aligned} \tag{2.208}$$

The conjugate relation in Eq. (2.195) (valid universally for fermions, in/out of equilibrium) ensures that the bottom set of equations are equivalent to the top set of equations in Eq. (2.208). This reduces computational effort: solving one determines the other.

Physically, the non-equilibrium dynamics captured by the $\mathcal{G} - \Sigma$ formalism is governed by the Kadanoff-Baym equations, which describe the coupled evolution of the Green's function $\mathcal{G}(t, t')$ and self-energy $\Sigma(t, t')$ in the two-time plane (t vs. t'). As shown in Fig. 2.2, this plane is divided into four quadrants, with a crucial physical interpretation:

1. Initial Equilibrium: Quadrant C (bottom-right) represents the *infinite* past, where the system resides in thermal equilibrium before any perturbation.
2. Non-Equilibrium Trigger: The origin ($t = t' = 0$) marks where a non-equilibrium perturbation (e.g., quench or drive) is applied, disrupting equilibrium.
3. Causal Propagation: The blue region shows the solution computed via the Kadanoff-Baym equations. Crucially, it incorporates:

- Memory Effects: Correlations from quadrants B (past times, $t > t'$) and D (past times, $t < t'$)

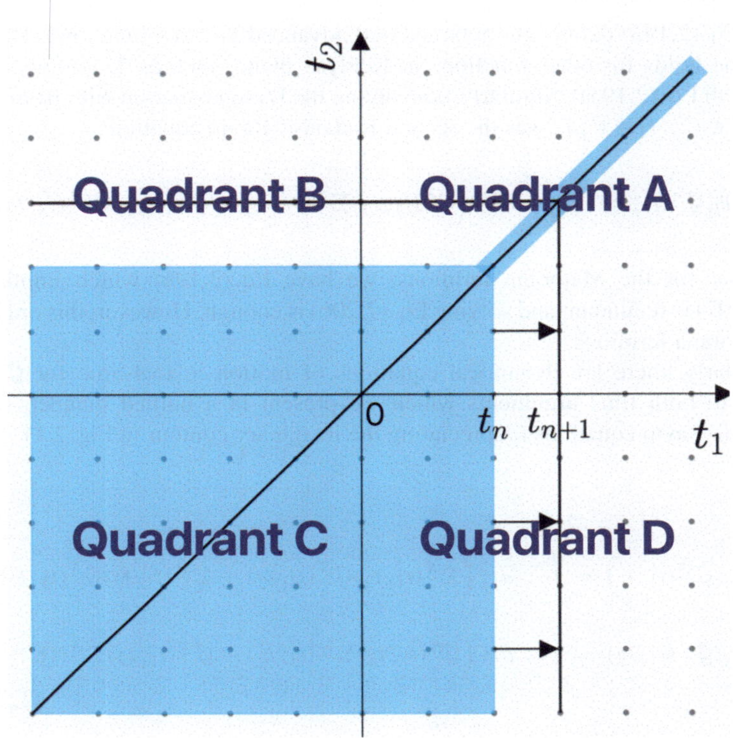

Fig. 2.2 Causal structure of the two-time plane (t_1-t_2) for Kadanoff-Baym equations (Eq. (2.208)). The system begins in equilibrium in quadrant C (distant past). A non-equilibrium perturbation at the origin induces time evolution. The blue region represents the solution computed via Eq. (2.208), where memory effects from quadrants B and D (historical correlations) propagate to quadrant A (far future), demonstrating non-Markovian dynamics

- Future Impact: These historical correlations propagate to quadrant A (top-left, far future), influencing long-time dynamics.

This formalism captures inherently non-Markovian behavior: the system's historical evolution (quadrants B/D) persistently influences its future dynamics (quadrant A), precluding any Markovian description where evolution would depend solely on instantaneous states. Compounding this complexity, the nonlinear structure of the Kadanoff-Baym equations creates formidable challenges for numerical solution.

2.9.4 Example of a Single Majorana SYK Dot

The Majorana SYK model with $q/2$-body interactions is governed by the Hamiltonian in Eq. (2.1). While we previously solved this model using imaginary-time path integrals (sufficient for equilibrium properties), we now revisit the same equilibrium problem using the real-time Keldysh formalism. This approach serves two key pedagogical purposes: (1) Redundancy check: explicitly solving in the large-q limit for the Green's function $\mathcal{G}(t)$ in real time provides an independent verification of our imaginary-time results (in particular the differential equation (2.136)), testing the consistency of our formalism; (2) calculating the system's energy $E = \langle \mathcal{H}_q \rangle$ via Keldysh techniques offers concrete practice with the Keldysh formalism.

2.9.4.1 Solving for the Green's Function at Large-q in Real-Time

We start with the Schwinger-Dyson equation for the self-energy in Eq. (2.242) using the Langreth rule in Appendix D (Eq. (D.8)) to get

$$\Sigma^{\gtrless}(t,t') = -(-1)^{q/2} J_q^2 \mathcal{G}^{\gtrless}(t,t')^{q-1}. \tag{2.209}$$

As mentioned above, the Majorana fermions satisfy the conjugate relation Eq. (2.196) that makes one of the Kadanoff-Baym equations (e.g., Eq. (2.207)) redundant and one needs to only solve Eq. (2.206) in order to solve the theory. We use the definition of retarded and advanced functions from Eq. (2.193) and plug into Eq. (2.206) to get

$$\begin{aligned} \imath \partial_{t_1} \mathcal{G}^>(t_1,t_2) = & \int_{-\infty}^{+\infty} dt_3 \Theta(t_1-t_3)\left[\Sigma^>(t_1,t_3) - \Sigma^<(t_1,t_3)\right]\mathcal{G}^>(t_3,t_2) \\ & + \int_{-\infty}^{+\infty} dt_3 \Theta(t_2-t_3)\Sigma^>(t_1,t_3)\left[\mathcal{G}^<(t_3,t_2) - \mathcal{G}^>(t_3,t_2)\right]. \end{aligned} \tag{2.210}$$

We now plug the expressions for the self-energies $\Sigma^{\gtrless}$ from Eq. (2.209) and simplify

$$\begin{aligned} & \imath \partial_{t_1} \mathcal{G}^>(t_1,t_2) \\ & = \int_{-\infty}^{t_1} dt_3 \left\{-(-1)^{q/2} J_q^2 \left[\mathcal{G}^>(t_1,t_3)^{q-1} - \mathcal{G}^<(t_1,t_3)^{q-1}\right]\mathcal{G}^>(t_3,t_2)\right\} \\ & + \int_{-\infty}^{t_2} dt_3 \left\{(-1)^{q/2} J_q^2 \mathcal{G}^>(t_1,t_3)^{q-1}\left[\mathcal{G}^>(t_3,t_2) - \mathcal{G}^<(t_3,t_2)\right]\right\}. \end{aligned} \tag{2.211}$$

This is the Kadanoff-Baym equation we need to solve.

We now take the large-q ansatz for the Green's function in real-time (this satisfies the Green's function identities in Eqs. (2.195) and (2.196))[32]

$$\mathcal{G}^>(t_1,t_2) = -\frac{\imath}{2} e^{g(t_1,t_2)/q} \tag{2.212}$$

where $g = \mathcal{O}(q^0)$. The boundary condition for g as well as the general conjugate relation for $\mathcal{G}^{\gtrless}$ (for all fermions, both in and out-of-equilibrium) in Eq. (2.195)

[32] We redirect the reader to Appendix F of Ref. [8] where an analytic continuation to the imaginary-time large-q ansatz, namely Eq. (2.124), is discussed.

translates to g as[33]

$$g(t,t) = 0, \quad g(t_1,t_2)^\star = g(t_2,t_1). \tag{2.213}$$

In equilibrium where there is a time-translational invariance, we have $g(-t) = g(t)$ (evenness) and $g^\star(t) = g(t)$ (reality) (these properties are the same as in imaginary-time formalism, see Eq. (2.124), as they should because both formalism are connected by analytical continuation). For Majorana fermions specifically, the conjugate relation (Eq. (2.196)) yields

$$\mathcal{G}^<(t_1,t_2) = +\frac{\imath}{2} e^{g(t_1,t_2)^\star/q}. \tag{2.214}$$

Then we employ the large-q limit where we expand $e^{g/q} = 1 + \frac{g}{q} + \mathcal{O}(1/q^2)$ and plug this in the Kadanoff-Baym equation in Eq. (2.211) and keeping up till the leading order in $1/q$. This gives (recall q is always even, therefore $q-1$ is odd)

$$\begin{aligned}
&\imath\left(\frac{-\imath}{2q}\right)\partial_{t_1} g\,(t_1,t_2) \\
&= \int_{-\infty}^{t_1} dt_3 \left\{ -(-1)^{\frac{q}{2}} J_q^2 \left[\left(\frac{-\imath}{2}\right)^{q-1} e^{g(t_1,t_3)} \right.\right. \\
&\qquad \left.\left. - \left(\frac{\imath}{2}\right)^{q-1} e^{g(t_1,t_3)^\star} \right] \left(\frac{-\imath}{2}\right) e^{\frac{g(t_3,t_2)}{q}} \right\} \\
&+ \int_{-\infty}^{t_2} dt_3 \left\{ (-1)^{\frac{q}{2}} J_q^2 \left(\frac{-\imath}{2}\right)^{q-1} e^{g(t_1,t_3)} \left[\left(\frac{-\imath}{2}\right) e^{g(t_3,t_2)/q} \right.\right. \\
&\qquad \left.\left. - \left(\frac{\imath}{2}\right) e^{g(t_3,t_2)^\star/q} \right] \right\} \\
&= -\int_{\infty}^{t_1} dt_3 \left\{ \frac{1}{2^{q-1}} \frac{1}{\imath} J_q^2 \left[e^{g(t_1,t_3)} + e^{g(t_1,t_3)^\star} \right] \left(\frac{\imath}{2}\right) e^{\frac{g(t_3,t_2)}{q}} \right\} \\
&+ \int_{-\infty}^{t_2} dt_3 \left\{ J_q^2 \left(\frac{1}{2^{q-1}\imath}\right) e^{g(t_1,t_3)} \left[\left(\frac{\imath}{2}\right) e^{g(t_3,t_2)/q} + \left(\frac{\imath}{2}\right) e^{g(t_3,t_2)^\star/q} \right] \right\} \\
&\Rightarrow \partial_{t_1} g\,(t_1,t_2) = -\int_{\infty}^{t_1} dt_3 \left\{ \frac{q}{2^{q-1}} J_q^2 \left[e^{g(t_1,t_3)} + e^{g(t_1,t_3)^\star} \right] e^{\frac{g(t_3,t_2)}{q}} \right\} \\
&+ \int_{-\infty}^{t_2} dt_3 \left\{ \frac{q}{2^{q-1}} J_q^2 e^{g(t_1,t_3)} \left[e^{g(t_3,t_2)/q} + e^{g(t_3,t_2)^\star/q} \right] \right\}
\end{aligned} \tag{2.215}$$

[33] The periodicity in imaginary-time (see Appendix A) translates to real-time at equilibrium as $\boxed{g(t) = g(-t - i\beta)}$ where we have assumed time-translational invariance (since it's in equilibrium) $g(t_1,t_2) = g(t_1 - t_2) = g(t)$ and β is the inverse temperature of the equilibrium state. This relation is known as the Kubo–Martin–Schwinger (KMS) relation (also see Appendix F). We will return to this later.

where we used the large-q limit in $e^{\frac{g}{q}(q-1)} \approx e^{g}$ and the simple manipulation $(-1)^{q/2}\imath^{q} = (-1)^{q/2}(\sqrt{-1})^{q} = (-1)^{q} = +1$. We identify $\mathcal{J}_q^2 \equiv \frac{q}{2^{q-1}}J_q^2$ that allows us to take the large-q limit that keeps $\mathcal{J}_q$ finite (see the discussion below Eq. (2.136)). Moreover, to keep the leading order in $1/q$, we see that all exponentials of the form $e^{g/q}$ is to be approximated by $1+g/q$ which to leading order in $\mathcal{O}(q^0)$ is simply 1. Therefore we finally get

$$\Rightarrow \partial_{t_1} g\,(t_1, t_2) = -\mathcal{J}_q^2 \int_{\infty}^{t_1} dt_3 \left[e^{g(t_1,t_3)} + e^{g(t_1,t_3)^\star}\right] + 2\mathcal{J}_q^2 \int_{-\infty}^{t_2} dt_3 e^{g(t_1,t_3)}. \tag{2.216}$$

Then we take derivative with respect to t_2 where we notice that the first term on the right-hand side has no t_2 dependence. Using Leibniz integral rule, namely $\frac{d}{dx}\left(\int_{a(x)}^{b(x)} f(x,t)dt\right) = f(x,b(x)) \cdot \frac{d}{dx}b(x) - f(x,a(x)) \cdot \frac{d}{dx}a(x) + \int_{a(x)}^{b(x)} \frac{\partial}{\partial x} f(x,t)dt$, we get (recall that partial derivatives commute)

$$\partial_{t_2}\partial_{t_1} g\,(t_1, t_2) = 2\mathcal{J}_q^2 e^{g(t_1,t_2)}. \tag{2.217}$$

Now we impose equilibrium where $g(t_1,t_2) = g(t_1-t_2) = g(t)$. Then using chain rule, we have $\partial_{t_2} g = \frac{dg}{dt} \cdot \partial_{t_2}(t) = \frac{dg}{dt} \cdot (-1) = -g'(t)$, followed by $\partial_{t_1}\partial_{t_2} g = \partial_{t_1}\left[-g'(t)\right]$. Using chain rule again, $\partial_{t_1}\left[-g'(t)\right] = -\frac{d}{dt}\left[g'(t)\right]\cdot\partial_{t_1} t = -g''(t)\cdot 1 = -g''(t)$. Therefore, the left-hand side becomes $-\frac{d^2g(t)}{dt^2}$ and we get

$$\frac{d^2 g(t)}{dt^2} = -2\mathcal{J}_q^2 e^{g(t)}. \tag{2.218}$$

In order to connect to the imaginary-time formalism, we perform the Wick's rotation $t \to -\imath\tau$ and continue $g(t)$ analytically to $g(\tau)$. This leads to the same differential equation (therefore, the same solution) we obtained in Eq. (2.136), thereby reaffirming the redundancy check for the real-time formalism.

We can check directly by plugging the following solution (including the boundary condition $g(t=0)=0$ and periodicity as mentioned in footnote 33) that the equilibrium Green's function in real-time is exactly given by (at any temperature $1/\beta$)

$$g(t) = \ln\left\{\frac{c_1}{4\mathcal{J}_q^2}\left(1 - \tanh^2\left(\frac{1}{2}\sqrt{c_1(t+c_2)^2}\right)\right)\right\} \tag{2.219}$$

where c_1 and c_2 are constants of integration that are determined by the initial condition. Without loss of generality, we can always chose to re-define

$$\sigma \equiv \sqrt{c_1}/2, \quad \theta \equiv \frac{-\imath c_2\sqrt{c_1}}{2} \tag{2.220}$$

such that the initial condition $g(0) = 0$ becomes

$$\cos\theta = \frac{\sigma}{\mathcal{J}_q} \tag{2.221}$$

which implies the bounds of σ, namely $-\mathcal{J}_q \leq \sigma \leq +\mathcal{J}_q$. Accordingly, we get

$$g(t) = 2\log\left\{\frac{\sigma}{\mathcal{J}_q}\,\mathrm{sech}(\imath\theta + \sigma t)\right\}. \tag{2.222}$$

from which we deduce that the initial state is completely determined by the ratio $\sigma/\mathcal{J}_q$. We can determine the temperature by imposing the KMS relation (see footnote 33) which gives the relation

$$\mathcal{J}_q\beta = 2\frac{\theta}{\sigma/\mathcal{J}_q} \tag{2.223}$$

where β is the inverse temperature of the state. Accordingly, we restrict to $0 \leq \sigma \leq \mathcal{J}_q$ for temperature to remain positive because $\theta = \cos^{-1}(\sigma/\mathcal{J}_q)$. The identity $\cos^{-1}(-x) = \pi - \cos(x)$ has been used where $x \in [-1, 1]$.

2.9.4.2 Energy in the Keldysh Plane

We already discussed the form of energy and its relation to Green's function in imaginary time in Sect. 2.8.1. We mentioned about the relation that it is a simpler version of a much more powerful (generalized) Galitskii-Migdal sum rule. We will return to this in later chapters. Here we wish to provide a hands-on experience in calculating the energy in the Keldysh plane. As we did in imaginary-time formalism in Sect. 2.2.1 by introducing a source field, we perform a similar manipulation in the Keldysh formalism. We are interested in calculating the energy $U_q = \langle H_q(t_1)\rangle$ where $H_q(t_1) = \int dt\mathcal{H}_q(t)$. Therefore, we use Eq. (2.189) where we identify $\hat{O}(t) = H_q = \int \mathcal{H}_q$. Accordingly, the generating functional takes the form

$$\mathcal{Z}[\eta] = \int D\psi_i \exp\{iS + \imath S_\eta\}$$

where we have introduced the source field $S_\eta = -\int dt\eta(t)\mathcal{H}(t)$. The disorder-averaged generating functional is given by

$$\overline{\mathcal{Z}}[\eta] = \int \mathcal{D}j_q\mathcal{P}_q[j_q]\mathcal{Z}[\eta] \tag{2.224}$$

where we use Eq. (2.5) and repeat the calculation as done above to get

$$\overline{\mathcal{Z}}[\eta] = \exp\left[\imath S_{\text{eff}}[\mathcal{G}, \Sigma] + \imath S_{\eta,\text{eff}}[\mathcal{G}, \Sigma, \eta]\right], \tag{2.225}$$

where $S_{\text{eff}}[\mathcal{G}, \Sigma]$ is given in Eq. (2.200) and $S_{\eta,\text{eff}}[\mathcal{G}, \Sigma, \eta]$ contains the source field dependency $\eta(t)$ given by

$$\frac{S_{\eta,\text{eff}}[\mathcal{G}, \Sigma, \eta]}{N} = \frac{\imath(-1)^{q/2}}{4q^2 2^{-q}} \mathcal{J}_q^2 \int_{\mathcal{C}} dt dt' \left[\eta(t) + \eta(t') + \eta(t)\eta(t')\right] \mathcal{G}(t, t')^q, \tag{2.226}$$

where we again identified the definition of $\mathcal{J}_q^2 \equiv J_q^2 \frac{q}{2^{q-1}}$. Using Eqs. (2.189) (where we taken $t_1 \in \mathcal{C}_-$ without loss of generality) and (2.225), we get

$$\begin{aligned} U_q(t_1) = \langle H_q(t_1) \rangle = \langle \int \mathcal{H}_q dt \rangle &= \imath \left. \frac{\delta \overline{\mathcal{Z}}[\eta]}{\delta \eta(t_1)} \right|_{\eta=0} \\ &= \imath \overline{\mathcal{Z}}[\eta] \imath \left. \frac{\delta S_{\eta,\text{eff}}[\mathcal{G}, \Sigma, \eta]}{\delta \eta(t_1)} \right|_{\eta=0} \\ &= -\imath \frac{N}{q^2} \frac{(-1)^{q/2}}{2^{2-q}} \mathcal{J}_q^2 \int_{\mathcal{C}} dt \left(\mathcal{G}(t_1, t)^q + \mathcal{G}(t, t_1)^q \right) \end{aligned} \tag{2.227}$$

where we used the fact that $\frac{\delta \eta(t')}{\delta \eta(t)} = \delta(t - t')$ and that the Keldysh partition function is normalized $\overline{\mathcal{Z}}[\eta = 0] = 1$. Note that unlike Sect. 2.2.1, we did not have to take any logarithm of the Keldysh generating functional before taking derivatives.

We can now proceed to simplify this integral over Keldysh contour using $\mathcal{C} = \mathcal{C}_+ + \mathcal{C}_-$, namely $\int_{\mathcal{C}} dt(\ldots) = \int_{-\infty}^{+\infty} dt(\ldots) + \int_{+\infty}^{-\infty} dt(\ldots)$. So we get

$$\begin{aligned} U_q(t_1) = -\imath \frac{N}{q^2} \frac{(-1)^{q/2}}{2^{2-q}} \mathcal{J}_q^2 \Bigg[&\int_{-\infty}^{+\infty} \mathcal{G}^>(t_1, t)^q dt + \int_{+\infty}^{-\infty} \mathcal{G}(t_1, t)^q dt \\ &+ \int_{-\infty}^{+\infty} \mathcal{G}^<(t, t_1)^q dt + \int_{+\infty}^{-\infty} \mathcal{G}(t, t_1)^q dt \Bigg] \end{aligned} \tag{2.228}$$

Invoking the Majorana conjugate relation from Eq. (2.196), we express $\mathcal{G}^<(t, t_1)^q = [-\mathcal{G}^>(t_1, t)]^q = \mathcal{G}^>(t_1, t)^q$ (since q is even) and we identify that the first and the third terms are the same. We next associate $\mathcal{G}(t, t_1)^q$ in the second and fourth terms with the anti-time-ordered Green's function, as both $t, t_1 \in \mathcal{C}_-$ (backward contour) (see Eq. (2.190)). We use Eq. (2.191) for the anti-time-ordered Green's function

$$\begin{aligned} \mathcal{G}(t, t_1)^q &= \left(\Theta(t_1 - t) \mathcal{G}^>(t, t_1) + \Theta(t - t_1) \mathcal{G}^<(t, t_1) \right)^q \\ &= \Theta(t_1 - t) \mathcal{G}^>(t, t_1)^q + \Theta(t - t_1) \mathcal{G}^<(t, t_1)^q, \text{ and} \\ \mathcal{G}(t_1, t)^q &= \Theta(t - t_1) \mathcal{G}^>(t_1, t)^q + \Theta(t_1 - t) \mathcal{G}^<(t_1, t)^q, \end{aligned} \tag{2.229}$$

where cross-terms vanish due to the properties of the Heaviside step function. We convert all functions in the form where their arguments follow the ordering (t_1, t).

$$U_q(t_1) = -\imath \frac{N}{q^2} \frac{(-1)^{q/2}}{2^{2-q}} \mathcal{J}_q^2 \times \Bigg[\int_{-\infty}^{+\infty} 2\mathcal{G}^>(t_1, t)^q dt - \int_{-\infty}^{+\infty} \Theta(t - t_1)\mathcal{G}^>(t_1, t)^q dt - \int_{-\infty}^{+\infty} \Theta(t_1 - t)\mathcal{G}^<(t_1, t)^q dt - \int_{-\infty}^{+\infty} \Theta(t_1 - t) \underbrace{\mathcal{G}^>(t, t_1)^q}_{=\mathcal{G}^<(t_1,t)^q} dt - \int_{-\infty}^{+\infty} \Theta(t - t_1) \underbrace{\mathcal{G}^<(t, t_1)^q}_{=\mathcal{G}^>(t_1,t)^q} dt \Bigg] \tag{2.230}$$

$$\Rightarrow U_q(t_1) = -\imath \frac{N}{q^2} \frac{(-1)^{q/2}}{2^{2-q}} \mathcal{J}_q^2 \Bigg[\int_{-\infty}^{+\infty} 2\mathcal{G}^>(t_1, t)^q dt - 2\int_{-\infty}^{+\infty} \Theta(t - t_1)\mathcal{G}^>(t_1, t)^q dt - 2\int_{-\infty}^{+\infty} \Theta(t_1 - t)\mathcal{G}^<(t_1, t)^q dt \Bigg] \tag{2.231}$$

$$\begin{aligned}
\Rightarrow U_q(t_1) &= -\imath \frac{N}{q^2} \frac{(-1)^{q/2}}{2^{2-q}} \mathcal{J}_q^2 \Bigg[\int_{-\infty}^{+\infty} 2\mathcal{G}^>(t_1, t)^q dt - 2\int_{t_1}^{+\infty} \mathcal{G}^>(t_1, t)^q dt - 2\int_{-\infty}^{t_1} \mathcal{G}^<(t_1, t)^q dt \Bigg] \\
&= -\imath \frac{N}{q^2} \frac{(-1)^{q/2}}{2^{2-q}} \mathcal{J}_q^2 \Bigg[\int_{-\infty}^{t_1} 2\mathcal{G}^>(t_1, t)^q dt - 2\int_{-\infty}^{t_1} \mathcal{G}^<(t_1, t)^q dt \Bigg] \\
&= -\imath \frac{N}{q^2} \frac{(-1)^{q/2}}{2^{1-q}} \mathcal{J}_q^2 \int_{-\infty}^{t_1} \Big(\mathcal{G}^>(t_1, t)^q - \mathcal{G}^<(t_1, t)^q \Big) dt.
\end{aligned} \tag{2.232}$$

Now we use the large-q ansatz from Eqs. (2.212) and (2.214) to get (recall that q is even)

$$
\begin{aligned}
U_q(t_1) =& -\imath \frac{N}{q^2}\frac{(-1)^{q/2}}{2^{1-q}}\mathcal{J}_q^2 \int_{-\infty}^{t_1} \left(\frac{\imath}{2}\right)^q \left(e^{g(t_1,t)} - e^{g^\star(t_1,t)}\right)dt \\
=& -\imath \frac{N}{2q^2}\mathcal{J}_q^2 \int_{-\infty}^{t_1} \left(e^{g(t_1,t)} - e^{g^\star(t_1,t)}\right)dt \\
=& \frac{N\mathcal{J}_q^2}{q^2}\mathrm{Im}\int_{-\infty}^{t_1} e^{g(t_1,t)}dt \\
u_q(t_1) \equiv \frac{U_q(t_1)}{N} =& \frac{\mathcal{J}_q^2}{q^2}\mathrm{Im}\int_{-\infty}^{t_1} e^{g(t_1,t)}dt \quad \text{(energy density)},
\end{aligned}
\tag{2.233}
$$

where Im[...] is the imaginary part of its argument. We already have the differential equation for $g(t_1, t_2)$ in Eq. (2.218) which can be solved for different non-equilibrium conditions and plugged here to get the energy density. At equilibrium, $g(t_1, t_2) = g(t_1 - t_2) = g(t)$ and we already know the solution in real-time from previous section, namely Eq. (2.222). Plugging here and performing the integral gives the equilibrium energy density as[34]

$$
u_q = -\frac{\sigma}{q^2}\tan\theta = -\frac{\mathcal{J}_q}{q^2}\sqrt{1-\left(\frac{\sigma}{\mathcal{J}_q}\right)^2}, \tag{2.234}
$$

where θ and σ are defined in Eq. (2.221) while the associated temperature is given by the KMS constraint in Eq. (2.223).

2.9.5 Example of a Simple Quench

We consider the Hamiltonian consisting of $\mathrm{SYK}_q + \mathrm{SYK}_2$ terms where we keep their interaction strengths time-dependent.

$$
\begin{aligned}
\mathcal{H}(t) =& \imath^{q/2}\sum_{\{i_q\}_\leq} j_{q;\{i_q\}}(t)\psi_{i_1}\ldots\psi_{i_q} + \imath\sum_{\{l_2\}_\leq} j_{2;\{l_2\}}(t)\psi_{l_1}\psi_{l_2} \\
=& \mathrm{SYK}_q + \mathrm{SYK}_2,
\end{aligned}
\tag{2.235}
$$

[34] Note the presence of q^2 in the denominator. Therefore, to have the correct scaling at large-N and at large-q, generally the energy is defined as $E \equiv \frac{q^2}{N}\langle H_q\rangle$ in the literature.

where the notation is already explained above in Eq. (2.3). The couplings are time-dependent[35] random variables drawn from Gaussian ensembles (see Eq. (2.4)) which we reproduce here for convenience

$$\begin{aligned}\mathcal{P}_q[j_{q;\{i_q\}}] &= \sqrt{\frac{N^{q-1}q^2 2^{-q}}{\pi \mathcal{J}_q^2 q!}}\exp\left(-\frac{1}{2\langle j_q^2\rangle}\sum_{\{i_q\}_\leq} j_{q;\{i_q\}}^2\right)\\ \mathcal{P}_2[j_{2;\{l_2\}}] &= \sqrt{\frac{Nq}{4\pi\mathcal{J}_2^2}}\exp\left(-\frac{1}{2\langle j_2^2\rangle}\sum_{\{l_2\}_\leq} j_{2;\{l_q\}}^2\right).\end{aligned} \tag{2.236}$$

from which we deduce the mean

$$\langle j_q\rangle = 0, \quad \langle j_2\rangle = 0,$$

and variances

$$\langle j_q^2\rangle = \frac{J_q^2(q-1)!}{N^{q-1}} = \frac{\mathcal{J}_q^2 q!}{N^{q-1}2^{1-q}q^2}, \quad \langle j_2^2\rangle = \frac{J_2^2}{N} = \frac{2\mathcal{J}_2^2}{Nq}$$

where we define $\mathcal{J}_q^2 \equiv 2^{1-q}qJ_q^2$ and $\mathcal{J}_2^2 \equiv 2^{-1}qJ_2^2$ as above.

Coherence Temperature

Note that SYK_2 term acts as the kinetic term while SYK_q contributes as the interacting Hamiltonian. A crucial property of SYK_2 is that there exists quasiparticles (in the sense of Fermi liquid theory, see Ref. [5]) in the system while there are no stable quasiparticles for SYK_q Hamiltonian. Therefore the mixed Hamiltonian provides a natural playground for competing interests of model with and without quasiparticles. In non-equilibrium dynamics at low temperatures, the renormalization group argument suggests that the Hamiltonian term with lesser number of fermions dominate. This argument is true only below a characteristic temperature scale, known as the *coherence temperature*, below which the kinetic term dominates. The *coherence temperature* $T_{\text{coherence}}$ defines the characteristic energy scale for the crossover regime where SYK_2 and SYK_q interactions become comparable. This competition is necessary for Planckian dynamics [7] to emerge, as the pure SYK_q model lacks such dynamics. Below $T_{\text{coherence}}$, the SYK_2

(continued)

[35] Until now, we only had Majorana fields as time-dependent while the coupling strengths were kept time-independent. Here, we put time-dependence in both the Majorana fields (as above) as well as the coupling strengths.

term dominates, driving the system toward Fermi liquid-like quasiparticle behavior. Above $T_{\text{coherence}}$, SYK_q interactions prevail, producing non-Fermi liquid (strange metal) physics. We have provided a detailed derivation of the coherence temperature in Appendix E whose result we state here

$$T_{\text{coherence}} = \left(\frac{J_2^q}{J_q^2}\right)^{\frac{1}{q-2}}, \tag{2.237}$$

where J_q and J_2 denote the coupling strengths of SYK_q and SYK_2 terms, respectively. Accordingly, in the large-q limit, $T_{\text{coherence}} = J_2$ is solely decided by the interaction strength of the SYK_2 term.

The quench protocol we wish to perform is as follows

$$\boxed{\mathcal{J}_q(t) = \mathcal{J}_q\Theta(-t), \quad \mathcal{J}_2(t) = \mathcal{J}_2\Theta(t)} \tag{2.238}$$

where we have the SYK_q term prepared in equilibrium in infinite past and at time $t = 0$, we switch off the SYK_q term and switch on the SYK_2 term. This sudden change is referred to as *quench* and is one of the useful techniques and quite often used in the literature for inducing non-equilibrium dynamics in a system. Quenches can be quite complex and analytically infeasible to track but this one is simpler and the purpose is to solve the quench for the post-quench Green's function.

We can think in terms of the $t - t'$ plane in Fig. 2.2 where the pre-quench Hamiltonian is just the SYK_q term whose solution we know from above. This implies that we already know the Green's function in quadrant C in Fig. 2.2. Now the quench happens at the origin and if we successfully solve for the Green's function in the far future (quadrant A) as well as quadrants B and D, we have successfully solved the non-equilibrium dynamics.

Since we are dealing with non-equilibrium dynamics, we must use the Keldysh formalism and solve for the Kadanoff-Baym equations. As we saw in Sect. 2.9.3, we need the Schwinger-Dyson equations in real-time. So we start with disorder-averaged partition function that will give us the Schwinger-Dyson equations, followed by writing down the Kadanoff-Baym equations which we will simplify using the large-q ansatz and finally solve the simplified equations. All steps are exactly the same as presented in detail above. However, if the reader needs further help in reproducing the calculations, we refer them to Ref. [8] (especially the appendices) where all mathematical steps pertaining to this model are presented in detail.

We start with the partition function in the Keldysh plane (substituting the Keldysh action (2.186) in the Keldysh partition function (2.185))

$$\mathcal{Z} = \int \mathcal{D}\psi_i \exp\left[-\frac{1}{2} \int dt \sum_{i=1} \psi_i \partial_t \psi_i - \imath \int dt \mathcal{H} \right] \tag{2.239}$$

where integration is performed over the Keldysh contour $\mathcal{C}$. Then the disorder-averaged partition function is given by

$$\begin{aligned} \overline{\mathcal{Z}} &= \iint \mathcal{D} j_{q;\{i_q\}} \mathcal{D} j_{2;\{l_2\}} \mathcal{P}_q[j_{q;\{i_q\}}] \mathcal{P}_2[j_{2;\{l_2\}}] \mathcal{Z} \\ &= \iint \mathcal{D}\mathcal{G}\mathcal{D}\Sigma e^{\imath S_{\text{eff}}[\mathcal{G},\Sigma]}, \end{aligned} \tag{2.240}$$

where we introduced the bi-local fields $\mathcal{G}$ and Σ using Eq. (2.198). The effective action is given by

$$\begin{aligned} \frac{S_{\text{eff}}[\mathcal{G},\Sigma]}{N} =& \frac{-\imath}{2} \log\det(\partial_t + \imath\Sigma) + \frac{\imath(-1)^{q/2}}{4q^2 2^{-q}} \int dt dt' \mathcal{J}_q(t) \mathcal{J}_q(t') \mathcal{G}(t,t')^q \\ &- \frac{\imath}{2q} \int dt dt' \mathcal{J}_2(t) \mathcal{J}_2(t') \mathcal{G}(t,t')^2 + \frac{\imath}{2} \int dt dt' \Sigma(t,t') \mathcal{G}(t,t'). \end{aligned} \tag{2.241}$$

Then the corresponding Euler-Lagrange equations by extremizing the action $\frac{\delta S_{\text{eff}}[\mathcal{G},\Sigma]}{\delta\Sigma} \stackrel{!}{=} 0$ and $\frac{\delta S_{\text{eff}}[\mathcal{G},\Sigma]}{\delta\mathcal{G}} \stackrel{!}{=} 0$ give the Schwinger-Dyson equations

$$\begin{aligned} \mathcal{G}_0(t,t')^{-1} =& \mathcal{G}(t,t')^{-1} + \Sigma(t,t') \\ \Sigma(t,t') =& -\frac{(-1)^{q/2}}{2^{1-q}q} \mathcal{J}_q(t) \mathcal{J}_q(t') \mathcal{G}^{q-1}(t,t') + \frac{2}{q} \mathcal{J}_2(t) \mathcal{J}_2(t') \mathcal{G}(t,t'). \end{aligned} \tag{2.242}$$

As per the steps detailed in Sect. 2.9.3, we go to the Keldysh contour, use the definitions highlighted in Sect. 2.9.2, and get the Kadanoff-Baym equation (as highlighted in Sect. 2.9.3, we need to only solve one Kadanoff-Baym equations for Majorana fermions to solve the system). We consider Eq. (2.206) which we reproduce here for convenience

$$\imath \partial_{t_1} \mathcal{G}^>(t_1,t_2) = \int_{-\infty}^{\infty} dt_3 \left(\Sigma^R(t_1,t_3) \mathcal{G}^>(t_3,t_2) + \Sigma^>(t_1,t_3) \mathcal{G}^A(t_3,t_2) \right).$$

Following the same steps as in Sect. 2.9.4 for our setup here, we get the full Kadanoff-Baym equation

$$\begin{aligned}
\imath\partial_{t_1}\mathcal{G}^{>}(t_1,t_2) = \int_{-\infty}^{t_1} dt_3 \Bigg\{ & -\frac{(-1)^{q/2}}{2^{1-q}q}\mathcal{J}_q(t_1)\mathcal{J}_q(t_3) \\
& \left[G^{>}(t_1,t_3)^{q-1} - G^{<}(t_1,t_3)^{q-1}\right] G^{>}(t_3,t_2) \\
& + \frac{2}{q}\mathcal{J}_2(t_1)\mathcal{J}_2(t_3)\left[G^{>}(t_1,t_3) - G^{<}(t_1,t_3)\right] G^{>}(t_3,t_2) \Bigg\} \\
& + \int_{-\infty}^{t_2} dt_3 \Bigg\{ \frac{(-1)^{q/2}}{2^{1-q}q}\mathcal{J}_q(t_1)\mathcal{J}_q(t_3) G^{>}(t_1,t_3)^{q-1} \\
& \times \left[G^{>}(t_3,t_2) - G^{<}(t_3,t_2)\right] \\
& - \frac{2}{q}\mathcal{J}_2(t_1)\mathcal{J}_2(t_3) G^{>}(t_1,t_3)\left[G^{>}(t_3,t_2) - G^{<}(t_3,t_2)\right] \Bigg\} .
\end{aligned} \tag{2.243}$$

which we intend to solve at large-q. So we plug in the large-q ansatz for the Green's function in Eqs. (2.212) and (2.214) and simplify to leading order in $1/q$ to get

$$\begin{aligned}
\partial_{t_1} g(t_1,t_2) = & -\int_{-\infty}^{t_1} dt_3 \left\{ \mathcal{J}_q(t_1)\mathcal{J}_q(t_3)\left[e^{g(t_1,t_3)} + e^{g^\star(t_1,t_3)}\right] + 2\mathcal{J}_2(t_1)\mathcal{J}_2(t_3) \right\} \\
& + \int_{-\infty}^{t_2} dt_3 \left\{ 2\mathcal{J}_q(t_1)\mathcal{J}_q(t_3) e^{g(t_1,t_3)} + 2\mathcal{J}_2(t_1)\mathcal{J}_2(t_3) \right\} .
\end{aligned} \tag{2.244}$$

We have kept things general (in terms of time-dependence) and exact to leading order in N and q. Now we impose the quench protocol in Eq. (2.238) on the coupling strengths. The quantum quench requires partitioning the Kadanoff-Baym equations into four causal quadrants in the two-time plane (refer Fig. 2.2):

- Quadrant A: $t_1, t_2 \geq 0$ (both times post-quench)
- Quadrant B: $t_1 \leq 0, t_2 \geq 0$ (pre- and post-quench)
- Quadrant C: $t_1, t_2 \leq 0$ (both pre-quench)
- Quadrant D: $t_1 \geq 0, t_2 \leq 0$ (post- and pre-quench).

Each quadrant must be solved independently, with integration constants subsequently determined through boundary consistency conditions. The causal structure is visualized in Fig. 2.2.

We now solve the system quadrant by quadrant, beginning with quadrant C. As quadrant C represents the initial equilibrium large-q SYK state, its solution $g_c(t_1, t_2) = g(t_1 - t_2) = g(t)$ follows directly from Eq. (2.222) whose initial temperature $1/\beta_0$ is given by Eq. (2.223).

Unlike equilibrium quadrant C, quadrants B and D represent non-equilibrium dynamics where Eq. (2.244) remains a full integro-differential equation that cannot be simplified to an ordinary differential equation. Crucially, the conjugate relation for $g(t_1, t_2)$ in Eq. (2.195) (which translates for g to Eq. (2.213)) enables a symmetric solution strategy: solving either quadrant B or D provides the solution for its conjugate partner through complex conjugation. We, therefore, focus on quadrant D ($t_1 \geq 0$, $t_2 \leq 0$), where the discontinuous coupling constants, governed by step functions in Eq. (2.238), permit direct integration of Eq. (2.244).

$$\begin{aligned}\partial_{t_1} g(t_1, t_2) = &- \int_{-\infty}^{0} dt_3 \left\{ \mathcal{J}_q(t_1)\, \mathcal{J}_q(t_3) \left[e^{g(t_1,t_3)} + e^{g^\star(t_1,t_3)} \right] + 2\mathcal{J}_2(t_1)\mathcal{J}_2(t_3) \right\} \\ &- \int_{0}^{t_1} dt_3 \left\{ \mathcal{J}_q(t_1)\, \mathcal{J}_q(t_3) \left[e^{g(t_1,t_3)} + e^{g^\star(t_1,t_3)} \right] + 2\mathcal{J}_2(t_1)\mathcal{J}_2(t_3) \right\} \\ &+ 2 \int_{-\infty}^{t_2} dt_3 \left\{ \mathcal{J}_q(t_1)\, \mathcal{J}_q(t_3)\, e^{g(t_1,t_2)} + \mathcal{J}_2(t_1)\mathcal{J}_2(t_3) \right\},\end{aligned} \tag{2.245}$$

where we substitute Eq. (2.238) for the couplings and use $t_1 \geq 0$, $t_2 \leq 0$ (since we are in quadrant D) to get

$$\partial_{t_1} g_D(t_1, t_2) = -2\mathcal{J}_2^2 t_1 \tag{2.246}$$

that gives

$$g_D(t_1, t_2) = -\mathcal{J}_2^2 t_1^2 + D(t_2). \tag{2.247}$$

The integration constant is found by the continuity requirement of the Green's function at the boundary of quadrants, namely

$$g_D(0, t_2) = g_C(0, t_2) \tag{2.248}$$

where we know the solution for g in quadrant C from Eq. (2.222). Using that, we get for quadrant D

$$\boxed{g_D(t_1, t_2) = -\mathcal{J}_2^2 t_1^2 + 2\ln\left[\frac{\sigma}{\mathcal{J}_q \cosh(\imath\theta - \sigma t_2)}\right]}. \tag{2.249}$$

which immediately gives via complex conjugation the result for quadrant B

$$\boxed{g_B(t_1, t_2) = g_D(t_2, t_1)^\star = -\mathcal{J}_2^2 t_2^2 + 2\ln\left[\frac{\sigma}{\mathcal{J}_q \cosh(\imath\theta + \sigma t_1)}\right]}. \tag{2.250}$$

Recall that σ and θ are related to each other and to the initial temperature $1/\beta_0$ via Eqs. (2.221) and (2.223).

The last quadrant that remains is quadrant A where the non-equilibrium dynamics settle down in the far future ($t_1, t_2 \gg 0$) and the system reaches some form of stationarity and equilibrium. Here we directly integrate the Kadanoff-Baym equation subject to boundary conditions. The first-order derivative equation derived from Eq. (2.245) while using Eq. (2.238) for the couplings is

$$\partial_{t_1} g_A(t_1, t_2) = -2\mathcal{J}_2^2 t_1 + 2\mathcal{J}_2^2 t_2 \quad (t_1, t_2 > 0) \tag{2.251}$$

which is integrated to

$$g_A(t_1, t_2) = -\mathcal{J}_2^2 t_1^2 + 2\mathcal{J}_2^2 t_1 t_2 + A(t_2) \tag{2.252}$$

where we find the integration constant $A(t_2)$[36] via the same-time boundary condition $g(t, t) = 0$) (see Eq. (2.213)). This gives

$$A(t_2) = \mathcal{J}_2^2 t_2^2 - 2\mathcal{J}_2^2 t_2^2 = -\mathcal{J}_2^2 t_2^2. \tag{2.253}$$

Therefore we have solved for quadrant A too:

$$\boxed{g_A(t_1, t_2) = -\mathcal{J}_2^2 t_1^2 + 2\mathcal{J}_2^2 t_1 t_2 - \mathcal{J}_2^2 t_2^2 = -\mathcal{J}_2^2 (t_1 - t_2)^2} \tag{2.254}$$

where time-translational invariance emerges on its own (without imposing) right after the quench. Therefore the system reaches stationary conditions (where time-translational invariance holds) instantaneously after the quench. We also note that g_A obtained is always real.

We have successfully solved the quench dynamics analytically by evaluating the Green's functions in all four quadrants of Fig. 2.2.

2.9.5.1 Aside: Energy

Just like we evaluated the energy density for a single Majorana SYK dot in Eq. (2.233), we can evaluate along the same lines the energy for our setup where the generating functional is given by

$$\overline{\mathcal{Z}}[\eta] = \exp\left[\imath S_{\text{eff}}[\mathcal{G}, \Sigma] + \imath S_{\eta,\text{eff}}[\mathcal{G}, \Sigma]\right]. \tag{2.255}$$

The effective action $S_{\text{eff}}[\mathcal{G}, \Sigma]$ is the same as Eq. (2.241) while $S_{\eta,\text{eff}}[\mathcal{G}, \Sigma]$ is evaluated the same as in Eq. (2.226) to get

$$\begin{aligned}\frac{S_{\eta,\text{eff}}[\mathcal{G}, \Sigma]}{N} = &\frac{\imath(-1)^{q/2}}{4q^2 2^{-q}} \int_C dt dt' \mathcal{J}_q(t)\mathcal{J}_q(t') \left[\eta(t) + \eta(t') + \eta(t)\eta(t')\right] \mathcal{G}(t, t')^q \\ &- \frac{\imath}{2q} \int_C dt dt' \mathcal{J}_2(t)\mathcal{J}_2(t') \left[\eta(t) + \eta(t') + \eta(t)\eta(t')\right] \mathcal{G}(t, t')^2.\end{aligned} \tag{2.256}$$

[36] Since we integrate with respect to t_1, the integration constant cannot depend on t_1.

Then the energy is calculated the same as in Eq. (2.227) ($\mathcal{H}$ is given in Eq. (2.235))

$$U(t_1) = \langle H(t_1) \rangle = \langle \int \mathcal{H} dt \rangle = \imath \left. \frac{\delta \overline{\mathcal{Z}}[\eta]}{\delta \eta(t_1)} \right|_{\eta=0} \tag{2.257}$$

to get

$$\begin{aligned} U(t_1) = -\imath \frac{N}{q^2} \Big[\frac{(-1)^{q/2}}{2^{2-q}} \int_C dt \mathcal{J}_q(t_1) \mathcal{J}_q(t) \left(\mathcal{G}(t_1, t)^q + \mathcal{G}(t, t_1)^q \right) \\ - \frac{q}{2} \int_C dt \mathcal{J}_2(t_1) \mathcal{J}_2(t) \left(\mathcal{G}(t_1, t)^2 + \mathcal{G}(t, t_1)^2 \right) \Big] \end{aligned} \tag{2.258}$$

where we see that if we put $\mathcal{J}_2 = 0$, we re-derive Eq. (2.227) (with time-independent couplings). Accordingly the energy density is given by using large-q ansatz Eqs. (2.212) and (2.214) at leading order in $1/q$

$$\boxed{u(t_1) \equiv \frac{U(t_1)}{N} = \frac{1}{q^2} \mathrm{Im} \Big\{ \int_{-\infty}^{t_1} dt_2 \Big(\mathcal{J}_q(t_1) \mathcal{J}_q(t_2) e^{g(t_1, t_2)} + \mathcal{J}_2(t_1) \mathcal{J}_2(t_2) g(t_1, t_2) \Big) \Big\}} \tag{2.259}$$

which reduces to Eq. (2.233) for time-independent couplings and $\mathcal{J}_2 = 0$.

We can use this relation to calculate the energy density post-quench (quadrant A)[37] using Eq. (2.254) which we found to be time-translational invariant instantaneously after the quench and always real. Since we are in quadrant A, the integration must start from the origin. Therefore the energy density right after quench is

$$u(t_1 \geq 0) = \frac{1}{q^2} \mathcal{J}_2^2 \mathrm{Im} \int_{-\infty}^{t_1} dt_2 \, g_A(t_1, t_2) = 0. \tag{2.260}$$

Consequently, the kinetic energy (which constitutes the total energy here) adjusts instantaneously to its post-quench value. The pre-quench energy density in quadrant C is determined by the initial inverse temperature β_0. Remarkably, the post-quench state always reaches zero energy instantaneously regardless of β_0, corresponding to an infinite-temperature thermal state.

References

1. Bhattacharya, R., Jatkar, D.P., Sorokhaibam, N.: Quantum quenches and thermalization in SYK models. J. High Energy Phys. **2019**(7), 66 (2019)

[37] Energy density pre-quench (quadrant C) is just the same as the single dot Majorana SYK at large-q given by Eq. (2.233).

2. Bruus, H., Flensberg, K.: Many-Body Quantum Theory in Condensed Matter Physics. Oxford University Press, Oxford (2004)
3. Castellani, T., Cavagna, A.: Spin-glass theory for pedestrians. J. Stat. Mech.: Theory Exp. **2005**(05), P05012 (2005)
4. Chowdhury, D., Georges, A., Parcollet, O., Sachdev, S.: Sachdev-Ye-Kitaev models and beyond: a window into non-Fermi liquids. Rev. Mod. Phys. **94**(3), 035004 (2022). arXiv:2109.05037 [cond-mat, physics:hep-th, physics:quant-ph]
5. Coleman, P.: Introduction to Many-Body Physics. Cambridge University Press, Cambridge (2015)
6. Eberlein, A., Kasper, V., Sachdev, S., Steinberg, J.: Quantum quench of the Sachdev-Ye-Kitaev model. Phys. Rev. B **96**(20), 205123 (2017)
7. Hartnoll, S.A., Mackenzie, A.P.: Colloquium: planckian dissipation in metals. Rev. Mod. Phys. **94**(4), 041002 (2022)
8. Jaramillo, S.S., Jha, R., Kehrein, S.: Thermalization of a closed Sachdev-Ye-Kitaev system in the thermodynamic limit. Phys. Rev. B **111**(19), 195153 (2025)
9. Kamenev, A.: Field Theory of Non-Equilibrium Systems. Cambridge University Press, Cambridge (2023)
10. A. Kitaev, A simple model of quantum holography. Talks given at "Entanglement in Strongly-Correlated Quantum Matter,". (Part1,. Part2), KITP (2015)
11. Maldacena, J., Stanford, D.: Remarks on the Sachdev-Ye-Kitaev model. Phys. Rev. D **94**(10), 106002 (2016)
12. Maldacena, J., Shenker, S.H., Stanford, D.: A bound on chaos. J. High Energy Phys. **2016**(8), 1–17 (2016)
13. Milekhin, A.: Non-local reparametrization action in coupled Sachdev-Ye-Kitaev models. J. High Energy Phys. **2021**(12), 1–45 (2021)
14. Sarosi, G.: Ads$_2$ holography and the syk model. In: Proceedings of XIII Modave Summer School in Mathematical Physics — PoS(Modave2017). Sissa Medialab (2018)
15. Stefanucci, G., van Leeuwen, R.: Nonequilibrium Many-Body Theory of Quantum Systems: A Modern Introduction. Cambridge University Press, Cambridge (2013)
16. Zinn-Justin, J., Zinn-Justin, J.: Quantum Field Theory and Critical Phenomena. Oxford University Press, Oxford (2021)

Complex Generalization

3

Abstract

This chapter continues the construction of the SYK framework by introducing the complex fermion generalization, which supplements the Majorana model with a conserved $U(1)$ charge and an associated chemical potential. Starting from an all to all $q/2$ body Hamiltonian of spinless complex fermions with random Gaussian couplings, the chapter defines the grand canonical setup, the charge density, and the precise ensemble for the disordered interactions. The Euclidean path integral is then constructed, disorder averaged, and rewritten in terms of bilocal Green's function and self energy fields, yielding an effective action whose large N saddle gives the Schwinger Dyson equations for both free and interacting complex SYK models at general q. Special attention is paid to the modified structure of the free Green's function in the presence of $U(1)$ symmetry, its Matsubara representation, and the resulting distinction between $\mathcal{G}(\tau)$ and $\mathcal{G}(-\tau)$. The chapter further analyzes the residual $U(1)$ symmetry that survives the breaking of $U(N)$ by random couplings, develops the infrared and conformal limit with $\text{Diff}(\mathbb{R}) \times U(1)$ invariance, and derives conformal Green's functions and self energies including spectral and particle hole asymmetry. Finally, a large-q expansion provides closed analytic solutions for the Green's function and self energy at finite temperature, establishing the complex SYK model as a charged, solvable non Fermi liquid laboratory for thermodynamics and transport in later chapters.

Having established the foundational Majorana SYK model as a minimal framework for non-Fermi liquid physics with hints of holography, we now turn to its complex fermion generalization. This extension replaces Majorana operators with conventional complex fermions—a seemingly simple modification that profoundly enriches the model's physical scope. Crucially, the complex SYK model exhibits a conserved $U(1)$ charge $\mathcal{Q}$ corresponding to the charged fermions, bringing it

R. Jha, *Theory of Strongly Correlated Systems*, Lecture Notes in Physics 1051,
https://doi.org/10.1007/978-3-032-20910-8_3

closer to electronic systems in condensed matter where such conservation laws are ubiquitous. This symmetry unlocks new frontiers inaccessible to the Majorana variant:

- Transport Phenomena: Charge conservation enables the study of electrical/thermal conductivity and diffusivity in strongly correlated non-Fermi liquids.
- Thermalization Dynamics: The emergent time re-parameterization and conserved $U(1)$ symmetries govern thermalization processes in charge sectors.
- Finite-Density States: Access to finite chemical potential regimes relevant for accessing insulating, strange metal and Fermi-liquid type phases.

These features make the complex SYK an indispensable tool for exploring quantum critical transport, many-body quantum chaos, and holographic mappings to charged black holes. In this chapter, we develop this generalization before delving into advanced (aforementioned) topics in later chapters. Our focus while introducing the complex SYK model in this chapter will be mainly coming from the applications in condensed matter theory and we refrain from divulging towards holographic applications. Additionally, we will try to keep our setup in this chapter close to the Majorana SYK model, so that we can keep drawing parallels wherever it can be drawn. Most of the physics carries through, that's why we will not repeat everything here but refer the reader to Chap. 2. We also highlight the differences with the Majorana variant at all points where we encounter such physics.

3.1 Model

We consider a system of N spinless charged fermions where we have a conserved $U(1)$ symmetry in addition to the $O(N)$ symmetry as discussed in Sect. 2.4. Just like the Majorana SYK, we have all-to-all $q/2$-body interaction, effectively making the system spatially zero dimensional. The interacting Hamiltonian is given by (subscript q on the coupling strength is a label to denote $q/2$-body interaction, just like we did for the Majorana SYK)

$$\mathcal{H}_q = \sum_{\substack{\{i_{1:\frac{q}{2}}\}_{\leq} \\ \{i_{\frac{q}{2}+1:q}\}_{\leq}}} j_{q;\{i_{1:\frac{q}{2}}\},\{i_{\frac{q}{2}+1:q}\}} c^{\dagger}_{i_1} c^{\dagger}_{i_2} \dots c^{\dagger}_{i_{q/2}} c_{i_{q/2+1}} \dots c_{i_{q-1}} c_{i_q} \tag{3.1}$$

where we consider the coupling strengths to be time-independent but the fermionic operators may depend on time. We also borrow the notation from Eq. (2.3) with slight modification which we reproduce here for convenience

$$\{i_{1:q}\}_{\leq} \equiv 1 \leq i_1 < i_2 < \dots < i_q \leq N, \tag{3.2a}$$

$$\{i_{1:q}\} \equiv \{i_1, i_2, \dots, i_{q-1}, i_q\}. \tag{3.2b}$$

Therefore, for $q = 4$, the Hamiltonian looks like

$$\mathcal{H}_4 = \sum_{\substack{1\le i_1<i_2\le N\\ 1\le i_3<i_4\le N}} j_{4;i_1i_2,i_3i_4} c^{\dagger}_{i_1} c^{\dagger}_{i_2} c_{i_3} c_{i_4}. \tag{3.3}$$

Note that we have separated the two set of indices $\{i_{1:\frac{q}{2}}\}$ and $\{i_{\frac{q}{2}+1:q}\}$ separately by a comma in the coupling strength because one labels the fermionic creation operators $c_i^{\dagger}$ while the other set labels the fermionic annihilation operators c_i, which satisfy the following anti-commutation relations *at equal time*[1]

$$\left\{c_i, c_j^{\dagger}\right\} = \delta_{ij}, \quad \left\{c_i, c_j\right\} = 0 = \left\{c_i^{\dagger}, c_j^{\dagger}\right\}. \tag{3.4}$$

Due to the Hamiltonian being Hermitian, we get the following constraint for the coupling strength

$$j^{\star}_{q;\{i_{1:\frac{q}{2}}\},\{i_{\frac{q}{2}+1:q}\}} = j_{q;\{i_{\frac{q}{2}+1:q}\},\{i_{1:\frac{q}{2}}\}}. \tag{3.5}$$

As an illustration, for $q = 4$, this becomes $j^{\star}_{4;i_1i_2,i_3i_4} = j_{4;i_3i_4,i_1i_2}$.

Just like the Majorana SYK, the coupling constants j_q (other indices suppressed for brevity) are derived from Gaussian ensembles whose mean and variance are given by

$$\langle j_q\rangle = 0, \qquad \sigma_q^2 = \langle j_q^2\rangle = \frac{2(q/2!)^2 J_q^2}{(q/2)N^{q-1}}, \tag{3.6}$$

where J_q is a constant governing the strength of the interaction. The Gaussian ensemble is given by

$$\mathcal{P}_q\left[j_{q;\{i_{1:\frac{q}{2}}\},\{i_{\frac{q}{2}+1:q}\}}\right] = A\exp\left(-\frac{1}{2\sigma_q^2}\sum_{\{i_q\}_{\le}}\left|j_{q;\{i_{1:\frac{q}{2}}\},\{i_{\frac{q}{2}+1:q}\}}\right|^2\right), \tag{3.7}$$

where $A = \sqrt{\frac{1}{2\pi\sigma_q^2}}$ is calculated via the normalization condition. Note that we have the complex identity $|z|^2 = zz^{\star} = z^{\star}z$. Since the fermions are charged, there is

[1] This flows naturally from the choice of convention we made for the Majorana fermions anti-commutation relation in Eq. (2.2) (see footnote 2). In particular, every complex fermion can be decomposed into two real Majorana fermions as $\underbrace{c_i}_{\text{complex}} = \frac{\psi_{i1}+i\psi_{i2}}{\sqrt{2}}, \underbrace{c_i^{\dagger}}_{\text{complex}} = \frac{\psi_{i1}-i\psi_{i2}}{\sqrt{2}}$. Substituting this in Eq. (2.2), namely $\left\{\psi_{ia}, \psi_{jb}\right\} = \delta_{ij}\delta_{ab} \quad (a, b \in \{1, 2\})$, we get Eq. (3.4). In order to match with the other convention for the anti-commutation relation mentioned in footnote 2, the decomposition would be $c_i = \frac{\chi_{i1}+i\chi_{i2}}{2}, c_i^{\dagger} = \frac{\chi_{i1}-i\chi_{i2}}{2}$. We stick with our convention in Eq. (2.2) throughout this work.

an associated Chemical potential chemical potential μ whose contribution in the Hamiltonian is given by

$$\mathcal{H}_{\mu} = -\mu \sum_{i}^{N} c_i^{\dagger} c_i \tag{3.8}$$

where the conserved charge is given by the following charge density

$$\mathcal{Q} \equiv \frac{1}{N} \sum_{i=1}^{N} \left\langle c_i^{\dagger} c_i \right\rangle - \frac{1}{2}. \tag{3.9}$$

With this additional chemical potential term, the full Hamiltonian is given by

$$\mathcal{H} = \mathcal{H}_q + \mathcal{H}_{\mu}. \tag{3.10}$$

Much of the formalism (with the addition of the $U(1)$ symmetry) will be quite similar to the Majorana SYK model in the coming sections.

Caution on Conventions
As the reader might have noticed, there are a few conventions that goes in the formalism of the SYK model that might change the presence of factors here and there, depending on which literature one is referring. We briefly mention the relevant ones that the reader should be aware of and must check while referring to the literature. Needless to say, one must be consistent once a convention is decided upon and in general, no physics (should) depend on the nature of any particular chosen convention.

The first convention is that of choosing the variance of the Gaussian probability distributions as done in Eq. (3.6) or in Eq. (2.4) for the Majorana SYK model. The scaling of the theory is fundamentally rooted in how these Gaussian probability distributions are defined and one must be careful of always verifying the convention used. As an example, an alternative to the variance for the complex SYK model provided in Eq. (3.6) is the following (note that there is a presence of q^2 instead of q when compared to the variance in Eq. (3.6))

$$\sigma_q^2 = \frac{4\left(\frac{J_q}{q}\right)^2 \left[\left(\frac{q}{2}\right)!\right]^2}{\left(\frac{N}{2}\right)^{q-1}}. \tag{3.11}$$

We will get a hands-on experience with this variance in later chapters.

(continued)

The other crucial convention enters in the Green's function. The convention manifests itself as the prefactors that are chosen in the definition of the Green's function. For example, we choose previously in the real-time formalism a prefactor of $-\imath$ in Eq. (2.212). Other conventions include choosing a prefactor of -1. The way different conventions work is by re-adjusting the definition of the self-energy, which generally gets introduced as a Lagrange multiplier (e.g. Eq. (2.198)). Certain factors get absorbed in the self-energy so that different conventions balance out and no physical phenomenon depends on the convention chosen.

This work intentionally exposes the reader to different notational conventions used in SYK literature—an important skill for navigating research. While each chapter strictly adheres to one self-consistent convention (explicitly stated upfront), we vary conventions across chapters to give the reader a hands-on experience. We recommend to always check the convention declaration at the start of each chapter. This discipline mirrors real research practice, where convention-checking precedes engagement with any paper. We have already defined the convention for the variance in Eq. (3.6) above, while the convention for the Green's function will be introduced below in Eq. (3.12).

3.2 The Schwinger-Dyson Equations

3.2.1 The Free Case

We start by calculating the free Green's function where the interacting Hamiltonian is zero and $\mathcal{H}_0 = \mathcal{H}_\mu$ (subscript 0 denotes the free case). The reason we start from the free case in this chapter is to highlight the difference right from the outset that $U(1)$ symmetry brings in the structure of the Green's function when compared to the Majorana case. Since we are in equilibrium where there is time-translational invariance, we chose to work in the imaginary-time formalism τ obtained from the real time through Wick's rotation $t \to -\imath\tau$. We start by defining our Green's function in general (free or otherwise)

$$\boxed{\mathcal{G}_{ij}(\tau) = -\langle T c_i(\tau) c_j^\dagger(0)\rangle = -\left[\Theta(\tau)\langle c_i(\tau)c_j^\dagger(0)\rangle - \Theta(-\tau)\langle c_j^\dagger(0)c_i(\tau)\rangle\right]} \tag{3.12}$$

which sets our convention for this chapter. Here, T is the time-ordering operator $T c_i(\tau)c_j^\dagger(0) = \begin{cases} c_i(\tau)c_j^\dagger(0) & \text{if } \tau > 0 \\ -c_j^\dagger(0)c_i(\tau) & \text{if } \tau < 0 \end{cases}$ and $\Theta(\tau)$ is the Heaviside step function.

Now we specialize to the free case and take its derivative (subscript 0 denotes free case)

$$
\begin{aligned}
\partial_\tau \mathcal{G}_{ij}(\tau) =& -\partial_\tau(\Theta(\tau))\langle c_i(\tau)c_j^\dagger(0)\rangle - \Theta(\tau)\langle\partial_\tau(c_i(\tau))c_j^\dagger(0)\rangle \\
&+ \partial_\tau(\Theta(-\tau))\langle c_j^\dagger(0)c_i(\tau)\rangle + \Theta(-\tau)\langle c_j^\dagger(0)\partial_\tau(c_i(\tau))\rangle \\
=& -\delta(\tau)\langle c_i(\tau)c_j^\dagger(0)\rangle - \Theta(\tau)\langle\partial_\tau(c_i(\tau))c_j^\dagger(0)\rangle \\
&- \delta(\tau)\langle c_j^\dagger(0)c_i(\tau)\rangle + \Theta(-\tau)\langle c_j^\dagger(0)\partial_\tau(c_i(\tau))\rangle \\
=& -\delta(\tau)\langle\{c_i(\tau), c_j^\dagger(0)\}\rangle - \langle T\partial_\tau(c_i(\tau))c_j^\dagger(0)\rangle \\
=& -\delta(\tau)\delta_{ij} - \langle T\partial_\tau(c_i(\tau))c_j^\dagger(0)\rangle
\end{aligned}
\tag{3.13}
$$

where we used $\partial_\tau\Theta(\tau) = \delta(\tau)$, $\partial_\tau\Theta(-\tau) = -\delta(\tau)$ and the definition of time-ordering. Note that the presence of δ-function allowed us to enforce the anti-commutation relation in Eq. (3.4). We have not made use of the free theory yet. Now we have to evaluate the time derivative of the fermionic operators. We use the Heisenberg's equation of motion in imaginary-time using the free Hamiltonian $\mathcal{H}_0 = \mathcal{H}_\mu = -\mu\sum_i^N c_i^\dagger c_i$ to get[2]

$$
\partial_\tau c_j(\tau) = \left[\mathcal{H}_0, c_j\right](\tau) = \mu c_j(\tau) \tag{3.14}
$$

where we used

$$
\left[\mathcal{H}_0, c_j\right] = -\mu\sum_i\left[c_i^\dagger c_i, c_j\right] = -\mu\sum_i\left(-\delta_{ij}c_i\right) = \mu c_j, \tag{3.15}
$$

using $[AB, C] = A[B, C] + [A, C]B$.[3] Therefore we have for the free Green's function

$$
\partial_\tau \mathcal{G}_{0,ij}(\tau) = -\delta(\tau)\delta_{ij} - \mu\underbrace{\langle T c_i(\tau)c_j^\dagger(0)\rangle}_{=-\mathcal{G}_{0,ij}} = -\delta(\tau)\delta_{ij} + \mu\mathcal{G}_{0,ij}(\tau). \tag{3.16}
$$

[2] The real-time evolution of an operator in Heisenberg picture is given by (we are using natural units, so $\hbar = 1$) $A(\tau) = e^{\imath t\mathcal{H}}Ae^{-\imath t\mathcal{H}} \xrightarrow{\imath t\to\tau} A(\tau) = e^{\tau\mathcal{H}}Ae^{-\tau\mathcal{H}}$ which translates into $\frac{d}{d\tau}A(\tau) = [\mathcal{H}, A(\tau)]$. Equivalently, we could have also started from the real-time formulation of the Heisenberg's equation of motion $\frac{dA}{dt} = \imath[\mathcal{H}, A(t)]$ and use the chain rule $\frac{dA}{dt} = \frac{dA}{d\tau}\cdot\frac{d\tau}{dt}$ where $\tau = \imath t$ to get $\frac{dA}{dt} = \imath\frac{dA}{d\tau}$. Thus the Heisenberg's equation of motion in imaginary-time is given by $\frac{d}{d\tau}A(\tau) = [\mathcal{H}, A(\tau)]$.

[3] The way the commutation is calculated using the anti-commutation relations in Eq. (3.4)

$$
\left[c_i^\dagger c_i, c_j\right] = c_i^\dagger c_i c_j - \underbrace{c_j c_i^\dagger}_{=\delta_{ij}-c_i^\dagger c_j} c_i = c_i^\dagger c_i c_j - (\delta_{ij}c_i - c_i^\dagger\underbrace{c_j c_i}_{=-c_i c_j}) = -\delta_{ij}c_i.
$$

Taking the Fourier transform where $\partial_\tau \to -\imath\omega$ (since we are in imaginary-time formalism, ω denotes Matsubara frequencies as explained in Appendix F), we get

$$-i\omega\mathcal{G}_{0,ij}(i\omega) = -\delta_{ij} + \mu\mathcal{G}_{0,ij}(\imath\omega) \tag{3.17}$$

which can be re-arranged and solved as

$$\boxed{\mathcal{G}_{0,ij}(\omega) = \frac{\delta_{ij}}{\imath\omega + \mu}}. \tag{3.18}$$

Taking the inverse Fourier transform gives

$$\mathcal{G}_{0,ij}(\tau) = \begin{cases} -\delta_{ij}\frac{e^{\mu\tau}}{e^{\beta\mu}+1} & ; \quad 0 < \tau < \beta \\ \delta_{ij}\frac{e^{\mu\tau}}{e^{-\beta\mu}+1} & ; \quad -\beta < \tau < 0. \end{cases} \tag{3.19}$$

We can do a quick verification: using the Fourier transform in Eq. (F.2), we get for $0 < \tau < \beta$

$$\begin{aligned}\mathcal{G}_{0,ij}(\omega) = \int_0^\beta e^{\imath\omega\tau}\mathcal{G}_{0,ij}(\tau)d\tau &= -\delta_{ij}\int_0^\beta \frac{e^{(\imath\omega+\mu)\tau}}{e^{\beta\mu}+1}d\tau \\ &= -\delta_{ij}\frac{e^{(\imath\omega+\mu)\beta}-1}{(e^{\beta\mu}+1)(\imath\omega+\mu)}\end{aligned} \tag{3.20}$$

where we have removed the subscript n for Matsubara frequencies throughout (except Appendix F where this is introduced in detail). Now using the fermionic Matsubara frequency from Appendix F, namely $\omega = \frac{(2n+1)\pi}{\beta}$, where $n \in \mathbb{Z}$, we get

$$\mathcal{G}_{0,ij}(\omega) = -\delta_{ij}\frac{-e^{\mu\beta}-1}{(e^{\beta\mu}+1)(\imath\omega+\mu)} = \delta_{ij}\frac{e^{\mu\beta}+1}{(e^{\beta\mu}+1)(\imath\omega+\mu)} = \delta_{ij}\frac{1}{(\imath\omega+\mu)} \tag{3.21}$$

which matches with Eq. (3.18). Similarly for $-\beta < \tau < 0$, we have

$$\begin{aligned}\mathcal{G}_{0,ij}(\omega) &= \int_{-\beta}^0 e^{\imath\omega\tau}\mathcal{G}_{0,ij}(\tau)d\tau = \delta_{ij}\int_{-\beta}^0 \frac{e^{(\imath\omega+\mu)\tau}}{e^{-\beta\mu}+1}d\tau \\ &= \delta_{ij}\frac{1-e^{(\imath\omega+\mu)(-\beta)}}{(e^{-\beta\mu}+1)(\imath\omega+\mu)} \\ &= \delta_{ij}\frac{1-(-1)e^{-\beta\mu}}{(e^{-\beta\mu}+1)(\imath\omega+\mu)} = \delta_{ij}\frac{1}{(\imath\omega+\mu)} \quad \text{(same as Eq. (3.18))}.\end{aligned} \tag{3.22}$$

We can also verify that Eq. (3.19) satisfies the anti-periodicity, $\mathcal{G}_{ij}(\tau+\beta) = -\mathcal{G}_{ij}(\tau)$ (see Appendix A):

- For $-\beta < \tau < 0$: Here, $\tau' = \tau + \beta \in (0, \beta)$, so we use the functional form of $\mathcal{G}_{0,ij}(\tau')$ for $\tau' \in (0, \beta)$.

$$\mathcal{G}_{0,ij}(\tau+\beta) = -\delta_{ij}\frac{e^{\mu(\tau+\beta)}}{e^{\beta\mu}+1} = -\delta_{ij}e^{\mu\tau}\frac{e^{\beta\mu}}{e^{\beta\mu}+1} = -\left[\delta_{ij}\frac{e^{\mu\tau}}{e^{-\beta\mu}+1}\right]$$

  which matches $-\mathcal{G}_{0,ij}(\tau)$ for $\tau \in (-\beta, 0)$.
- For $0 < \tau < \beta$: Here, we have $\tau' = \tau + \beta \in (\beta, 2\beta)$ but to bring it back to the principal domain of $(-\beta, \beta)$, we subtract 2β as $\mathcal{G}_{0,ij}(\tau+\beta) = \mathcal{G}_{0,ij}(\tau+\beta-2\beta) = \mathcal{G}_{0,ij}(\tau-\beta)$. Since $\tau'' = \tau - \beta \in (-\beta, 0)$, we use the functional form of $\mathcal{G}_{0,ij}(\tau'')$ for $\tau'' \in (-\beta, 0)$.

$$\mathcal{G}_{0,ij}(\tau-\beta) = \delta_{ij}\frac{e^{\mu(\tau-\beta)}}{e^{-\beta\mu}+1} = -\left[-\delta_{ij}\frac{e^{\mu\tau}}{e^{\beta\mu}+1}\right]$$

  matching $-\mathcal{G}_{0,ij}(\tau)$ for $\tau \in (0, \beta)$.

  One final remark where we see a distinction from the Majorana case is

- Majorana Green's function (derived in Eq. (2.46)) satisfies $\mathcal{G}_{0,ij}(\tau) = -\mathcal{G}_{0,ij}(-\tau)$
- Complex Green's function, having $U(1)$ symmetry no longer satisfies this: $\boxed{\mathcal{G}_{0,ij}(\tau) \neq -\mathcal{G}_{0,ij}(-\tau)}$.

This will be of utmost importance in deriving the Schwinger-Dyson equations where, unlike the Majorana case, we need to understand the difference and keep both terms $\mathcal{G}(\tau)$ and $\mathcal{G}(-\tau)$, or in real time $\mathcal{G}(t_1, t_2)$ and $\mathcal{G}(t_2, t_1)$, wherever they emerge.

3.2.2 Interacting Case

We illustrate the calculations for $q = 4$, followed by a generalization to arbitrary q. Since we are in the imaginary-time formalism, we resort to the Euclidean path integral formalism for the partition function (Eq. (2.10)) where the Euclidean action is given in Eq. (2.11). We reproduce them here for convenience with modification for complex fermions:

$$\mathcal{Z}_E = \iint \mathcal{D}c_i \mathcal{D}c_i^\dagger e^{-S_E[\psi_i]}$$

where the measure is given by $\mathcal{D}c_i = \prod_{i=1}^N dc_i$ and $\mathcal{D}c_i^\dagger = \prod_{i=1}^N dc_i^\dagger$. The Euclidean action $S_E[\psi_i]$ is given by (where the chemical potential in Eq. (3.8) is also taken into account)

$$S_E[c_i, c_i^\dagger] = \int d\tau \left[\sum_i^N c_i^\dagger (\partial_\tau - \mu) c_i + \mathcal{H}_4 \right],$$

where integration is from 0 to β (see Appendixes A and F for properties of imaginary-time formalism), $\mathcal{H}_4$ is the Hamiltonian of complex SYK model given in Eq. (3.3). We can perform disorder-averaging of the partition function to get (subscript J_4 denotes averaging is being done for $q = 4$ case where J_4 enters the picture through the variance in Eq. (3.6))

$$\langle \mathcal{Z}_E \rangle_{J_4} = \iint \mathcal{D} j_4 \mathcal{D} j_4^\dagger \mathcal{P}_4 \left[j_{4;i_1 i_2, i_3 i_4} \right] \mathcal{Z}_E \tag{3.23}$$

where probability distribution is provided in Eq. (3.7) and the measure is given by (also see footnote 3) $\mathcal{D} j_4 \mathcal{D} j_4^\dagger = \prod_{\substack{1 \le i_1 < i_2 \le N \\ 1 \le i_3 < i_4 \le N}}^N d j_{4;i_1 i_2, i_3 i_4} d j_{4;i_1 i_2, i_3 i_4}^\dagger$. Since we are also integrating out the Hermitian conjugate of the coupling constants, it's appropriate to explicitly write the Hamiltonian in Eq. (3.3) as $\mathcal{H}_4 = \frac{1}{2}(\mathcal{H}_4 + \mathcal{H}_4^\dagger)$ (using the property in Eq. (3.5))

$$\begin{aligned} \mathcal{H}_4 &= \sum_{\substack{1 \le i_1 < i_2 \le N \\ 1 \le i_3 < i_4 \le N}} j_{4;i_1 i_2, i_3 i_4} c_{i_1}^\dagger c_{i_2}^\dagger c_{i_3} c_{i_4} \\ &= \frac{1}{2} \sum_{\substack{1 \le i_1 < i_2 \le N \\ 1 \le i_3 < i_4 \le N}} \left(j_{4;i_1 i_2, i_3 i_4} c_{i_1}^\dagger c_{i_2}^\dagger c_{i_3} c_{i_4} + c_{i_4}^\dagger c_{i_3}^\dagger c_{i_2} c_{i_1} j_{4;i_1 i_2, i_3 i_4}^\dagger \right). \end{aligned} \tag{3.24}$$

Collecting altogether, we have

$$\begin{aligned} \langle \mathcal{Z}_E \rangle_{J_4} &= A \int \mathcal{D} c_i \mathcal{D} c_i^\dagger \mathcal{D} j_4 \mathcal{D} j_4^\dagger \\ &\times \quad \exp\left(-\frac{1}{2\sigma_4^2} \sum_{\{i_4\}_\le} \left| j_{4;i_1 i_2, i_3 i_4} \right|^2 - \int d\tau \left[\sum_i^N c_i^\dagger (\partial_\tau - \mu) c_i + \mathcal{H}_4 \right] \right). \end{aligned} \tag{3.25}$$

We isolate the terms dependent on the interaction strength and use the Gaussian integration (same as the Majorana case), namely $\int dx e^{-ax^2 + bx} = \sqrt{\frac{\pi}{a}} e^{b^2/4a}$, to integrate out the strengths where the prefactor coming out of integration exactly cancels the factor of A. We are left with

$$\langle \mathcal{Z}_E \rangle_{J_4} = \int \mathcal{D}c_i \mathcal{D}c_i^\dagger \exp\left(- \iint d\tau d\tau' \sum_{i=1}^{N} c_i^\dagger(\tau)\delta\left(\tau' - \tau\right)\left(\partial_{\tau'} - \mu\right) c_i\left(\tau'\right) \right.$$

$$\left. + \sum_{\substack{1\le i_1<i_2\le N \\ 1\le i_3<i_4\le N}}^{N} \frac{\sigma_4^2}{2} \iint d\tau d\tau' c_{i_4}^\dagger c_{i_3}^\dagger c_{i_2} c_{i_1}(\tau) \cdot c_{i_1}^\dagger c_{i_2}^\dagger c_{i_3} c_{i_4}\left(\tau'\right) \right) \tag{3.26}$$

where the functional dependence on τ or τ' is the same for all the terms that precede the argument. We can further simplify

$$\sum_{\substack{i_1<i_2 \\ i_3<i_4}} c_{i_1}^\dagger c_{i_4}^\dagger c_{i_3}^\dagger c_{i_2} c_{i_1}(\tau) \cdot c_{i_1}^\dagger c_{i_2}^\dagger c_{i_3} c_{i_4}\left(\tau'\right)$$

$$= \frac{1}{(2!)^2} \sum_{\substack{i_1\neq i_2 \\ i_3\neq i_4}} c_{i_4}^\dagger c_{i_3}^\dagger c_{i_2} c_{i_1}(\tau) \cdot c_{i_1}^\dagger c_{i_2}^\dagger c_{i_3} c_{i_4}\left(\tau'\right)$$

$$= \frac{1}{(2!)^2} \sum_{\substack{i_1\neq i_2 \\ i_3\neq i_4}} c_{i_1}(\tau) c_{i_1}^\dagger(\tau')(-c_{i_4}(\tau') c_{i_4}^\dagger(\tau)) c_{i_2}(\tau) c_{i_2}^\dagger(\tau')(-c_{i_3}(\tau') c_{i_3}^\dagger(\tau)) \tag{3.27}$$

where we introduce the ansatz (as in the Majorana case) $\mathcal{G}(\tau, \tau') = -\frac{1}{N}\sum_{i=1}^{N} c_i(\tau) c_i^\dagger(\tau')$ which we will later identify as the averaged Green's function.[4] Since $\mathcal{G}$ is site independent (see footnote 4), we can group terms with indices i_1 and i_2 together, and same for terms with indices i_3 and i_4. Thus, we get

$$\sum_{\substack{i_1<i_2 \\ i_3<i_4}} c_{i_1}^\dagger c_{i_4}^\dagger c_{i_3}^\dagger c_{i_2} c_{i_1}(\tau) \cdot c_{i_1}^\dagger c_{i_2}^\dagger c_{i_3} c_{i_4}\left(\tau'\right) = \frac{N^4}{(2!)^2}\Big[- \mathcal{G}(\tau, \tau')\mathcal{G}(\tau', \tau)\Big]^2 \tag{3.28}$$

where we recall our lesson from the free case that for complex fermions that $\mathcal{G}(\tau', \tau) \neq -\mathcal{G}(\tau, \tau')$ as in the Majorana case. So, we have to retain both

[4] $\mathcal{G}(\tau, \tau')$ is the averaged Green's function obtained from Eq. (3.12). As for the Majorana case, this is a general property of SYK-like systems that disorder averaging ensures the Green's function is site-diagonal and uniform, i.e., $\mathcal{G}_{ij} \propto \delta_{ij}$. Prior to disorder averaging, the Green's function for a fixed disorder realization is defined as $\mathcal{G}_{ij}\left(\tau, \tau'\right) = -\left\langle T c_i(\tau) c_j^\dagger\left(\tau'\right)\right\rangle$ as in Eq. (3.12). After disorder averaging, the Green's function $\mathcal{G}\left(\tau, \tau'\right)$ is obtained by averaging the site-diagonal components $\mathcal{G}\left(\tau, \tau'\right) = -\frac{1}{N}\sum_{j=1}^{N} \mathcal{G}_{jj}\left(\tau, \tau'\right)$, implying uniformity $\mathcal{G}_{ij}\left(\tau, \tau'\right) = \mathcal{G}\left(\tau, \tau'\right)\delta_{ij}$ where $\mathcal{G}$ is independent of site indices i and j. Therefore, we can club terms with indices i_1 and i_2 together, and same for terms with indices i_3 and i_4.

arguments throughout our analysis. Hence, we have for our disorder-averaged partition function

$$\langle \mathcal{Z}_E \rangle_{J_4} = \int \mathcal{D}c_i \mathcal{D}c_i^\dagger \exp\left(- \iint d\tau d\tau' \sum_{i=1}^{N} c_i^\dagger(\tau)\delta\left(\tau' - \tau\right)\left(\partial_{\tau'} - \mu\right) c_i\left(\tau'\right) \right.$$
$$\left. + \frac{\sigma_4^2 N^4}{2(2!)^2} \iint d\tau d\tau' \Big[- \mathcal{G}(\tau, \tau')\mathcal{G}(\tau', \tau)\Big]^2 \right). \tag{3.29}$$

Like the Majorana case, we insert the bi-local fields $\mathcal{G}(\tau, \tau')$ and $\Sigma(\tau, \tau')$ via the identity[5]

$$\iint \mathcal{D}\mathcal{G}\mathcal{D}\Sigma \exp\left\{ - \iint d\tau d\tau' N\Sigma\left(\tau, \tau'\right)\left(\mathcal{G}\left(\tau', \tau\right) + \frac{1}{N}\sum_{i}^{N} c_i\left(\tau'\right) c_i^\dagger(\tau) \right) \right\} = 1 \tag{3.30}$$

which enforces the definition of $\mathcal{G}$[6] and introduces Σ as a Lagrange multiplier (later to be identified as the self-energy). We have chosen Σ such that the imaginary unit is absorbed in its definition and is assumed to be analytically continued to the complex plane. We refer the reader to the discussion about the measures of integration in the paragraph below Eq. (2.21). Then the partition function becomes

$$\langle \mathcal{Z}_E \rangle_{J_4} = \int \mathcal{D}c_i \mathcal{D}c_i^\dagger \mathcal{D}\mathcal{G}\mathcal{D}\Sigma \exp\left(- \iint d\tau d\tau' \sum_{i=1}^{N} c_i^\dagger(\tau) \right.$$
$$\left. \times \quad \left[\delta\left(\tau' - \tau\right)\left(\partial_{\tau'} - \mu\right) - \Sigma\left(\tau, \tau'\right)\right] c_i\left(\tau'\right) \right)$$
$$\times \exp\left(\int d\tau d\tau' \left[\frac{N^4 \sigma_4^2}{2(2!)^2} \left(-\mathcal{G}\left(\tau, \tau'\right)\mathcal{G}\left(\tau', \tau\right)\right)^2 - N\Sigma\left(\tau, \tau'\right)\mathcal{G}\left(\tau', \tau\right) \right] \right). \tag{3.31}$$

We have isolated the integral over fermionic fields c_i and $c_i^\dagger$ in the first line which we integrate out using the identity

$$\int \mathcal{D}c_i \mathcal{D}c_i^\dagger \exp\left(- \iint d\tau d\tau' \sum_{i=1}^{N} c_i^\dagger(\tau)\left[\delta\left(\tau' - \tau\right)\left(\partial_{\tau'} - \mu\right) - \Sigma\left(\tau, \tau'\right)\right] c_i\left(\tau'\right) \right)$$
$$= \exp\left(N \ln \det\left[\mathcal{G}_0^{-1} - \Sigma \right] \right) \tag{3.32}$$

[5] Under $\mathcal{G} \to -\mathcal{G}$, we must take $\Sigma \to -\Sigma$ to preserve: (1) the Dyson equation structure, (2) saddle-point equations, and (3) the invariance of the interaction term. This accommodates different conventions for the Green's function.

[6] Recall the identity $\int dx e^{-\imath kx} \sim \delta(k)$.

to finally get (where we now use $\sigma_4^2 = \frac{4J_4^2}{N^3}$ from Eq. (3.6))

$$\boxed{\langle \mathcal{Z}_E \rangle_{J_4} = \iint \mathcal{D}\mathcal{G}\mathcal{D}\Sigma e^{-S_{E,\text{eff}}[\mathcal{G},\Sigma]}} \tag{3.33}$$

where $S_{\text{eff}}[\mathcal{G}, \Sigma]$ is the effective action

$$\begin{aligned}\frac{S_{E,\text{eff}}[\mathcal{G}, \Sigma]}{N} \equiv &- \ln \det \left[\mathcal{G}_0^{-1} - \Sigma\right] \\ &+ \iint d\tau d\tau' \Big(\Sigma(\tau, \tau')\mathcal{G}(\tau', \tau) - \frac{J_4^2}{2} \left(-\mathcal{G}\left(\tau, \tau'\right) \mathcal{G}\left(\tau', \tau\right)\right)^2 \Big).\end{aligned} \tag{3.34}$$

The ordering of imaginary time is of crucial importance due to fermions being charged. In the large-N, the action becomes (semi-)classical and the saddle point solutions dominate which are given by the Euler-Lagrange equations (ordering of time matters!):

$$\frac{\delta S_{\text{eff}}[\mathcal{G}, \Sigma]}{\delta \Sigma} \overset{!}{=} 0, \qquad \frac{\delta S_{\text{eff}}[\mathcal{G}, \Sigma]}{\delta \mathcal{G}} \overset{!}{=} 0. \tag{3.35}$$

The equations are

$$\boxed{\mathcal{G}^{-1} = \mathcal{G}_0^{-1} - \Sigma, \quad \Sigma(\tau) = -J_4^2 \mathcal{G}(\tau)^2 \mathcal{G}(-\tau)}, \tag{3.36}$$

where $\mathcal{G}_0$ is the free Green's function.[7] These are the Schwinger-Dyson equations of the theory. We are in equilibrium, so we have time-translational invariance where we substitute $\mathcal{G}(\tau, \tau') = \mathcal{G}(\tau - \tau') \to \mathcal{G}(\tau)$. Accordingly, $\mathcal{G}(\tau', \tau) = \mathcal{G}(\tau' - \tau) = \mathcal{G}(-\tau)$. Note that the first equation in the Dyson equation which is the reason we identify $\mathcal{G}$ and Σ as the Green's function and the self-energy, respectively.

We can generalize our results to $q/2$-body interactions where the disorder-averaged partition function is given by

$$\boxed{\langle \mathcal{Z}_q \rangle_{J_q} = \iint \mathcal{D}\mathcal{G}\mathcal{D}\Sigma e^{-S_{E,\text{eff}}[\mathcal{G},\Sigma]}} \tag{3.37}$$

[7] Our convention for the Green's function in Chap. 2 had a prefactor of $1/N$, while in this chapter, our convention is $-1/N$. Accordingly, the free Green's function in Fourier space in Eq. (3.18) has an overall positive sign. Upon taking the inverse Fourier transform based on the convention defined in Appendix F (Eq. (F.2)), we get $\mathcal{G}_0^{-1}(\omega) = \imath\omega + \mu \to \mathcal{G}_0^{-1} = -\partial_\tau + \mu$. If the convention would be $+1/N$, that would have changed the integral in Eq. (3.30) and inverse of the free Green's function would be $\mathcal{G}_0^{-1}(\omega) = -\imath\omega - \mu \to \mathcal{G}_0^{-1} = \partial_\tau - \mu$.

and the effective action is given by

$$\begin{aligned}\frac{S_{E,\text{eff}}[\mathcal{G},\Sigma]}{N} &\equiv -\ln\det\left[\mathcal{G}_0^{-1}-\Sigma\right]\\ &+\iint d\tau d\tau'\Big(\Sigma(\tau,\tau')\mathcal{G}(\tau',\tau)-\frac{J_q^2}{(q/2)}\left(-\mathcal{G}(\tau,\tau')\,\mathcal{G}(\tau',\tau)\right)^{q/2}\Big).\end{aligned} \tag{3.38}$$

The associated saddle point solutions in the large-N are the following Schwinger-Dyson equations:

$$\boxed{\mathcal{G}=\mathcal{G}_0^{-1}-\Sigma,\quad \Sigma(\tau)=-(-1)^{\frac{q}{2}}J_q^2\mathcal{G}(\tau)^{\frac{q}{2}}\mathcal{G}(-\tau)^{\frac{q}{2}-1}}. \tag{3.39}$$

Finally, we can connect the effective action in the large-N limit to the free energy F of the theory at temperature $T=1/\beta$ as

$$\beta F = S_{E,\text{eff}}[\mathcal{G},\Sigma] \tag{3.40}$$

where the scaling is linear in N as it should (since $S_{E,\text{eff}}[\mathcal{G},\Sigma]$ scales as N).

3.3 $U(N)$ vs. $U(1)$ Symmetry

The complex fermions do not satisfy the reality condition as Majorana fermions do ($\psi^\dagger=\psi$). Accordingly, the generalization of Sect. 2.4 is the $U(N)$ symmetry. We start with the action

$$S_E[c_i,c_i^\dagger]=\int d\tau\left[\sum_i^N c_i^\dagger\left(\partial_\tau-\mu\right)c_i+\mathcal{H}_4\right],$$

where $\mathcal{H}_4$ is the $q=4$ complex SYK Hamiltonian given in Eq. (3.3). We start by focusing on the kinetic term in the action. The free (kinetic) part of the action

$$S_{\text{kin}}=\int d\tau\sum_i^N c_i^\dagger\left(\partial_\tau-\mu\right)c_i$$

is invariant under $U(N)$ transformations (Einstein summation convention implied)

$$c_i\to U_{ij}c_j,\quad c_i^\dagger\to c_j^\dagger U_{ji}^\dagger,\quad U\in \mathrm{U}(N),$$

where U matrices satisfy $U^\dagger U = UU^\dagger = \mathbb{1}$. The matrices $U\in U(N)$ are inherently time-independent objects that allows for the kinetic term to have the $U(N)$ symmetry.

Now we switch on the interaction where the interaction term is

$$\mathcal{H}_4 = \sum_{\substack{1\leq i_1<i_2\leq N\\ 1\leq i_3<i_4\leq N}} j_{4;i_1i_2,i_3i_4} c^\dagger_{i_1} c^\dagger_{i_2} c_{i_3} c_{i_4}$$

transforms under $U(N)$ as:

$$c^\dagger_{i_1} c^\dagger_{i_2} c_{i_3} c_{i_4} \to U_{i_1a} U_{i_2b} U^\dagger_{i_3c} U^\dagger_{i_4d} c^\dagger_a c^\dagger_b c_c c_d.$$

For $\mathcal{H}_4$ to be $U(N)$-invariant, the couplings must satisfy

$$j_{4;i_1i_2,i_3i_4} = j_{4;ab,cd} U_{i_1a} U_{i_2b} U^\dagger_{i_3c} U^\dagger_{i_4d}.$$

However, in the complex SYK model, the couplings $j_{4;i_1i_2,i_3i_4}$ are independent random variables with no special symmetry. They are not invariant under $U(N)$ transformations, so $\mathcal{H}_4$ breaks $U(N)$ symmetry.

The only U(N) transformation that is preserved is the U(1) subgroup:

$$U_{ij} = e^{\imath\theta}\delta_{ij}, \quad \theta \in \mathbb{R}.$$

Under this:

$$c^\dagger_{i_1} c^\dagger_{i_2} c_{i_3} c_{i_4} \to e^{-\imath\theta} e^{-\imath\theta} e^{\imath\theta} e^{\imath\theta} c^\dagger_{i_1} c^\dagger_{i_2} c_{i_3} c_{i_4} = c^\dagger_{i_1} c^\dagger_{i_2} c_{i_3} c_{i_4}.$$

Thus, $\mathcal{H}_4$ is invariant under $U(1)$ but not under the full $U(N)$. Since the kinetic term is symmetric under $U(N)$, that makes it symmetric under $U(1)$ too.

Accordingly, the kinetic term respects the full $U(N)$ symmetry but the interactions of complex SYK Hamiltonian break the $U(N)$ symmetry, leaving $U(1)$ symmetry as the symmetry for the full action as a residual symmetry. This is the same $U(1)$ symmetry we mentioned at the start of the chapter. This is associated with the charge density in Eq. (3.9), namely $\mathcal{Q} \equiv \frac{1}{N}\sum_{i=1}^{N}\left\langle c^\dagger_i c_i\right\rangle - \frac{1}{2}$, which one can show to commute with the SYK Hamiltonian.

3.4 The IR Limit

We start with the Schwinger-Dyson equations in Eq. (3.39) where the Fourier transform gives (recall ω are the Matsubara frequencies as we are in imaginary-time formalism; see Appendix F for more details)

$$\mathcal{G}(\omega)^{-1} = \imath\omega + \mu - \Sigma(\omega) = \mathcal{G}_0(\omega)^{-1} - \Sigma(\omega), \quad \Sigma(\omega) = -J_q^2 \mathcal{G}^2(\omega)\mathcal{G}(-\omega). \tag{3.41}$$

The IR limit is defined for $\omega \to 0$ limit where we can ignore the free Green's function (exactly as we did in the Majorana case, see footnote 7). So we are left with

$$\mathcal{G}(\omega)^{-1} \simeq -\Sigma(\omega), \quad \Sigma(\omega) = -J_q^2 \mathcal{G}^2(\omega)\,\mathcal{G}(-\omega) \qquad \text{(IR limit)} \tag{3.42}$$

3.4.1 Effective Action and the Diff($\mathbb{R}$) Symmetry

As for the Majorana case, we impose the time-reparameterization $\tau \to f(\tau)$ (Diff($\mathbb{R}$)) in addition to the $U(1)$ symmetry. Therefore, the complete Diff($\mathbb{R}$) $\times\, U(1)$ transformation in the IR limit is given by (recall in 1D, Conf($\mathbb{R}$) $\cong$ Diff($\mathbb{R}$))

$$\begin{aligned} \mathcal{G}(\tau,\tau') &\to \left[f'(\tau)f'(\tau')\right]^{\Delta} e^{(\Lambda(\tau)-\Lambda(\tau'))}\mathcal{G}\left(f(\tau), f(\tau')\right) \\ \Sigma(\tau,\tau') &\to \left[f'(\tau)f'(\tau')\right]^{\Delta(q-1)} e^{(\Lambda(\tau)-\Lambda(\tau'))}\Sigma\left(f(\tau), f(\tau')\right) \end{aligned} \tag{3.43}$$

where $'$ denotes derivative with respect to its argument. We show that the Schwinger-Dyson equations inherit this symmetry[8] in the IR limit where the Schwinger-Dyson equations become

$$\begin{aligned} &\int d\tau'' \mathcal{G}(\tau,\tau'')\Sigma(\tau''\tau') = -\delta(\tau-\tau'), \\ &\Sigma(\tau,\tau') = -(-1)^{\frac{q}{2}} J_q^2 \mathcal{G}(\tau,\tau')^{\frac{q}{2}}\mathcal{G}(\tau',\tau)^{\frac{q}{2}-1}. \end{aligned} \tag{3.44}$$

The second equation is simple to observe for the symmetry:

$$\begin{aligned} &\left[f'(\tau)f'(\tau')\right]^{\Delta(q-1)} e^{(\Lambda(\tau)-\Lambda(\tau'))}\Sigma\left(f(\tau), f(\tau')\right) \\ &= -(-1)^{\frac{q}{2}} J_q^2 \Big[\left[f'(\tau)f'(\tau')\right]^{\Delta} e^{(\Lambda(\tau)-\Lambda(\tau'))}\mathcal{G}\left(f(\tau), f(\tau')\right)\Big]^{q/2} \\ &\times \Big[\left[f'(\tau')f'(\tau)\right]^{\Delta} e^{(\Lambda(\tau')-\Lambda(\tau))}\mathcal{G}\left(f(\tau'), f(\tau)\right)\Big]^{\frac{q}{2}-1} \end{aligned} \tag{3.45}$$

where all terms cancel perfectly if $\boxed{\Delta = \frac{1}{q}}$ (same as the Majorana case), leaving us with

$$\Sigma\left(f(\tau), f(\tau')\right) = -(-1)^{\frac{q}{2}} J_q^2 \mathcal{G}\left(f(\tau), f(\tau')\right)^{\frac{q}{2}} \mathcal{G}\left(f(\tau'), f(\tau)\right)^{\frac{q}{2}-1}$$

[8] There isn't a perfect cancellation of the Diff($\mathbb{R}$) $\times$ $U(1)$ transformation at the level of the full (IR + UV) effective action, however, the IR limit causes the free Green's function to be ignored, making the effective action in the IR limit invariant. This is in essence the same physics as we did for the Majorana case in Sect. 2.5.2. For more advanced discussions, we refer the reader to Ref. [2].

which shows the invariance. Similarly, for the Dyson equation, we get

$$
\begin{aligned}
&\int \frac{1}{\frac{df}{d\tau''}} df(\tau'') \left[f'(\tau) f'(\tau'')\right]^{\Delta} e^{(\Lambda(\tau)-\Lambda(\tau''))} \mathcal{G}\left(f(\tau), f(\tau'')\right) \\
&\quad \times \left[f'(\tau'') f'(\tau')\right]^{\Delta(q-1)} e^{(\Lambda(\tau'')-\Lambda(\tau'))} \Sigma\left(f(\tau''), f(\tau')\right) \\
&\quad = \int df \left(\frac{f'(\tau)}{f'(\tau')}\right)^{\frac{1}{q}} e^{(\Lambda(\tau)-\Lambda(\tau'))} \mathcal{G}\left(f(\tau), f(\tau'')\right) \Sigma\left(f(\tau''), f(\tau')\right) \\
&\quad (\Delta = 1/q) \\
&\quad = -\left(\frac{f'(\tau)}{f'(\tau')}\right)^{\frac{1}{q}} e^{(\Lambda(\tau)-\Lambda(\tau'))} \delta(f(\tau) - f(\tau')) \\
&\quad = -\left(\frac{f'(\tau)}{f'(\tau')}\right)^{\frac{1}{q}} e^{(\Lambda(\tau)-\Lambda(\tau'))} \frac{1}{|f'(\tau)|} \delta(\tau - \tau') \xrightarrow{\tau \to \tau'} -\frac{1}{|f'(\tau)|} \delta(\tau - \tau') \\
&\quad = -\delta(f(\tau) - f(\tau'))
\end{aligned}
\tag{3.46}
$$

where we treat f' and Λ as independent of f and used the Dyson equation to get the δ-function in the second to last equality. We used the identity $\delta(f(x) - f(x_0)) = \frac{1}{|f'(x_0)|}\delta(x - x_0)$ where the δ-function enforced $\tau \to \tau'$ that canceled all the prefactors and we were left with the Dyson equation, showing its invariance.

3.4.2 Conformal Green's Function

We use the spectral representation of the Euclidean Green's function that we analytically continue to the complex plane, namely $\mathcal{G}(\tau) \to \mathcal{G}(z)$ where $z = \imath\omega$, ω being the Matsubara frequency (see Appendix F). Recall that we have consistently not used the subscript n in Matsubara frequencies for brevity.

$$
\mathcal{G}(z) = \int \frac{d\Omega}{\pi} \frac{\rho(\Omega)}{z - \Omega} \qquad \text{(spectral representation)}, \tag{3.47}
$$

where $\rho(\Omega)$ is given by

$$
\rho(\Omega) = -\operatorname{Im}[\mathcal{G}(\Omega + \imath\eta)] = -\frac{1}{2\imath}\operatorname{Im}\Big[\mathcal{G}(\Omega + \imath\eta) - \mathcal{G}(\Omega + \imath\eta)^{\star}\Big]. \tag{3.48}
$$

We now take the ansatz[9] (the same power law in the IR limit as we derived for the Majorana case)

[9] We recommend the reader to a wonderfully written Ref. [1] for the motivation behind this ansatz.

$$\mathcal{G}(z) = c\frac{e^{-\imath(\pi\Delta+\theta)}}{z^{1-2\Delta}} \tag{3.49}$$

where $c, \Delta, \theta \in \mathbb{R}$ are constants and $z \in \mathbb{C}$ as well as $\text{Im}\{z\} > 0$. Substituting the ansatz in the spectral function, we get

$$\rho(\Omega) = c\frac{\sin(\pi\Delta+\theta)}{\Omega^{1-2\Delta}}. \tag{3.50}$$

The purpose of this section is to evaluate the conformal Green's function $\mathcal{G}(\tau)$ using the spectral representation $\mathcal{G}(z)$ by evaluating $\rho(\Omega)$ and then taking an inverse Fourier transform.

We start with the Matsubara formulation where Fourier transform is defined in Eq. (F.2) to get

$$\mathcal{G}(\tau) = \frac{1}{\beta}\sum_{\imath\omega} \mathcal{G}(\imath\omega)\, e^{-\imath\omega\tau} = \frac{1}{\beta}\sum_{\imath\omega}\int \frac{d\Omega}{\pi}\frac{\rho(\Omega)}{z-\Omega}e^{-\imath\omega\tau}. \tag{3.51}$$

In the IR limit, we are tending to zero temperature, therefore $\beta \to \infty$. Accordingly, $\frac{1}{\beta}\sum_{\imath\omega} \to \int \frac{dz}{2\pi\imath}$ $(z = \imath\omega)$ to get

$$\begin{aligned}\mathcal{G}(\tau) &= \int \frac{dz}{2\pi\imath}\int \frac{d\Omega}{\pi}\frac{\rho(\Omega)}{z-\Omega}e^{-z\tau} \\ &= \int \frac{d\Omega}{\pi}\rho(\Omega)\int \frac{dz}{2\pi\imath}\frac{e^{-z\tau}}{z-\Omega} = \int \frac{d\Omega}{\pi}\rho(\Omega)\mathcal{I}(\Omega).\end{aligned} \tag{3.52}$$

We have defined $\mathcal{I}(\Omega) \equiv \int \frac{dz}{2\pi\imath}\frac{e^{-z\tau}}{z-\Omega}$. We can break the integral into two components $\int \frac{d\Omega}{\pi} = \int_{-\infty}^{0}\frac{d\Omega}{\pi} + \int_{0}^{\infty}\frac{d\Omega}{\pi}$ where we can substitute $\Omega \to -\Omega$ to make the first integration limit positive. We get

$$\mathcal{G}(\tau) = \int_0^\infty \frac{d\Omega}{\pi}\rho(-\Omega)\mathcal{I}(-\Omega) + \int_0^\infty \frac{d\Omega}{\pi}\rho(\Omega)\mathcal{I}(\Omega) \tag{3.53}$$

where $\Omega > 0$ as enforced by the integration limits.

We first evaluate $\mathcal{I}(\pm\Omega)$ where we have two cases $\tau < 0$ and $\tau > 0$. Starting with $\mathcal{I}(-\Omega)$, we have

$$\mathcal{I}(-\Omega) = \int \frac{dz}{2\pi\imath}\frac{e^{-z\tau}}{z+\Omega} \tag{3.54}$$

whose complex contour is represented in Fig. 3.1. Note that there is a pole at $z_0 = -\Omega < 0$, as also shown in the Figure. Now the choice of contour is decided such

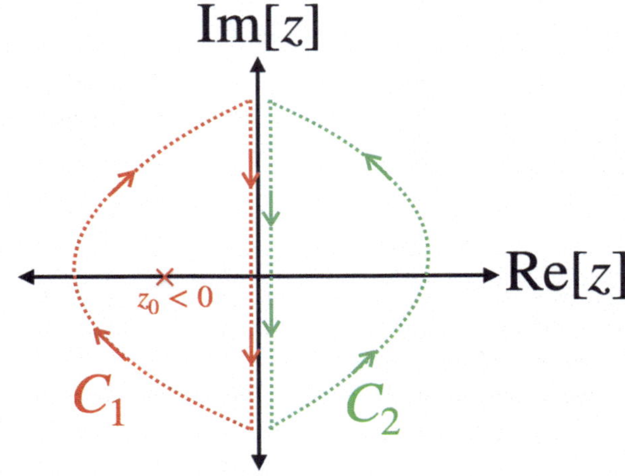

Fig. 3.1 The complex contour to perform complex integration in Eq. (3.54). There are two contours C_1 and C_2 where the selection is made by demanding the exponential in the integrand should go to zero at infinity. There is a pole at $z_0 = -\Omega < 0$

that the exponential vanishes at infinity. With this in mind, we have

$$\mathcal{I}(-\Omega) = \begin{cases} \frac{1}{2\pi\imath} \oint_{C_2} \frac{e^{-z\tau}}{z+\Omega} dz & ; \tau > 0 \\ \frac{1}{2\pi\imath} \oint_{C_1} \frac{e^{-z\tau}}{z+\Omega} dz & ; \tau < 0 \end{cases} \tag{3.55}$$

which gives

$$\mathcal{I}(-\Omega) = \begin{cases} 0 & ; \tau > 0 \\ -\operatorname{Res}\left\{\frac{e^{-z\tau}}{z+\Omega}\right\}_{z=-\Omega} = -e^{\Omega\tau} & ; \tau < 0 \end{cases} \tag{3.56}$$

where $\operatorname{Res}\left\{\frac{e^{-z\tau}}{z+\Omega}\right\}_{z=-\Omega}$ denotes the residue being evaluated at the pole $z = -\Omega$. The negative sign comes in because fo the clock-wise nature of C_1 contour.

We next evaluate $\mathcal{I}(\Omega)$

$$\mathcal{I}(\Omega) = \int \frac{dz}{2\pi\imath} \frac{e^{-z\tau}}{z - \Omega} \tag{3.57}$$

whose contour is given in Fig. 3.2.

Accordingly, the integral is evaluated to

$$\mathcal{I}(\Omega) = \begin{cases} \frac{1}{2\pi\imath} \oint_{C_2} \frac{e^{-z\tau}}{z-\Omega} dz & ; \tau > 0 \\ \frac{1}{2\pi\imath} \oint_{C_1} \frac{e^{-z\tau}}{z-\Omega} dz & ; \tau < 0 \end{cases} \tag{3.58}$$

which gives

$$\mathcal{I}(-\Omega) = \begin{cases} \operatorname{Res}\left\{\frac{e^{-z\tau}}{z-\Omega}\right\}_{z=\Omega} = e^{-\Omega\tau} & ; \tau > 0 \\ 0 \quad ; \tau < 0 \end{cases} \tag{3.59}$$

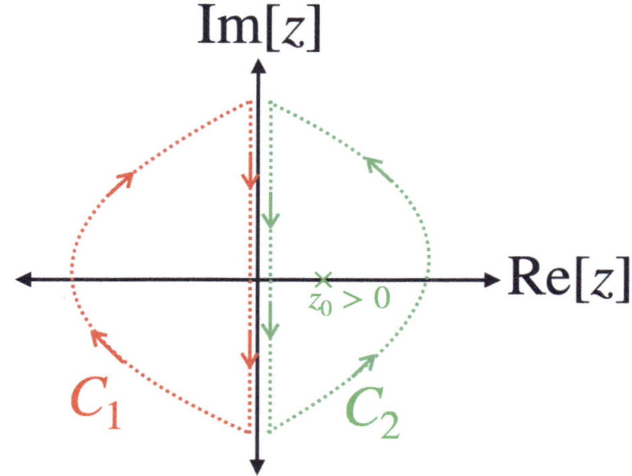

Fig. 3.2 The complex contour to perform complex integration in Eq. (3.57). There are two contours C_1 and C_2 where the selection is made by demanding the exponential in the integrand should go to zero at infinity. There is a pole at $z_0 = +\Omega > 0$

where the residue has positive sign in front because the contour C_2 is anti-clockwise. Plugging back in Eq. (3.53), we get

$$\mathcal{G}(\tau) = \begin{cases} \int_0^\infty \frac{d\Omega}{\pi} \rho(\Omega) e^{-\Omega\tau} & ; \tau > 0 \\ -\int_0^\infty \frac{d\Omega}{\pi} \rho(-\Omega) e^{\Omega\tau} & ; \tau < 0 \end{cases} \tag{3.60}$$

where

$$\rho(\Omega) = c \frac{\sin(\pi\Delta + \theta)}{\Omega^{1-2\Delta}} \tag{3.61}$$

and

$$\begin{aligned} \rho(-\Omega) = c\frac{\sin(\pi\Delta + \theta)}{(-\Omega)^{1-2\Delta}} &= \frac{c\sin(\pi\Delta + \theta)e^{\pm\pi i(2\Delta-1)}}{\Omega^{1-2\Delta}} \\ &= \frac{c\,\mathrm{Im}\left\{e^{i(\pi\Delta+\theta)}\right\} e^{\pm\pi i(2\Delta-1)}}{\Omega^{1-2\Delta}} \\ &= -\frac{c}{\Omega^{1-2\Delta}}\sin(\pi + \pi\Delta - \theta) = \frac{c}{\Omega^{1-2\Delta}}\sin(\pi\Delta - \theta). \end{aligned} \tag{3.62}$$

Thus, we get

$$\mathcal{G}(\tau) = \begin{cases} c\frac{\sin(\pi\Delta+\theta)}{\pi}\int_0^\infty \Omega^{2\Delta-1}e^{-\Omega|\tau|}d\Omega & ; \tau > 0 \\ -c\frac{\sin(\pi\Delta-\theta)}{\pi}\int_0^\infty \Omega^{2\Delta-1}e^{-\Omega|\tau|}d\Omega & ; \tau < 0 \end{cases} \tag{3.63}$$

where we substitute $\Omega = x/|\tau|$ and identify $\Gamma(x) = \int_0^\infty t^{x-1}e^{-t}dt$ to get

$$\boxed{\mathcal{G}(\tau) = \begin{cases} c\frac{\sin(\pi\Delta+\theta)}{\pi|\tau|^{2\Delta}}\Gamma(2\Delta) & ; \tau > 0 \\ -c\frac{\sin(\pi\Delta-\theta)}{\pi|\tau|^{2\Delta}}\Gamma(2\Delta) & ; \tau < 0 \end{cases}}. \tag{3.64}$$

We now take the ansatz in Eq. (3.49) and use the Dyson equation in the Fourier space in the IR limit (Eq. (3.42)) to get the ansatz for the self-energy (recall $z = \imath\omega$ in the Fourier space):

$$\Sigma(z) = -\mathcal{G}(z)^{-1} = -\frac{z^{1-2\Delta}}{ce^{-\imath(\pi\Delta+\theta)}}. \tag{3.65}$$

Then we use the spectral representation of the self-energy as

$$\Sigma(z) = \int \frac{d\Omega}{\pi} \frac{\sigma(\Omega)}{z-\Omega} \tag{3.66}$$

where the spectral function $\sigma(\Omega)$ is given by

$$\sigma(\Omega) = -\,\mathrm{Im}[\Sigma(\Omega + \imath\eta)] = -\frac{1}{2\imath}\mathrm{Im}\Big[\Sigma(\Omega+\imath\eta) - \Sigma(\Omega+\imath\eta)^{\star}\Big]. \tag{3.67}$$

Then repeating the same procedure as the Green's function gives

$$\boxed{\Sigma(\tau) = \begin{cases} -\frac{\sin(\pi\Delta+\theta)}{c\pi|\tau|^{2-2\Delta}}\Gamma(2-2\Delta) & ;\tau > 0 \\ +\frac{\sin(\pi\Delta-\theta)}{c\pi|\tau|^{2-2\Delta}}\Gamma(2-2\Delta) & ;\tau < 0 \end{cases}.} \tag{3.68}$$

Now we use the second Schwinger-Dyson equation in Eq. (3.44) where we plug the expressions for $\mathcal{G}$ and Σ from Eqs. (3.64) and (3.68), respectively to determine the constant c as follows:

$$\boxed{c = \left[\frac{\Gamma(2-2\Delta)}{\pi J_q^2}\right]^{\Delta}\left(\frac{\pi}{\Gamma(2\Delta)}\right)^{1-\Delta}\{\sin(\pi\Delta+\theta)\sin(\pi\Delta-\theta)\}^{\Delta-\frac{1}{2}}}. \tag{3.69}$$

We already have derived $\Delta = 1/q$ and with this we have found the conformal solutions for the Green's function as well as the self-energy in the IR limit for arbitrary q. Note that the physical meaning of θ is the particle-hole asymmetry where $\theta = 0$ is the point where particle-hole symmetry is established (coinciding with the Majorana case). Since the spectral function has to be positive, that sets the limit $-\pi\Delta < \theta < \pi\Delta$. We again see a power-law behavior and that the Diff($\mathbb{R}$) symmetry is reduced to the $SL(2,\mathbb{R})$ symmetry on top of the $U(1)$ symmetry. The discussions are essentially the same as for the Majorana case (Sect. 2.5.4). Just like the Majorana case where the collective modes in the IR limit are governed by a Schwarzian action (see Sect. 2.6, an analogous Schwarzian theory also exists for the complex SYK where $U(1)$ symmetry is included as well. Surprisingly, the $U(1)$ symmetry and the $SL(2,\mathbb{R})$ symmetry are decoupled. The reader seeking advanced (at the same time, pedagogical) discussions on the Schwarzian theory in complex SYK model is referred to Ref. [1].

3.4.3 A Brief Note on Interpolation to Finite Temperature*

All calculations are done in the zero temperature limit. As noticed for the Majorana case, the conformal symmetry allows us to go to finite temperature. We will not go in details here but briefly outline the basic idea behind. We refer the reader to Refs. [2] and [1] for advanced deep dive. The discussions are essentially the same as done for the Majorana case in Sect. 2.5.5. There is one ambiguity for the complex SYK model, namely the presence of $U(1)$ symmetry. This is encoded in terms such as $e^{\Lambda(\tau)}$ (see Eq. (3.43) where time re-parameterization Diff($\mathbb{R}$) and $U(1)$ symmetries are encoded) which gets fixed by using the KMS relation (see Appendix A and footnote 33)

$$\mathcal{G}(\tau+\beta) = -\mathcal{G}(\tau) \tag{3.70}$$

and choosing the normalization $e^{\Lambda(0)} = 1$. Without loss of generality, we choose $\tau < 0$ and $\tau + \beta > 0$ where we use Eq. (3.64) such that we cancel the denominator by approximating $\tau + \beta \approx \tau$ to get

$$e^{-\Lambda(\tau+\beta)} \sin(\pi\Delta + \theta) = e^{-\Lambda(\tau)} \sin(\pi\Delta - \theta). \tag{3.71}$$

We can now capture the "spectral asymmetry" (see Ref. [1]) via the parameter $\mathcal{E}$ defined as

$$e^{2\pi\mathcal{E}} \equiv \frac{\sin(\pi\Delta+\theta)}{\sin(\pi\Delta-\theta)} \tag{3.72}$$

and we (linearly) approximate $\Lambda(\tau+\beta) - \Lambda(\tau) \approx \Lambda\beta/\tau$ to get (keeping normalization in mind)

$$\Lambda(\tau) = \frac{2\pi\mathcal{E}\tau}{\beta}. \tag{3.73}$$

Now we use this combined with the time re-parameterization

$$\tau \longmapsto f(\tau) = \frac{\beta}{\pi}\sin\left(\frac{\pi|\tau|}{\beta}\right) \tag{3.74}$$

to get the finite temperature Green's function (the limitations of this approach restricted to the low-temperatures is the same as the Majorana case where we refer the reader to Sect. 2.5.5)

$$\boxed{\mathcal{G}(\tau)_\beta = \begin{cases} c\,\dfrac{\sin(\pi\Delta+\theta)\Gamma(2\Delta)}{\pi} e^{\frac{2\pi\mathcal{E}\tau}{\beta}} \left(\dfrac{\pi}{\beta\sin\left(\frac{\pi|z|}{\beta}\right)}\right)^{2\Delta} & ; 0 < \tau < \beta \\ -c\,\dfrac{\sin(\pi\Delta-\theta)\Gamma(2\Delta)}{\pi} e^{\frac{2\pi\mathcal{E}\tau}{\beta}} \left(\dfrac{\pi}{\beta\sin\left(\frac{\pi|z|}{\beta}\right)}\right)^{2\Delta} & ; -\beta < \tau < 0 \end{cases}.} \tag{3.75}$$

where $\Delta = \frac{1}{q}$, $\theta \in \mathbb{R}$ measures particle-hole asymmetry, $\mathcal{E}$ and c are defined in Eqs. (3.72) and (3.69), respectively.

3.5 The Large-q Limit

Just like the ansatz for the Majorana case in the large-q limit in Eq. (2.124), we have the following large-q ansatz for the complex SYK:

$$\boxed{\mathcal{G}(\tau) = \mathcal{G}_0(\tau) e^{\frac{g(\tau)}{q}} \xrightarrow{\text{large-}q} \mathcal{G}_0(\tau)\left[1 + \frac{g(\tau)}{q}\right]}, \tag{3.76}$$

where $g = \mathcal{O}(q^0)$.

The Schwinger-Dyson equations are given in Eq. (3.39) which we reproduce for convenience

$$\mathcal{G}^{-1} = \mathcal{G}_0^{-1} - \Sigma, \quad \Sigma(\tau) = -(-1)^{\frac{q}{2}} J_q^2 \mathcal{G}(\tau)^{\frac{q}{2}} \mathcal{G}(-\tau)^{\frac{q}{2}-1}.$$

Using the second equation, we get

$$\Sigma(\tau) = -(-1)^{q/2} J_q^2 \mathcal{G}_0(\tau)^{\frac{q}{2}} \left[1 + \frac{g(\tau)}{q}\right]^{\frac{q}{2}} \mathcal{G}_0(-\tau)^{\frac{q}{2}-1} \left[1 + \frac{g(-\tau)}{q}\right]^{\frac{q}{2}-1}. \tag{3.77}$$

Then taking $0 < \tau < \beta$[10] and using the form of free Green's function from Eq. (3.19), we get (recall q is even)

$$\begin{aligned}
\Rightarrow \Sigma(\tau) &= -(-1)^{\frac{q}{2}} J_q^2 (-1)^{\frac{q}{2}} \left(\frac{e^{\mu\tau}}{e^{\beta\mu}+1}\right)^{\frac{q}{2}} \left[1 + \frac{g(\tau)}{q}\right]^{\frac{q}{2}} \\
&\quad \times \left(\frac{e^{\mu(-\tau)}}{e^{-\beta\mu}+1}\right)^{\frac{q}{2}-1} \left[1 + \frac{g(-\tau)}{q}\right]^{\frac{q}{2}-1} \\
&= -\underbrace{\left(\frac{e^{\mu\tau}}{e^{\beta\mu}+1}\right)}_{\mathcal{G}_0(\tau)} \frac{J_q^2}{(e^{\mu\beta}+1)^{\frac{q}{2}-1}(e^{-\mu\beta}+1)^{\frac{q}{2}-1}} \\
&\quad \times \left[1 + \frac{g(\tau)}{q}\right]^{\frac{q}{2}} \left[1 + \frac{g(-\tau)}{q}\right]^{\frac{q}{2}-1} \\
&= \frac{J_q^2 \mathcal{G}_0(\tau)}{(2 + 2\cosh(\mu\beta))^{\frac{q}{2}-1}} \left[1 + \frac{g(\tau)}{q}\right]^{\frac{q}{2}} \left[1 + \frac{g(-\tau)}{q}\right]^{\frac{q}{2}-1},
\end{aligned} \tag{3.78}$$

[10] We will see below by deriving Eq. (3.85) that this choice does not matter as the differential equation for $g(\tau)$ is symmetric in $\tau \leftrightarrow -\tau$.

where we used $\left(e^{\mu\beta}+1\right)\left(e^{-\mu\beta}+1\right) = 1+e^{\mu\beta}+e^{-\mu\beta}+1 = 2+2\cosh(\mu\beta)$. Then taking large-$q$ limit, we get (where we approximate, for instance, $\left[1+\frac{g(-\tau)}{q}\right]^{\frac{q}{2}-1} \simeq e^{\frac{g(-\tau)}{q}(\frac{q}{2}-1)} \xrightarrow{q\gg 1} e^{\frac{g(-\tau)}{2}}$)

$$\boxed{\Sigma(\tau) = \frac{2}{q}\ell_q^2 \mathcal{G}_0(\tau) e^{\frac{1}{2}[g(\tau)+g(-\tau)]}} \quad \left(\ell_q^2 \equiv \frac{qJ_q^2}{2(2+2\cosh(\mu\beta))^{\frac{q}{2}-1}}\right). \tag{3.79}$$

Now we consider the first (Dyson) equation to obtain in Fourier space (Eq. (3.41), also see footnote 4) which we reproduce here for convenience

$$\mathcal{G}(\omega)^{-1} = \imath\omega + \mu - \Sigma(\omega) = \mathcal{G}_0(\omega)^{-1} - \Sigma(\omega).$$

Then we use the Fourier transform of the large-q ansatz in Eq. (3.76) to get

$$\mathcal{G}^{-1} = \mathcal{G}_0^{-1}\left[1+\frac{g}{q}\right]^{-1} \simeq \mathcal{G}_0^{-1}\left[1-\frac{g}{q}\right]$$

which we plug on the left-hand side of the Dyson equation, using $\mathcal{G}_0^{-1} = \imath\omega + \mu$ (Eq. (3.18)), to get

$$\begin{aligned} (\imath\omega+\mu)\left[1-\frac{g}{q}\right] &= \imath\omega+\mu-\Sigma(\omega) \\ \Rightarrow \Sigma = (\imath\omega+\mu)\frac{g}{q} &= (\imath\omega+\mu)^2 \frac{1}{\imath\omega+\mu}\frac{g}{q} = (\imath\omega+\mu)^2\mathcal{G}_0\frac{g}{q} \end{aligned} \tag{3.80}$$

whose inverse Fourier transform gives

$$\Sigma(\tau) = \frac{1}{q}(\partial_\tau - \mu)^2\left(\mathcal{G}_0(\tau)\frac{g(\tau)}{q}\right). \tag{3.81}$$

Expanding the right-hand side, we get (μ is time-independent)

$$\begin{aligned} \text{Right-hand side} &= \frac{1}{q}(\partial_\tau^2 + \mu^2 - 2\mu\partial_\tau)\left(\mathcal{G}_0(\tau)\frac{g(\tau)}{q}\right) \\ &= \frac{1}{q}\left[\left(\partial_\tau^2\mathcal{G}_0\right)g + \left(\partial_\tau^2 g\right)\mathcal{G}_0 + 2\partial_\tau\mathcal{G}_0\partial_\tau g + \mu^2\mathcal{G}_0 g - 2\mu(\partial_\tau\mathcal{G}_0)g - 2\mu\mathcal{G}_0\partial_\tau g\right], \end{aligned} \tag{3.82}$$

where we used $\partial_\tau^2(\mathcal{G}_0 g) = \partial_\tau(g\partial_\tau\mathcal{G}_0 + \mathcal{G}_0\partial_\tau g) = (\partial_\tau^2\mathcal{G}_0)g + \mathcal{G}_0\partial_\tau^g + 2\partial_\tau\mathcal{G}_0\partial_\tau g$. Next, we use the form of free Green's function in Eq. (3.19) to get for both $\tau > 0$ and $\tau < 0$:

$$\partial_\tau\mathcal{G}_0(\tau) = +\mu\mathcal{G}_0(\tau), \quad \partial_\tau^2\mathcal{G}_0(\tau) = +\mu^2\mathcal{G}_0(\tau) \tag{3.83}$$

which we use in the right-hand side to find all terms cancel except one and we are left with

$$\boxed{\Sigma(\tau) = \frac{1}{q}(\partial_\tau^2 g)\mathcal{G}_0}. \tag{3.84}$$

Therefore, we finally have two expressions for $\Sigma(\tau)$ using both equations of Schwinger-Dyson equations in Eqs. (3.79) and (3.84) which we equate to get the differential equation for $g(\tau)$:

$$\partial_\tau^2 g(\tau) = 2\ell_q^2 e^{\frac{1}{2}[g(\tau)+g(-\tau)]} \tag{3.85}$$

where recall from Eq. (3.79) that $\ell_q^2 \equiv \frac{qJ_q^2}{2(2+2\cosh(\mu\beta))^{\frac{q}{2}-1}}$. Clearly, the equation is symmetric in $\tau \leftrightarrow -\tau$, implying that the solutions for $g(\tau)$ and $g(-\tau)$ are the same. Accordingly, we can use this symmetric property of $\boxed{g(\tau) = g(-\tau)}$ to get the final differential equation for $g(\tau)$ for complex SYK model

$$\boxed{\partial_\tau^2 g(\tau) = 2\ell_q^2 e^{g(\tau)}} \quad \left(\ell_q^2 \equiv \frac{qJ_q^2}{2(2+2\cosh(\mu\beta))^{\frac{q}{2}-1}}\right). \tag{3.86}$$

As we saw in Eq. (2.136) in the Majorana case, the large-q limit is well-defined if ℓ_q^2 remain finite and constant. Recall that $q \to \infty$ limit needs to be taken only after the large-N limit ($N \to \infty$) has been taken (the limits don't commute).

Now, our life becomes easier when it comes to solving for the Green's function in Eq. (3.86) because we see that this equation is mathematically (including the boundary conditions $g(0) = 0$) identical to the Majorana case in Eq. (2.136) with $\mathcal{J}_q^2 \to \ell_q^2$. Therefore, the entirety of Sect. 2.7.2 carries through with this identification in mind, all the way up to the final solution in Eq. (2.150) with Eq. (2.149). Therefore, the solution for the complex SYK case is

$$\boxed{g(\tau) = 2\ln\left\{\frac{\cos\left(\frac{\pi\nu}{2}\right)}{\cos\left[\pi\nu\left(\frac{1}{2} - \frac{|\tau|}{\beta}\right)\right]}\right\}} \tag{3.87}$$

where

$$\boxed{\beta\ell_q = \frac{\pi\nu}{\cos(\pi\nu/2)}}. \tag{3.88}$$

The physical content of ν carries forward too, in the sense that ν runs from 0 to 1 as the coupling runs from 0 to ∞. The case of $\nu = 0$ corresponds to the free case ($g(\tau)|_{\nu=0} = 0$) while $\nu = 1$ corresponds to infinitely strong coupling as the temperature $T \to 0$.

Therefore, we have solved the model completely by determining $g(\tau)$ which determines the Green's function (via Eq. (3.76)) and self-energy (either via Eq. (3.79) or Eq. (3.84)) for all temperature. We observe from the expression of self-energy that there is an exponential suppression at low temperatures. Accordingly, the theory becomes free (dominated by the free Green's function) at zero temperature, no matter the value of the chemical potential μ. Conversely, when the chemical potential diverges $\mu \to \infty$, the theory again becomes free, independent of the value of temperature.

The thermodynamics of the complex SYK model is similar to that of the Majorana case except that we now have a chemical potential and we deal with the grand canonical ensemble. The partition function $\mathcal{Z}$ is treated as the grand partition function and the connection to the grand potential is (in natural units where the Boltzmann constant is set to unity) $\beta N \Omega = -\ln \mathcal{Z}$ where the $\mathcal{Z} = \text{Tr}\left[e^{-\beta N(H_q/N - \mu \mathcal{Q})}\right]$. A good reference is Ref. [1]. We will study the thermodynamics of the complex SYK model in detail in the next Chap. 4, where equilibrium properties are considered.

We have exclusively focused on the imaginary-time formalism. The real-time dynamics is captured by the Keldysh formalism which has been introduced properly in Sect. 2.9. The formalism remains exactly the same for the complex SYK model where the only difference lies in the properties of the Green's function due to the charged nature of complex fermions. We will consider the non-equilibrium and transport properties of the complex SYK model in Chap. 5, where we will deal with the real-time formalism in depth.

References

1. Davison, R.A., Fu, W., Georges, A., Gu, Y., Jensen, K., Sachdev, S.: Thermoelectric transport in disordered metals without quasiparticles: the Sachdev-Ye-Kitaev models and holography. Phys. Rev. B **95**(15), 155131 (2017)
2. Gu, Y., Kitaev, A., Sachdev, S., Tarnopolsky, G.: Notes on the complex Sachdev-Ye-Kitaev model. J. High Energy Phys. **2020**(2), 1–74 (2020). Publisher: Springer Berlin Heidelberg

Equilibrium Properties

4

Abstract

This chapter develops a real-time Keldysh formulation of the complex Sachdev-Ye-Kitaev (SYK) model and applies it to equilibrium thermodynamics, chaos, and chains. Starting from a random all-to-all $q/2$-body Hamiltonian for charged fermions with Gaussian distributed couplings, the chapter constructs the real-time path integral on the Keldysh contour, performs disorder averaging, and introduces bilocal Green's function and self energy fields to obtain an effective action whose large-N saddle yields Schwinger-Dyson and Kadanoff-Baym equations. A large-q ansatz then reduces these integro-differential equations to solvable equations for symmetric and antisymmetric Green's functions, leading to explicit solutions that obey KMS relations and encode charge density and chemical potential dependence. These solutions determine the equation of state and grand potential, revealing a first order phase transition in charge density that terminates at a critical point with Landau-Ginzburg mean field critical exponents, closely analogous to van der Waals fluids and Anti de Sitter black holes. The same formalism yields a generalized Galitskii-Migdal sum rule, closed expressions for the quantum Lyapunov exponent that saturates the Maldacena-Shenker-Stanford bound on quantum chaos at low temperatures, and a ladder kernel description of four-point functions. Finally, the chapter extends the construction to one-dimensional SYK chains with random on-site interactions and random hopping, deriving effective actions and Schwinger-Dyson equations which serve as the equilibrium starting point for non equilibrium and transport analyses in the following chapter.

We investigate equilibrium properties of the complex SYK model, beginning with thermodynamics. The equation of state and grand potential for a single SYK dot reveal a first-order phase transition terminating at a critical point (continuous transition) at low temperatures. Strikingly, the critical exponents belong to the

R. Jha, *Theory of Strongly Correlated Systems*, Lecture Notes in Physics 1051,
https://doi.org/10.1007/978-3-032-20910-8_4

Landau-Ginzburg (mean-field) universality class. This aligns with critical exponents of black hole phase transitions in holographic duals, offering a non-integrable, conformally symmetric testbed for gauge-gravity duality.

We then outline quantum chaos, bridging classical Lyapunov exponents (unbounded) to their quantum analogs (bounded). The SYK model saturates the universal upper bound for quantum Lyapunov exponents $\lambda_L = 2\pi k_B T/\hbar$ (T is the temperature), establishing it as maximally chaotic.

Finally, we generalize the SYK dot to 1D chains with nearest-neighbor hopping. These chains—essential for probing transport, dimensional scaling of chaos, and thermodynamics—form the foundation for non-equilibrium studies in Chap. 5.

4.1 Real-Time Formalism

We are interested in the thermodynamics of the model which presumes equilibrium. As we showed for the Majorana case in Sect. 2.9.4 that the real-time formalism leads to the same equilibrium physics as the imaginary-time formalism,[1] we will resort to the real-time formalism in this chapter, which we develop in this section. The purpose is to present a full non-equilibrium picture in real-time which will serve as the basis for all work to be presented and then specialize to equilibrium properties from the next section.

4.1.1 Model

As mentioned in boxed frame in Sect. 3.1 as caution that the SYK model has different conventions for the variances of the distribution considered and in the definition of the Green's function (we recommend the reader to revisit the boxed frame). That's why, we will start from the model with all details provided and be consistent with it throughout this chapter. We are considering the thermodynamics of a single dot complex SYK model whose Hamiltonian is given by

$$\mathcal{H}_q = J_q \sum_{\substack{\{\boldsymbol{\mu}\}_\leq \\ \{\boldsymbol{\nu}\}_\leq}} X^{\boldsymbol{\mu}}_{\boldsymbol{\nu}} c^\dagger_{\mu_1} \cdots c^\dagger_{\mu_{q/2}} c_{\nu_{q/2}} \cdots c_{\nu_1} \tag{4.1}$$

where the notation in Eq. (3.2) is slightly modified, namely $\{\boldsymbol{\mu}\}_\leq \equiv 1 \leq \mu_1 < \mu_2 < \cdots < \mu_{\frac{q}{2}-1} < \mu_{\frac{q}{2}} \leq N$ and $\boldsymbol{\mu} \equiv \{\mu_1, \mu_2, \mu_3, \ldots, \mu_{\frac{q}{2}}\}$ (where the label running from 1 to $q/2$ is assumed and will be consistent throughout this chapter). We have extracted the strength of the interaction J_q in front while the interactions themselves are random which is controlled by the matrix $X^{\boldsymbol{\mu}}_{\boldsymbol{\nu}}$, derived from a Gaussian ensemble

[1] However, the converse is not true: non-equilibrium dynamics require the real-time Keldysh formalism which is introduced in detail in Sect. 2.9.

with the following mean and variance:[2]

$$\overline{X} = 0, \quad \overline{|X|^2} = \sigma_q^2 = \frac{4\left[\left(\frac{q}{2}\right)!\right]^2}{q^2\left(\frac{N}{2}\right)^{q-1}}. \tag{4.2}$$

The Gaussian probability distribution itself is given by

$$\mathcal{P}_q\left[X_\nu^\mu\right] = A\exp\Big(-\frac{1}{2\sigma_q^2}\sum_{\substack{\{\mu\}_\leq \\ \{\nu\}_\leq}}\left|X_\nu^\mu\right|^2\Big), \tag{4.3}$$

where $A = \sqrt{\frac{1}{2\pi\sigma_q^2}}$ is the normalization factor. The fermionic creation and annihilation operators satisfy the anti-commutation algebra as in Eq. (3.4) and are, in general, time-dependent.

4.1.2 Schwinger-Dyson Equations

The Keldysh (real-time) partition function is given in Eq. (2.185) (also see footnote 29) which we reproduce here for the charged fermions

$$\mathcal{Z} = \int \mathcal{D}c_i \mathcal{D}c_i^\dagger e^{\imath S[c_i, c_i^\dagger]}, \tag{4.4}$$

where the measure is provided by $\mathcal{D}c_i = \prod_{i=1}^N dc_i$ and $\mathcal{D}c_i^\dagger = \prod_{i=1}^N dc_i^\dagger$. The action $S[\psi_i]$ is given by

$$S[c_i, c_i^\dagger] = \int dt\left(\imath c_i^\dagger(t)\partial_t c_i(t) - \mathcal{H}_q(t)\right). \tag{4.5}$$

Integration is carried over the Keldysh contour $\mathcal{C}$, as shown in Fig. 2.1. Then we follow the same steps as done in real-time for Majorana fermions in Sect. 2.9 and charged fermions but in imaginary-time in Sect. 3.2.2 to evaluate the disorder-averaged partition function, starting from Eqs. (4.4) and (4.5). As the disorder-averaging involves introducing the bi-local Green's function and the self-energy, we define (and set our convention) for the Green's function as follows:

$$\mathcal{G}(t_1, t_2) \equiv \frac{-1}{N}\sum_{j=1}^{N}\left\langle T_{\mathcal{C}} c_j(t_1)\, c_j^\dagger(t_2)\right\rangle \tag{4.6}$$

[2] We discussed this convention already in Eq. (3.11) which we now use to get a hands-on experience with this one.

where T_C is the Keldysh time-ordering operator that leads to various types of Green's functions as defined in Sect. 2.9.2. Then, the disorder-averaged partition function is given by[3]

$$\langle \mathcal{Z} \rangle = \iint \mathcal{D}\mathcal{G}\mathcal{D}\Sigma e^{\imath S_{\text{eff}}[\mathcal{G},\Sigma]} \tag{4.7}$$

where the effective action is given by

$$\imath \frac{S_{\text{eff}}[\mathcal{G},\Sigma]}{N} = \ln \det[\mathcal{G}_0^{-1} - \Sigma] + \iint dt_1 dt_2 \left(\Sigma\left(t_1,t_2\right) \mathcal{G}\left(t_2,t_1\right) \right.$$
$$\left. + \frac{J_q^2}{q^2} \left(-4\mathcal{G}\left(t_1,t_2\right)\mathcal{G}\left(t_2,t_1\right)\right)^{\frac{q}{2}} \right) \tag{4.8}$$

where $\mathcal{G}_0^{-1}$ is the free Green's function (see footnote 7). Then the saddle point solutions dominate in the large-N limit where the effective action becomes (semi-)classical and the saddle point solutions are calculated by extremizing the action (Euler-Lagrange equations) that result in the Schwinger-Dyson equations of the theory, namely

$$\boxed{\mathcal{G}^{-1} = \mathcal{G}_0^{-1} - \Sigma, \qquad \Sigma(t_1,t_2) = \frac{2}{q} J_q^2 \mathcal{G}\left(t_1,t_2\right)\left[-4\mathcal{G}\left(t_1,t_2\right)\mathcal{G}\left(t_2,t_1\right)\right]^{\frac{q}{2}-1}}. \tag{4.9}$$

Since we are in the two-time Keldysh plane, we can invoke the definitions of functions from Sect. 2.9.2 and invoke the Langreth rule in Appendix D (Eq. (D.8)) to get for the self-energy

$$\Sigma^{\gtrless}(t_1,t_2) = \frac{2}{q} J_q^2 \mathcal{G}^{\gtrless}\left(t_1,t_2\right)\left[-4\mathcal{G}^{\gtrless}\left(t_1,t_2\right)\mathcal{G}^{\lessgtr}\left(t_2,t_1\right)\right]^{\frac{q}{2}-1} \tag{4.10}$$

where we note that for the current convention of the Green's function, namely $\mathcal{G}\left(t_1,t_2\right) \equiv \frac{-1}{N}\sum_{j=1}^{N}\left\langle T_C c_j\left(t_1\right) c_j^\dagger\left(t_2\right)\right\rangle$, the general conjugate relation becomes

$$\left[\mathcal{G}^{\gtrless}(t_1,t_2)\right]^\star = +\mathcal{G}^{\gtrless}(t_2,t_1). \tag{4.11}$$

Contrast this with Eq. (2.195) where a different convention was used for Chap. 2. We repeat what we have stated after Eq. (2.195) that we stick to a convention consistently throughout the chapter and every chapter starts by definition clearly what the convention will be for the remainder of the chapter. Unfortunately, different conventions can be found in the literature and that's why we wish to show how to deal with them in different chapters.

[3] See the discussion below Eq. (2.21) on measures of integration.

Now we take the large-q limit where we posit the ansatz for the Green's function (as done in previous chapters) in real-time. We start with the convention chosen for the Green's function in Eq. (4.6) at the level of greater and lesser functions

$$\mathcal{G}^{>}(t_1,t_2)=\frac{-1}{N}\sum_{j=1}^{N}\left\langle c_j(t_1)\,c_j^{\dagger}(t_2)\right\rangle,\qquad \mathcal{G}^{<}(t_1,t_2)=\frac{+1}{N}\sum_{j=1}^{N}\left\langle c_j^{\dagger}(t_2)\,c_j(t_1)\right\rangle \tag{4.12}$$

which at equal time becomes

$$\begin{aligned}\mathcal{G}^{>}(t,t)&=-\frac{1}{N}\sum_j\left\langle c_jc_j^{\dagger}\right\rangle=-\frac{1}{N}\sum_j\left\langle 1-c_j^{\dagger}c_j\right\rangle=-(1-\langle n\rangle),\\ \mathcal{G}^{<}(t,t)&=+\frac{1}{N}\sum_j\left\langle c_j^{\dagger}c_j\right\rangle=\langle n\rangle,\end{aligned} \tag{4.13}$$

where $\langle n\rangle\equiv\frac{1}{N}\sum_j\left\langle c_j^{\dagger}c_j\right\rangle$ is the average particle density and we used the (equal-time) anti-commutation relation $\left\{c_j,c_j^{\dagger}\right\}=1$ to get $c_j(t)c_j^{\dagger}(t)=1-c_j^{\dagger}c_j$. From the definition of $\langle n\rangle$ and the $U(1)$ conserved charged density given in Eq. (3.9), we have

$$\mathcal{Q}\equiv\langle n\rangle-\frac{1}{2} \tag{4.14}$$

which implies $\mathcal{Q}=0$ at half-filling $\langle n\rangle=\frac{1}{2}$ (Majorana limit of charged fermions), as expected. Thus, $\mathcal{Q}$ is a measure of deviation from half-filling. Therefore, we have

$$\langle n\rangle=\mathcal{Q}+\frac{1}{2},\quad 1-\langle n\rangle=\frac{1}{2}-\mathcal{Q}, \tag{4.15}$$

which leads to the following equal-time greater and lesser Green's function:

$$\boxed{\mathcal{G}^{>}(t,t)=\mathcal{Q}-\frac{1}{2},\quad \mathcal{G}^{<}(t,t)=\mathcal{Q}+\frac{1}{2}}. \tag{4.16}$$

This fixes the equal-time value of the Green's functions, and the ansatz must satisfy this boundary condition. This is further verified by having a redundancy check $\mathcal{G}^{>}(t,t)-\mathcal{G}^{<}(t,t)=\left(\mathcal{Q}-\frac{1}{2}\right)-\left(\mathcal{Q}+\frac{1}{2}\right)=-1$, which matches $-\frac{1}{N}\sum_j\left\langle\left\{c_j,c_j^{\dagger}\right\}\right\rangle=-1$ from anti-commutation. This further validates the prefactor choice.

In the large-q limit, correlations between fermions weaken. The exponential form decouples the two-time correlation into a static charge factor $(\mathcal{Q}\mp\frac{1}{2})$ and a dynamical factor which we denote as $g^{\gtrless}(t_1,t_2)$, governed by the SYK interaction.

Accordingly, the large-q ansatz for all $\mathcal{Q}$ is

$$\boxed{\mathcal{G}^{\gtrless}(t_1,t_2) = \left(\mathcal{Q} \mp \frac{1}{2}\right) e^{g^{\gtrless}(t_1,t_2)/q}}, \tag{4.17}$$

with $g^{\gtrless}(t_1,t_2)$ satisfying the equal-time boundary condition $g^{\gtrless}(t,t) = 0$ and $g^{\gtrless} = \mathcal{O}(q^0)$. Using the general conjugate relation in Eq. (4.11) for the convention of Green's function being followed in this chapter, we get

$$g^{\gtrless}(t_1,t_2)^\star = g^{\gtrless}(t_2,t_1) \quad \text{(general conjugate relation)}. \tag{4.18}$$

We find that this is the same as with the other convention for the Green's function (Eq. (2.213)) that is followed in Chap. 2 (see Eq. (2.195)).

We now plug in the self-energy expression in Eq. (4.10) to get

$$\begin{aligned}
\Sigma^{\gtrless}(t_1,t_2) &= \frac{2}{q} J_q^2 \mathcal{G}^{\gtrless}(t_1,t_2) \left[-4\mathcal{G}^{\gtrless}(t_1,t_2)\,\mathcal{G}^{\lessgtr}(t_2,t_1)\right]^{\frac{q}{2}-1} \\
&= \frac{2}{q} J_q^2 \mathcal{G}^{\gtrless}(t_1,t_2) \left[-4\left(\mathcal{Q}^2 - \frac{1}{4}\right) e^{\frac{g^{\gtrless}(t_1,t_2)+g^{\lessgtr}(t_2,t_1)}{q}}\right]^{\frac{q}{2}-1} \\
&= \frac{2}{q} J_q^2 (1-4\mathcal{Q}^2)^{\frac{q}{2}-1} \mathcal{G}^{\gtrless}(t_1,t_2) \left[e^{\frac{g^{\gtrless}(t_1,t_2)+g^{\lessgtr}(t_2,t_1)}{2}}\right] \quad \text{(large-}q\text{ limit)}
\end{aligned} \tag{4.19}$$

where we define

$$\boxed{\mathcal{J}_q^2 \equiv J_q^2(1-4\mathcal{Q}^2)^{\frac{q}{2}-1}, \quad g^{\pm}(t_1,t_2) \equiv \frac{g^>(t_1,t_2) \pm g^<(t_2,t_1)}{2}}. \tag{4.20}$$

Note the time ordering in the arguments which can be confusing if one if not aware of. We briefly discuss the physical intuition behind $g^{\pm}(t_1,t_2)$. In the Majorana limit, the Green's function also satisfies $\mathcal{G}^>(t_1,t_2) = +\mathcal{G}^<(t_2,t_1)$ (contrast with Eq. (2.196) where a different convention for the Green's function is being followed), on top of the general conjugate relation in Eq. (4.11) that led to Eq. (4.18), that translates to (recall that $\mathcal{Q} = 0$ in the Majorana limit)

$$g^>(t_1,t_2) = g^<(t_2,t_1) \quad \text{(Majorana condition)}. \tag{4.21}$$

We find this relation to be the same as with the convention of the Green's function in Chap. 2. Accordingly, we have $g^-(t_1,t_2) = 0$ in the Majorana limit for all time t_1 and t_2. Therefore, $g^-(t_1,t_2)$ is a measure of charge fluctuations away from the Majorana limit (similar to $\mathcal{Q}$ as we showed above). Also $g^+(t_1,t_2) = g^>(t_1,t_2) = g^<(t_2,t_1)$ for the Majorana case. That's why $g^{\pm}(t_1,t_2)$ are also referred to as "symmetric" and "anti-symmetric" (little) Green's function.

So, the self-energy takes the form

$$\boxed{\Sigma_q^{\gtrless}(t_1, t_2) = \frac{1}{q}\mathcal{L}^{\gtrless}(t_1, t_2)\mathcal{G}^{\gtrless}(t_1, t_2)}, \tag{4.22}$$

where

$$\mathcal{L}^{>}(t_1, t_2) \equiv 2\mathcal{J}_q^2 e^{g_+(t_1,t_2)}, \quad \mathcal{L}^{<}(t_1, t_2) = \mathcal{L}^{>}(t_1, t_2)^{\star}. \tag{4.23}$$

4.1.3 The Kadanoff-Baym Equations

Now, we repeat the same steps of Sect. 2.9.3 to get the Kadanoff-Baym equations using our convention for the Green's function in Eq. (4.12) to get (recall that we depreciate the imaginary contour due to Bogoliubov's principle)[4]

$$\partial_{t_1}\mathcal{G}^{\gtrless}(t_1, t_2) = \int_{-\infty}^{\infty} dt_3 \Big[\Sigma^{\gtrless}(t_1, t_3)\,\mathcal{G}^{A}(t_3, t_2) + \Sigma^{R}(t_1, t_3)\,\mathcal{G}^{\gtrless}(t_3, t_2)\Big]. \tag{4.24}$$

We use the definition of retarded and advanced functions from Eq. (2.193) to have

$$\begin{aligned}\mathcal{G}^{A}(t_3, t_2) &= \Theta(t_2 - t_3)\left[\mathcal{G}^{<}(t_3, t_2) - \mathcal{G}^{>}(t_3, t_2)\right]\\ \Sigma^{R}(t_1, t_3) &= \Theta(t_1 - t_3)\left[\Sigma^{>}(t_1, t_3) - \Sigma^{<}(t_1, t_3)\right].\end{aligned}$$

which gives

$$\begin{aligned}\partial_{t_1}\mathcal{G}^{\gtrless}(t_1, t_2) = &\int_{-\infty}^{t_2} dt_3\,\Sigma^{\gtrless}(t_1, t_3)\left[\mathcal{G}^{<}(t_3, t_2) - \mathcal{G}^{>}(t_3, t_2)\right]\\ &+\int_{-\infty}^{t_1} dt_3\left[\Sigma^{>}(t_1, t_3) - \Sigma^{<}(t_1, t_3)\right]\mathcal{G}^{\gtrless}(t_3, t_2)\end{aligned} \tag{4.25}$$

[4] There is an implicit dependence on the chemical potential which controls the charge density of the system and is set by the initial conditions. Strictly speaking, the μ dependence is captured by a phase transformation of the Green's function $\mathcal{G}^{\gtrless}(t_1, t_2) = e^{\imath\mu(t_1-t_2)}\widetilde{\mathcal{G}}^{\gtrless}(t_1, t_2)$ (accordingly, the advanced function transforms the same). Since $\Sigma^{\gtrless}(t_1, t_2) \propto \mathcal{G}^{\gtrless}(t_1, t_2)^{\frac{q}{2}}\mathcal{G}^{\gtrless}(t_2, t_1)^{\frac{q}{2}-1}$ (see Eq. (4.10)), we have $\Sigma^{\gtrless}(t_1, t_2) = e^{\imath\mu(t_1-t_2)}\widetilde{\Sigma}^{\gtrless}(t_1, t_2)$ (accordingly, the retarded function transforms the same). Then Eq. (4.24) is satisfied by $\widetilde{\mathcal{G}}$ and $\widetilde{\Sigma}$. By plugging the phase transformations, we recover the explicit μ dependence in the Kadanoff-Baym equations: $(\partial_{t_1} - \imath\mu)\mathcal{G}^{\gtrless}(t_1, t_2) = \int_{-\infty}^{\infty} dt_3\Big[\Sigma^{\gtrless}(t_1, t_3)\,\mathcal{G}^{A}(t_3, t_2) + \Sigma^{R}(t_1, t_3)\,\mathcal{G}^{\gtrless}(t_3, t_2)\Big]$. Therefore, one needs to solve Eq. (4.24) for $\widetilde{\mathcal{G}}$ and transform back to $\mathcal{G}$ to extract physics. However, the scaling limit, chaos exponent, or entropy growth are independent of μ. The phase $e^{-\imath\mu(t_1-t_2)}$ is a trivial shift. Also, all equal-time properties such as particle density or charge density are also universal as the chemical potential phase factor vanishes. Therefore, we drop the tilde symbol and continue with Eq. (4.24) for the rest of this work while keeping this footnote in mind. We also refer to Ref. [9] for a systematic analysis by keeping the mass term in the Green's function.

where we re-write formally the first line as

$$\int_{-\infty}^{t_2} \mathrm{d}t_3 \, \Sigma^{\gtrless}(t_1, t_3) \, \mathcal{G}^A(t_3, t_2)$$
$$= \int_{t_1}^{t_2} \mathrm{d}t_3 \, \Sigma^{\gtrless}(t_1, t_3) \, \mathcal{G}^A(t_3, t_2) + \int_{-\infty}^{t_1} \mathrm{d}t_3 \, \Sigma^{\gtrless}(t_1, t_3) \, \mathcal{G}^A(t_3, t_2) \tag{4.26}$$

to get in the Kadanoff-Baym equation

$$\partial_{t_1} \mathcal{G}^{\gtrless}(t_1, t_2) = \int_{t_1}^{t_2} \mathrm{d}t_3 \, \Sigma^{\gtrless}(t_1, t_3) \, \mathcal{G}^A(t_3, t_2) + \int_{-\infty}^{t_1} \mathrm{d}t_3 \, \Sigma^{\gtrless}(t_1, t_3) \, \mathcal{G}^A(t_3, t_2)$$
$$+ \int_{-\infty}^{t_1} dt_3 \left[\Sigma^{>}(t_1, t_3) - \Sigma^{<}(t_1, t_3)\right] \mathcal{G}^{\gtrless}(t_3, t_2) \, . \tag{4.27}$$

We combine the last two terms on the right-hand side and again use the definition of $\mathcal{G}^A(t_3, t_2)$ to get

$$\int_{-\infty}^{t_1} dt_3 \, \Sigma^{\gtrless}(t_1, t_3) \, \mathcal{G}^A(t_3, t_2) + \left[\Sigma^{>}(t_1, t_3) - \Sigma^{<}(t_1, t_3)\right] \mathcal{G}^{\gtrless}(t_3, t_2)$$
$$= \int_{-\infty}^{t_1} dt_3 \, \Sigma^{\gtrless}(t_1, t_3) \left[\mathcal{G}^{<}(t_3, t_2) - \mathcal{G}^{>}(t_3, t_2)\right] \tag{4.28}$$
$$+ \left[\Sigma^{>}(t_1, t_3) - \Sigma^{<}(t_1, t_3)\right] \mathcal{G}^{\gtrless}(t_3, t_2) \, .$$

If we take all the top signs, we get in the integrand

$$\Sigma^{>}(t_1, t_3) \left[\mathcal{G}^{<}(t_3, t_2) - \mathcal{G}^{>}(t_3, t_2)\right] + \left[\Sigma^{>}(t_1, t_3) - \Sigma^{<}(t_1, t_3)\right] \mathcal{G}^{>}(t_3, t_2)$$
$$= \Sigma^{>}(t_1, t_3) \, \mathcal{G}^{<}(t_3, t_2) - \Sigma^{<}(t_1, t_3) \, \mathcal{G}^{>}(t_3, t_2), \tag{4.29}$$

while the bottom sign gives the same as above

$$\Sigma^{<}(t_1, t_3) \left[\mathcal{G}^{<}(t_3, t_2) - \mathcal{G}^{>}(t_3, t_2)\right] + \left[\Sigma^{>}(t_1, t_3) - \Sigma^{<}(t_1, t_3)\right] \mathcal{G}^{<}(t_3, t_2)$$
$$= \Sigma^{>}(t_1, t_3) \, \mathcal{G}^{<}(t_3, t_2) - \Sigma^{<}(t_1, t_3) \, \mathcal{G}^{>}(t_3, t_2). \tag{4.30}$$

Therefore, the Kadanoff-Baym equation becomes

$$\boxed{\partial_{t_1} \mathcal{G}^{\gtrless}(t_1, t_2) = \int_{t_1}^{t_2} \mathrm{d}t_3 \, \Sigma^{\gtrless}(t_1, t_3) \left[\mathcal{G}^{<}(t_3, t_2) - \mathcal{G}^{>}(t_3, t_2)\right] + \mathcal{I}(t_1, t_2)} \tag{4.31}$$

where we define

$$\mathcal{I}(t_1, t_2) \equiv \int_{-\infty}^{t_1} dt_3 \Big[\Sigma^{>}(t_1, t_3)\, \mathcal{G}^{<}(t_3, t_2) - \Sigma^{<}(t_1, t_3)\, \mathcal{G}^{>}(t_3, t_2) \Big]. \tag{4.32}$$

Generalized Galitskii-Migdal Sum Rule

We already introduced the Galitskii-Migdal sum rule for the Majorana case in Eq. (2.175). Now we see a generalization to the complex SYK model. Here we generalize to charged fermions. Given the convention for the lesser Green's function in Eq. (4.12), we have

$$\imath \partial_t \mathcal{G}^{<}\left(t^{+}, t\right) \equiv \frac{\imath}{N} \sum_k \left\langle c_k^{\dagger}(t) \partial_t c_k\left(t^{+}\right)\right\rangle = \frac{1}{N} \sum_k \left\langle c_k^{\dagger}\left[c_k, \mathcal{H}\right]\right\rangle(t), \tag{4.33}$$

where $\lim_{t' \to t} \imath \partial_{t'} \mathcal{G}^{<}\left(t', t\right) = \imath \partial_t \mathcal{G}^{<}\left(t^{+}, t\right)$ and we used the real-time Heisenberg's equation of motion (see footnote 2). We provide a basic outline of the proof; we refer the reader to Appendix A of Ref. [5] (also see Ref. [9]). We evaluate a general commutator consisting of strings of creation and annihilation fermionic operators, namely (see

$$\begin{aligned}
\sum_{k=1}^{n} c_k^{\dagger}\left[c_k, c_1^{\dagger} c_2^{\dagger} \cdots c_n^{\dagger}\right] &= \sum_{k=1}^{n} \sum_{\nu=1}^{n} c_k^{\dagger}(-1)^{\nu-1} c_1^{\dagger} \cdots c_{\nu-1}^{\dagger}\left\{c_k, c_\nu^{\dagger}\right\} c_{\nu+1}^{\dagger} \cdots c_n^{\dagger} \\
&= \sum_{k=1}^{n} \sum_{\nu=1}^{n} c_k^{\dagger}(-1)^{\nu-1} c_1^{\dagger} \cdots c_{\nu-1}^{\dagger} \delta_{\nu,k} c_{\nu+1}^{\dagger} \cdots c_n^{\dagger} \\
&= \sum_{\nu=1}^{n} (-1)^{2\nu-2} c_1^{\dagger} \cdots c_{\nu-1}^{\dagger} c_\nu^{\dagger} c_{\nu+1}^{\dagger} \cdots c_n^{\dagger} \\
&= n c_1^{\dagger} \cdots c_n^{\dagger}.
\end{aligned} \tag{4.34}$$

Similarly, one can show trivially that

$$\sum_{k=1}^{n} c_k^{\dagger}\left[c_k, c_1 c_2 \cdots c_n\right] = 0. \tag{4.35}$$

Then specializing to the case of SYK Hamiltonian in Eq. (4.1) which can be symbolically written as $\mathcal{H}_q = \sum X C^{\dagger} C$, we have $\left[c_k, C^{\dagger} C\right] = C^{\dagger}\left[c_k, C\right] + \left[c_k, C^{\dagger}\right] C = \left[c_k, C^{\dagger}\right] C$. Here, $n = q/2$. Then plugging the commutator from Eq. (4.34) in Eq. (4.33) where average is taken over, we get the *equal-time Kadanoff-Baym equation* (also using Eq. (4.31))

(continued)

$$\boxed{\lim_{t'\to t} \partial_{t'} \mathcal{G}^{<}(t', t) = \mathcal{I}(t, t) = -\imath \frac{q}{2} \mathcal{E}_q(t)} \tag{4.36}$$

where we define $\mathcal{E}_q \equiv \frac{\langle \mathcal{H}_q \rangle}{N}$ as the energy density.[a]

[a]This can be generalized to a general SYK-like Hamiltonian $\mathcal{H} = \sum_q \mathcal{H}_q$ wherem sum is over all $q/2$-body interactions $\boxed{\lim_{t'\to t} \partial_{t'} \mathcal{G}^{<}(t', t) = \mathcal{I}(t, t) = -\imath \sum_{q>0} \frac{q}{2} \mathcal{E}_q(t)}$.

4.1.4 The Kadanoff-Baym Equations in the Large-q Limit

In this section, we apply the large-q ansatz equations and simplify the Kadanoff-Baym equations. In particular, we will use Eqs. (4.17), (4.22) and (4.23) in Eq. (4.31) (where $\mathcal{I}(t_1, t_2)$ is defined in Eq. (4.32)). Plugging and keeping everything to leading order in $1/q$, we get

$$\begin{aligned} &\left(\mathcal{Q} \mp \frac{1}{2}\right) \frac{1}{q} \partial_{t_1} g^{\gtrless}(t_1, t_2) = \frac{1}{q} \int_{t_1}^{t_2} dt_3 \mathcal{L}^{\gtrless}(t_t, t_3) \left(\mathcal{Q} \mp \frac{1}{2}\right) + \mathcal{I}(t_1, t_2) \\ &\Rightarrow \partial_{t_1} g^{\gtrless}(t_1, t_2) = \int_{t_1}^{t_2} dt_3 \mathcal{L}^{\gtrless}(t_t, t_3) + \frac{q}{\left(\mathcal{Q} \mp \frac{1}{2}\right)} \mathcal{I}(t_1, t_2) \end{aligned} \tag{4.37}$$

where we simplify the last term containing $\mathcal{I}(t_1, t_2)$ (defined in Eq. (4.32)) to get

$$\begin{aligned} &\frac{q}{\left(\mathcal{Q} \mp \frac{1}{2}\right)} \mathcal{I}(t_1, t_2) \\ &= \frac{q}{\left(\mathcal{Q} \mp \frac{1}{2}\right)} \int_{-\infty}^{t_1} dt_3 \frac{1}{q} \left(\mathcal{L}^{>}(t_1, t_3) \left(\mathcal{Q}^2 - \frac{1}{4}\right) - \mathcal{L}^{<}(t_1, t_3) \left(\mathcal{Q}^2 - \frac{1}{4}\right)\right) \\ &= 2\left(\mathcal{Q} \pm \frac{1}{2}\right) \underbrace{\int_{-\infty}^{t_1} dt_3 \frac{(\mathcal{L}^{>}(t_1, t_3) - \mathcal{L}^{<}(t_1, t_3))}{2}}_{\equiv \imath \alpha(t_1)} \\ &= 2\left(\mathcal{Q} \pm \frac{1}{2}\right) \imath \alpha(t_1). \end{aligned} \tag{4.38}$$

To determine $\alpha(t_1)$, we can use the equal-time property of the Kadanoff-Baym equation in Eq. (4.36) to get for the left-hand side $\frac{q}{\left(\mathcal{Q}\mp\frac{1}{2}\right)}(-\imath)\frac{q}{2}\mathcal{E}_q(t_1)$, where $\mathcal{E}_q \equiv \frac{\langle \mathcal{H}_q \rangle}{N}$. So, we get for $\alpha(t_1)$ the following:[5]

$$\boxed{(1-4\mathcal{Q}^2)\alpha(t_1) = q^2\mathcal{E}_q(t_1) \equiv \epsilon_q(t_1)} \tag{4.39}$$

where we introduced a re-scaled energy

$$\epsilon_q(t_1) \equiv q^2 \frac{\langle \mathcal{H}_q \rangle}{N}. \tag{4.40}$$

Putting things back, we get for the Kadanoff-Baym equation at leading order in $1/q$:

$$\partial_{t_1} g^{\gtrless}(t_1, t_2) = \int_{t_1}^{t_2} dt_3 \mathcal{L}^{\gtrless}(t_t, t_3) + 2\imath \left(\mathcal{Q} \pm \frac{1}{2}\right)\alpha(t_1) \tag{4.41}$$

where α is defined in Eq. (4.39). Therefore, the equation of motion for $g^>$ and $g^<$ are as follows:

$$\begin{aligned} \partial_{t_1} g^{>}(t_1, t_2) &= \int_{t_1}^{t_2} dt_3 \mathcal{L}^{>}(t_t, t_3) + 2\imath \left(\mathcal{Q} + \frac{1}{2}\right)\alpha(t_1) \\ \partial_{t_1} g^{<}(t_1, t_2) &= \int_{t_1}^{t_2} dt_3 \mathcal{L}^{<}(t_t, t_3) + 2\imath \left(\mathcal{Q} - \frac{1}{2}\right)\alpha(t_1). \end{aligned} \tag{4.42}$$

Taking the complex conjugate of the second equation and using $g^<(t_1, t_2)^\star = g^<(t_2, t_1)$ as well as $\mathcal{L}^<(t_1, t_3)^\star = \mathcal{L}^>(t_1, t_3)$ (see Eq. (4.23)) to get

$$\partial_{t_1} g^{<}(t_2, t_1) = \int_{t_1}^{t_2} dt_3 \mathcal{L}^{>}(t_t, t_3) - 2\imath \left(\mathcal{Q} - \frac{1}{2}\right)\alpha(t_1). \tag{4.43}$$

Then using the definition of symmetric and anti-symmetric Green's function from Eq. (4.20), we add and subtract the equations for $g^>(t_1, t_2)$ and $g^<(t_2, t_1)$ to get

$$\begin{aligned} \partial_{t_1} g^{+}(t_1, t_2) &= \int_{t_1}^{t_2} dt_3 \mathcal{L}^{>}(t_1, t_3) + \imath\alpha(t_1) \\ \partial_{t_1} g^{-}(t_1, t_2) &= 2\imath\mathcal{Q}\alpha(t_1). \end{aligned} \tag{4.44}$$

[5] We can generalize this relation to a sum of SYK-like Hamiltonians as mentioned in footnote a where it's given by $\boxed{\left(1-4\mathcal{Q}^2\right)\alpha(t) = q\mathcal{E}_2(t) + q^2\sum_{\kappa>0}\kappa\mathcal{E}_{\kappa q}(t)}$ where $\mathcal{E}_m \equiv \frac{\langle \mathcal{H}_m \rangle}{N}$.

The second equation is already in a solvable form at equilibrium. We take the derivative of the first equation with respect to t_2 where we use Leibniz rule (see above Eq. (2.217)) to get the following differential equations for the Green's functions

$$\boxed{\begin{aligned} \partial_{t_1}\partial_{t_2} g^+(t_1,t_2) &= \mathcal{L}^>(t_1,t_2) \\ \partial_{t_1} g^-(t_1,t_2) &= 2\imath \mathcal{Q}\alpha(t_1). \end{aligned}} \tag{4.45}$$

These are the equations that holds across equilibrium and non-equilibrium. They will form the basis for the rest of this work. We now specialize to the thermodynamics of the complex SYK model which, by construction, implies equilibrium.

4.2 Thermodynamics

We study the thermodynamics of a single dot complex SYK model whose details are given above in Sect. 4.1.1. The system is at equilibrium and therefore, there exists a time-translational invariance. We start by solving the for the Green's functions, followed by delving into the thermodynamics of the system.

4.2.1 Solving for the Green's Functions

We have time-translational differential equations for g^+ and g^- in Eq. (4.45) where $(t_1, t_2) \to (t_1 - t_2) = t$, which we write as (where we used the definition of $\mathcal{L}^>$ from Eq. (4.23))

$$\frac{\partial^2 g^+(t)}{\partial t^2} = -2\mathcal{J}_q^2 e^{g_+(t)}, \quad \partial_t g^-(t) = 2\imath \mathcal{Q}\alpha, \tag{4.46}$$

where α is time-independent because the energy density is a constant in equilibrium (see Eq. (4.39)). These equations admit a closed form solution as follows

$$\boxed{g^+(t) = 2\ln \frac{\pi\nu}{\beta\mathcal{J}_q \cos(\pi\nu(1/2 - \imath t/\beta))}, \quad g^-(t) = 2\imath \mathcal{Q}\alpha t}. \tag{4.47}$$

This matches with what we solved in the imaginary time formalism (see Eq. (3.87)). We have the closure relation (same as in Eq. (2.149)) coming from the requirement that $g^+(t=0) = 0$

$$\pi\nu = \beta\mathcal{J}_q \cos\left(\frac{\pi\nu}{2}\right). \tag{4.48}$$

We can directly verify the KMS relation[6]

$$\boxed{g^+(t) = g^+(-t - \imath\beta)}. \tag{4.49}$$

These relations allow us to determine the energy density of the system as captured by α in Eq. (4.39). By taking the equal-time limit of the first order differential equation for g^+ in Eq. (4.44) leads to

$$\boxed{\dot{g}^+(t = 0) = \imath\alpha} \tag{4.50}$$

where limit $t \to 0$ is taken after the derivative with respect to t has been performed. This leads to

$$\beta\alpha = -2\beta\mathcal{J}_q \sin\left(\frac{\pi\nu}{2}\right) \tag{4.51}$$

where the closure relation in Eq. (4.48) has been used. All properties of ν are the same as discussed below Eq. (2.149), which we refer the reader to, if required.

Finally, we note that the chemical potential term enters via the Boltzmann factor $e^{-\beta[\mathcal{H}-\mu N\mathcal{Q}]}$ and not directly via the Hamiltonian. So, the full Green's function (also read footnote 4) satisfy the following KMS relation

$$\boxed{\mathcal{G}^<(t + \imath\beta) = -e^{\beta\mu}\mathcal{G}^>(t)}. \tag{4.52}$$

We refer the reader to a very nice and detailed proof in Appendix C of Ref. [10], where a proof for the fluctuation-dissipation theorem in presence of chemical potential is also provided.

Finally, we also solve for $g^{\gtrless}$ that will determine the full Green's function in Eq. (4.17). Using the definition in Eq. (4.20), we get

$$g^>(t_1, t_2) = g^+(t_1, t_2) + g^-(t_1, t_2), \quad g^<(t_2, t_1) = g^+(t_1, t_2) - g^-(t_1, t_2), \tag{4.53}$$

where $g^<(t_2, t_1) = g^<(t_1, t_2)^\star$. This solves the theory completely.

4.2.2 Scaling Relations and Low-Temperature Limit

To find the scaling relations with respect to q, we need to first find the equation of state where we will see that there are contributions from the interacting Hamiltonian and a free term. Then demanding a competition between the two terms in the equation of state in the large-q limit will give us the scaling relations.

---

[6] The time dependence comes from the cosine factor, so we can check $\cos(\pi\nu(1/2 - \imath(-t - \imath\beta)/\beta)) = \cos(-\pi\nu/2 + \imath\pi\nu t/\beta)$ which is same as $\cos(\pi v(1/2 - \imath t/\beta))$ as $\cos(-x) = \cos(x)$.

We know that $\mathcal{G}^>$ and $\mathcal{G}^<$ satisfies the KMS relation as in Eq. (4.52). Now we use their large-q ansatz in Eq. (4.17) to get

$$\frac{\mathcal{G}^<(t+\imath\beta)}{-\mathcal{G}^>(t)} = \frac{\mathcal{Q}+\frac{1}{2}}{-(\mathcal{Q}-\frac{1}{2})} e^{\frac{1}{q}[g^<(t+\imath\beta)-g^>(t)]}, \tag{4.54}$$

where we focus on the term in the exponential where we use Eq. (4.53) and the KMS relation for g^+ from Eq. (4.49)

$$\begin{aligned} g^<(t+\imath\beta) - g^>(t) =& (g^+(-t-\imath\beta) - g^-(-t-\imath\beta)) - (g^+(t)+g^-(t)) \\ =& (g^+(t) - g^-(-t-\imath\beta)) - (g^+(t)+g^-(t)) \\ =& -g^-(-t-\imath\beta) - g^-(t) \\ =& -2\imath\mathcal{Q}\alpha(-t-\imath\beta) - 2\imath\mathcal{Q}\alpha t = -2\beta\mathcal{Q}\alpha \end{aligned} \tag{4.55}$$

where we used the solution for g^- at equilibrium in Eq. (4.47). Thus, we have

$$\frac{\mathcal{G}^<(t+\imath\beta)}{-\mathcal{G}^>(t)} = \frac{1+2\mathcal{Q}}{1-2\mathcal{Q}} e^{-\frac{2}{q}\beta\mathcal{Q}\alpha}. \tag{4.56}$$

But the left-hand side is equal to $e^{-\beta\mu}$ using the KMS relation in Eq. (4.52). Therefore, equating we get the equation of state

$$\boxed{\beta\mu = \ln\left[\frac{1+2\mathcal{Q}}{1-2\mathcal{Q}}\right] - \frac{2}{q}\mathcal{Q}\beta\alpha}. \tag{4.57}$$

Therefore, there is a competition between the free term and the interacting term (containing α) in the equation of state. We impose that both terms compete in the large-q limit. As an example, if $\beta\alpha = \mathcal{O}(q^0)$ and $\mathcal{Q} = \mathcal{O}(q^{-1/2})$, then the second term gets suppressed in the large-q limit, leaving us with the free contribution only. We don't want that and that's why this scaling is not what we desire.

We start with the definition of the coupling in Eq. (4.20), namely $\mathcal{J}_q^2 \equiv J_q^2(1-4\mathcal{Q}^2)^{\frac{q}{2}-1}$. In the large-$q$ limit, we get $\mathcal{J}_q^2 \sim e^{-2q\mathcal{Q}^2}$ which will get exponentially suppressed at large-q (effectively making the theory free) if $\mathcal{Q} = \mathcal{O}(q^0)$. That's why our first scaling relation comes from the charge density:

$$\mathcal{Q} = \frac{\tilde{\mathcal{Q}}}{q^{1/2}}, \qquad \tilde{\mathcal{Q}} = \mathcal{O}(q^0). \tag{4.58}$$

Then we focus on the equation of state and we re-scale the temperature as follows to keep everything to leading order in $1/q$:

$$T = \tilde{T}/q = \mathcal{O}(q^{-1}), \quad \beta = q\tilde{\beta} \equiv q/\tilde{T} \tag{4.59}$$

where $\tilde{T} = \mathcal{O}(q^0)$ and $\tilde{\beta} = \mathcal{O}(q^0)$. This defines our "low" temperature regime where $T = \tilde{T}/q$. Finally, we require

$$\mu = \tilde{\mu} q^{-3/2} \tag{4.60}$$

where $\tilde{\mu} = \mathcal{O}(q^0)$, to ensure competition between the interacting term and the free term in the equation of state in the large-q limit.

We are interested in the low temperature physics where the quantum fluctuations are dominant. We start by solving for ν in the closure relation Eq. (4.48) at low temperatures. This is important as ν determines our Green's function $g^+(t)$ in Eq. (4.47) which in turn decides the behavior of the system. Therefore, expanding ν to leading order in $1/q$ gives[7]

$$\nu = 1 - \frac{2\tilde{T}}{q\mathcal{J}_q} + \mathcal{O}\left(\frac{1}{q^2}\right) = 1 - \frac{2}{qU_q} + \mathcal{O}\left(\frac{1}{q^2}\right) \tag{4.61}$$

where $\tilde{T}/q = T$ is the low temperature regime and we have defined $U_q \equiv \tilde{\beta}\mathcal{J}_q$. We will later see that this saturates the upper bound of quantum Lyapunov exponent $\lambda_L \rightarrow 2\pi T$ at low temperatures—something known as the Maldacena-Shenker-Stanford (MSS) bound [6]. This is the reason why the SYK model is sometimes also referred to as the "maximally chaotic" model.

Now we have all ingredients to deep dive into the thermodynamics and see if there is any interesting phase diagram (such as phase transitions) in the model at low temperatures.

4.2.3 The Equation of State

We start by writing the closure relation in Eq. (4.48) as

$$\pi\nu = qU_q \cos\left(\frac{\pi\nu}{2}\right) \tag{4.62}$$

where we recall $U_q = \tilde{\beta}\mathcal{J}_q$. Accordingly, $\beta\alpha$ in Eq. (4.51) (which also appears in the equation of state (4.57)) becomes (since ν is 1 to leading order)

$$\beta\alpha = -2qU_q \quad \Rightarrow \tilde{\beta}\alpha = -2U_q. \tag{4.63}$$

[7] Software such as Mathematica can be quite handy in dealing with such manipulations where one needs to solve Eq. (4.48), namely $\pi\nu = q\tilde{\beta}\mathcal{J}_q \cos\left(\frac{\pi\nu}{2}\right)$ at leading order in q. We can take as ansatz $\nu = 1 - \delta$ and approximate $\cos(\pi\nu/2) = \sin(\pi\delta/2) \approx \pi\delta/2$ and solve for δ on both sides to get $\delta = \frac{2}{q\tilde{\beta}\mathcal{J}_q + 2}$ which is then expanded to leading order in $1/q$.

Then, the equation of state at leading order is given by[8]

$$\tilde{\beta}\tilde{\mu} = 4\tilde{\mathcal{Q}}(1 + U_q). \tag{4.64}$$

Now, we substitute $U_q = \tilde{\beta}\mathcal{J}_q = \tilde{\beta}J_q e^{-q\mathcal{Q}^2} = \tilde{\beta}J_q e^{-\tilde{\mathcal{Q}}^2}$ (see the discussion above Eq. (4.58)) to get

$$\boxed{\frac{\tilde{\mu}}{J_q} = 4\tilde{\mathcal{Q}}\left[\frac{\tilde{T}}{J_q} + e^{-\tilde{\mathcal{Q}}^2}\right]}. \tag{4.65}$$

We will return to this equation of state at low temperatures later when we analyze the phase diagram. Here we wish to give a brief essence of what's coming next. We can ask ourselves if there is any point of inflection in the system by analyzing $\tilde{\mu}(\tilde{\mathcal{Q}})$. We can find a critical value of the charge density $\tilde{\mathcal{Q}}_c$ such that $\left.\frac{\partial^2\tilde{\mu}(\tilde{\mathcal{Q}})}{\partial\tilde{\mathcal{Q}}^2}\right|_{\tilde{\mathcal{Q}}=\tilde{\mathcal{Q}}_c} = 0$. One can check that $\frac{\partial^2\tilde{\mu}(\tilde{\mathcal{Q}})}{\partial\tilde{\mathcal{Q}}^2} = e^{-\tilde{\mathcal{Q}}^2}(-24\tilde{\mathcal{Q}} + 16\tilde{\mathcal{Q}}^3)$ which gives $\tilde{\mathcal{Q}}_c = 0, \sqrt{3/2}$. Accordingly, we can get the critical temperature by demanding $\left.\frac{\partial\tilde{\mu}(\tilde{\mathcal{Q}})}{\partial\tilde{\mathcal{Q}}}\right|_{\tilde{\mathcal{Q}}=\tilde{\mathcal{Q}}_c} = 0$ which gives $\tilde{T}_c = J_q e^{-\tilde{\mathcal{Q}}_c^2}(2\tilde{\mathcal{Q}}_c^2 - 1)$. For $\tilde{\mathcal{Q}}_c = 0$, $\tilde{T}_c < 0$ which is clearly unphysical, thereby leaving us with one critical point, namely

$$\boxed{\tilde{\mathcal{Q}}_c = \sqrt{\frac{3}{2}}, \qquad \tilde{T}_c = J_q e^{-\tilde{\mathcal{Q}}_c^2}(2\tilde{\mathcal{Q}}_c^2 - 1) = 2J_q e^{-\frac{3}{2}}}. \tag{4.66}$$

The coupling J_q sets the temperature scale which we can set to unity without loss of generality. We visualize the equation of state in Fig. 4.1. We clearly see that there exists a phase transition in the system and looks very similar to van der Waals phase transition. Indeed, as we will see later, there exists a continuous phase transition whose critical exponents belong to the same universality class as van der Waals fluid (known as the mean-field universality class or, equivalently, Landau-Ginzburg universality class).

4.2.4 The Grand Potential and the Phase Diagram

The grand potential $\Omega = \mathcal{E} - \mu\mathcal{Q} - TS$ (where $\mathcal{E}$ and S are the energy and entropy densities, respectively) is a thermodynamic potential that governs the thermodynamics of the system and is of crucial importance to analyze the phase transitions and the corresponding phase diagram. With the benefit of hindsight, we

[8] We have used $\ln\left[\frac{1+2x}{1-2x}\right] \approx 4x$ where $x = \mathcal{Q}$ and recall that $\mathcal{Q} = \mathcal{O}\left(\frac{1}{\sqrt{q}}\right)$.

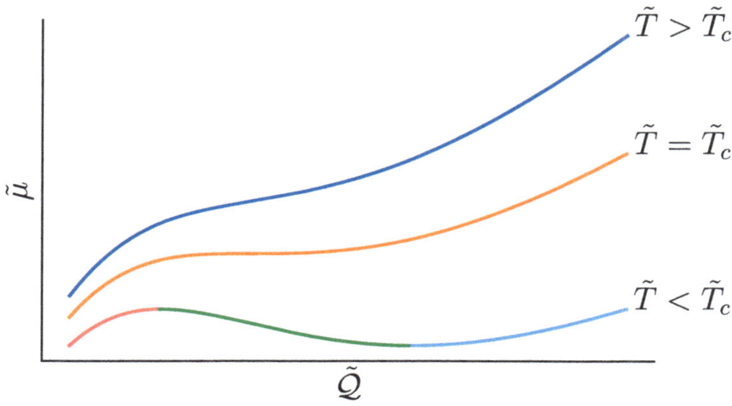

Fig. 4.1 The equation of state (Eq. (4.65)) at low temperatures where $\tilde{T}_c$ denotes the critical temperature as given by Eq. (4.66) where there exists a point of inflection. There exists a first-order phase transition in the system for $\tilde{T} < \tilde{T}_c$, similar to van der Waals phase transition.This culminates in a continuous order phase transition at the critical point $\tilde{T} = \tilde{T}_c$. Indeed, as shown later in Sect. 4.2.5, there is a continuous phase transition in the system which belongs to the same universality class as that of van der Waals fluids, known as the mean-field (or, Landau-Ginzburg) universality class. Color coding is done to ensure consistency with the phase diagram in Fig. 4.2 and the discussion in the main text after Eq. (4.72)

focus on the quantity $q\beta\Omega = q\beta\mathcal{E} - q\beta\mu\mathcal{Q} - qS$ such that the terms are of $\mathcal{O}(q^0)$ as shown below. We focus on each of the term separately, keeping terms at order $\mathcal{O}(q^0)$. We start with

$$q\beta\mathcal{E} = q^2\tilde{\beta}\tilde{\mathcal{E}} = \tilde{\beta}\left(1 - \frac{\tilde{\mathcal{Q}}^2}{q}\right)\alpha \simeq \tilde{\beta}\alpha \tag{4.67}$$

where we used Eq. (4.39). But we know $\tilde{\beta}\alpha$ from Eq. (4.51) that $\tilde{\beta}\alpha = -2\tilde{\beta}\mathcal{J}_q \sin(\pi\nu/2) \simeq -2U_q$ where we used the definition of $U_q \equiv \tilde{\beta}\mathcal{J}_q = \tilde{\beta}J_q e^{-\tilde{\mathcal{Q}}^2}$ and the leading order of ν was plugged from Eq. (4.61). Thus, for the first term, we have $-2U_q$.

The second term is

$$q\beta\mu\mathcal{Q} = qq\tilde{\beta}q^{-3/2}\tilde{\mu}q^{-1/2}\tilde{\mathcal{Q}} = \tilde{\beta}\tilde{\mu}\tilde{\mathcal{Q}} = 4\tilde{\mathcal{Q}}^2(1 + U_q), \tag{4.68}$$

where Eq. (4.64) was used.

Finally, we use the Maxwell's relation to get the entropy density

$$-\left(\frac{\partial S}{\partial \mathcal{Q}}\right)_{T, J_q, J_{q/2}} = \left(\frac{\partial \mu}{\partial T}\right)_{\mathcal{Q}, J_q, J_{q/2}} \tag{4.69}$$

and use the equation of state to get for the third term

$$qS = q\ln 2 - 2\tilde{Q}^2 + \mathcal{O}\left(\frac{1}{q}\right). \tag{4.70}$$

Therefore plugging back in the grand potential, we get at leading order in q^0

$$q\beta\Omega = q\beta\mathcal{E} - q\beta\mu Q - qS = -2U_q - 4\tilde{Q}^2(1+U_q) - q\ln 2 + 2\tilde{Q}^2. \tag{4.71}$$

We can define $\tilde{\beta}\tilde{\Omega} \equiv q\beta\Omega + q\ln 2$ where all terms in the expressions are at order $\mathcal{O}(q^0)$, to finally get for the potential

$$\boxed{\tilde{\beta}\tilde{\Omega} = -2U_q(1+2\tilde{Q}^2) - 2\tilde{Q}^2 = -2\tilde{\beta}J_q e^{-\tilde{Q}^2}(1+2\tilde{Q}^2) - 2\tilde{Q}^2}. \tag{4.72}$$

We now analyze the grand potential in Fig. 4.2 to identify phase transitions in the system and contrast against the equation of state in Fig. 4.1. Our analysis reveals two distinct regimes:

- First-order phase transitions occur for temperatures $\tilde{T} < \tilde{T}_c$, characterized by discontinuous jumps in the *order parameter* $\tilde{Q}$.

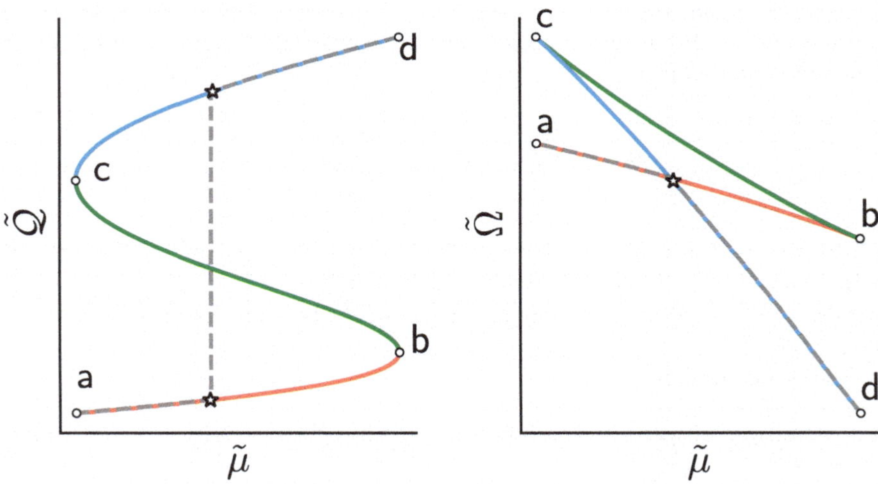

Fig. 4.2 The grand potential (Eq. (4.72)) is represented at low temperatures below the critical point $\tilde{T} < \tilde{T}_c$. Critical values are provided in Eq. (4.66). This reveals a first-order phase transition characterized by a discontinuous jump in the order parameter $\tilde{Q}$ (charge density). The dotted line indicates the stable phase path. Color coding is consistent with Fig. 4.1 and the main text discussion below Eq. (4.72). The phase transition is analogous to liquid-gas transitions in classical systems such as van der Waals fluids. We quantitatively analyze this transition in Sect. 4.2.5, where we establish its mean-field (Landau-Ginzburg) universality class

- These first-order phase transitions culminate at a critical point at $\tilde{T} = \tilde{T}_c$ where the phase transition becomes continuous (second-order).

Figure 4.2 illustrates a representative first-order transition ($\tilde{T} < \tilde{T}_c$). Key features include:

- An unstable phase region (green curve) that is thermodynamically inaccessible to the system
- Stable phases (dotted lines) formed by segments of the pink and blue branches
- A discontinuous jump in the $\tilde{Q}$-$\tilde{\mu}$ plane as the system avoids unstable states

This behavior is remarkably analogous to van der Waals fluids below their critical temperature:

- Pink Branch: "Gaseous" phase with lower charge density
- Blue Branch: "Liquid" phase with higher charge density

The correspondence extends naturally to our complex SYK model, where charge density plays the role of particle density in conventional fluids.

We now turn to the critical point at $\tilde{T} = \tilde{T}_c$, where the phase transition changes from first-order to continuous. This critical point represents a special regime where thermodynamic quantities exhibit singular behavior described by power laws. To quantitatively characterize these singularities and complete our thermodynamic analysis of the complex SYK model, we will calculate critical exponents—numerical indices that capture how quantities like specific heat and correlation length diverge near the critical point.

Remarkably, our analysis will reveal that the critical exponents of this quantum SYK model match those of the classical van der Waals fluid. This establishes that both systems belong to the same universality class—specifically, the mean-field (Landau-Ginzburg) universality class. Universality classes group diverse physical systems that share identical critical exponents, revealing that their critical behavior depends only on fundamental symmetries and dimensionality, not microscopic details. The emergence of mean-field exponents in both a quantum model and classical fluid demonstrates a profound connection between seemingly disparate systems at criticality.

4.2.5 Critical Exponents and the Universality Class

We already calculated the critical points in Eq. (4.66). Plugging in the chemical potential and the grand potential gives us the critical value of the chemical potential

$$\boxed{\tilde{\mu}_c = 12 J_q \sqrt{\frac{3}{2}} e^{-\frac{3}{2}}, \quad \tilde{\Omega}_c = -14 J_q e^{-\frac{3}{2}}}. \tag{4.73}$$

As we already saw in the phase diagram above (Fig. 4.2) that there exists a first-order phase transitions for temperatures $\tilde{T} < \tilde{T}_c$ that culminates at a continuous phase transition critical point at $\tilde{T} = \tilde{T}_c$. Critical exponents serve as the "fingerprints" of such continuous phase transitions at the critical point. Their universal values reveal deeper connections between seemingly disparate systems—from quantum condensed matter to black hole thermodynamics—demonstrating how collective behavior transcends microscopic physics at criticality. We first proceed to quantify the transition above by calculating the critical exponents and then later comment on the universality.

As is usual with the calculation of the critical exponents, we need to be in the vicinity of the critical point where there are power-law behaviors. That's why one resorts to the reduced variables, namely

$$m \equiv \left[\frac{\tilde{\mu}}{\tilde{\mu}_c} - 1\right], \quad \rho \equiv \frac{\tilde{Q}}{\tilde{Q}_c} - 1, \quad t \equiv \frac{\tilde{T}}{\tilde{T}_c} - 1 \tag{4.74}$$

so that the critical point corresponds to the coordinates $(m, \rho, t) = (0, 0, 0)$. We also rescale the grand potential in Eq. (4.72) around the critical point as

$$f \equiv \frac{\tilde{\Omega} - \tilde{\Omega}_c}{\tilde{\mu}_c \tilde{Q}_c} + m - \frac{t}{3}. \tag{4.75}$$

As we saw from the phase diagram in Fig. 4.2, the critical order parameter is ρ and we will focus in the vicinity of the critical point where power law behaviors arise.

The following manipulations, while theoretically possible with pen and paper, are complex; we strongly recommend using software such as Mathematica. We have provided an implementation in Mathematica for all expressions used below in Appendix G. We encourage readers to use this code to explore the technical details involved in calculating the critical exponents. We now begin by expressing the chemical potential and the grand potential in terms of the reduced variables.

$$\begin{aligned} \tilde{\mu} &= \frac{2\sqrt{6}J_q(\rho+1)\left(e^{\frac{1}{2}(-3)\rho(\rho+2)} + 2t + 2\right)}{e^{3/2}}, \\ \tilde{\Omega} &= -2J_q e^{\frac{1}{2}(-3)(\rho+1)^2}(3\rho(\rho+2)+4) - \frac{6J_q(\rho+1)^2(t+1)}{e^{3/2}}. \end{aligned} \tag{4.76}$$

Accordingly, the reduced variables in Eq. (4.74) become

$$\begin{aligned} m &= \frac{1}{3}J_q(\rho+1)\left(e^{-\frac{3}{2}\rho(\rho+2)} + 2t + 2\right) - 1, \\ f &= \frac{1}{18}\left(-2J_q e^{-\frac{3}{2}\rho(\rho+2)}(3\rho(\rho+1)+1) - 6J_q\left(\rho^2 - 1\right)(t+1) - 6t - 4\right). \end{aligned} \tag{4.77}$$

Then we expand f and m around the critical order parameter $\rho = 0$ to get (we remind the reader that Appendix G provides a Mathematica implementation for all expressions being dealt in this section). Since the coupling strength J_q provides a reference for the energy scale in the system (accordingly, for the temperature), without loss of generality, we set $J_q = 1.0$ below.[9] The expansions are given by

$$f = -\rho^2 \frac{t + (3\rho/2)^2}{3} + \mathcal{O}(\rho^5), \quad m = \frac{2t}{3} + \rho\left(\frac{2t}{3} + \rho^2\right) + \mathcal{O}(\rho^4). \tag{4.78}$$

Consequently, field mixing must be incorporated to obtain the correct scaling function and critical exponents as explained in Ref. [11]. This means the true ordering field h is not solely the chemical potential, but includes a linear combination with the temperature field: $h \equiv m - \frac{2}{3}t$. Our goal is to express both the reduced grand potential f and reduced charge density ρ in terms of t and h. To achieve this, we

1. solve the cubic equation (4.78) for $\rho(t, m)$,
2. apply the field transformation $h \equiv m - \frac{2}{3}t$ to eliminate m,
3. substitute $\rho(t, h)$ into Eq. (4.78) for f, and
4. expand $f(t, h)$ in asymptotic limits (e.g., small t or small h).

Again, while theoretically possible with pen and paper, software programs such as Mathematica are highly recommended. We have implemented all these steps in Mathematica whose implementation is given in Appendix G. We get for the reduced grand potential f and the reduced order parameter ρ around $h = 0$ for small values of h

$$\begin{aligned} f_{h=0}(t, h) &= -\frac{|t|^2}{9} + |h|\left|\frac{2t}{3}\right|^{1/2} - \frac{3h^2}{8}|t|^{-1} + \mathcal{O}\left(h^3|t|^{-5/2}\right), \\ \rho_{h=0}(t, h) &= -\operatorname{sgn}(h)\left|\frac{2t}{3}\right|^{1/2} + 2h\left|\frac{8t}{3}\right|^{-1} + \mathcal{O}\left(h^2|t|^{-5/2}\right), \end{aligned} \tag{4.79}$$

respectively. While the same quantities when expanded around $t = 0$ for small values of t give

$$\begin{aligned} f_{t=0}(t, h) &= -\frac{3}{4}|h|^{4/3}\left[1 + \mathcal{O}\left(h^{-2/3}|t|\right)\right], \\ \rho_{t=0}(t, h) &= -\partial_h f_{t=0}(t, h). \end{aligned} \tag{4.80}$$

Finally, we can calculate the critical exponents using these relations in the vicinity of the critical point. We define and calculate the exponents

[9] As you can see in Appendix G where the expressions are evaluated for arbitrary J_q, the critical exponents are the same for all J_q.

Table 4.1 Critical exponents for the phase transition of complex SYK model

α	β	γ	δ
0	$\frac{1}{2}$	1	3

- Exponent α is defined via the specific heat $C_h \propto -\partial_t^2 f_{h=0}(t,0) \propto |t|^{-\alpha}$ that gives us $upalpha = 0$.
- Exponent β is defined through the order parameter $\rho_{h=0}(t,0) \propto |t|^{\beta}$ that gives us $\beta = \frac{1}{2}$.
- Exponent γ is defined via the susceptibility $\chi_h \propto \partial_h^2 f_{h=0}(t,h)\big|_{h=0} \propto |t|^{-\gamma}$ that gives us $\gamma = 1$.
- Finally, the exponent δ is defined through the critical isotherm $\rho_{t=0}(0,h) \propto h^{1/\delta}$ that gives us $\delta = 3$.

We summarize the critical exponents in Table 4.1 where we conclude that the continuous phase transition of a complex SYK model belongs to the same universality class as the Landau-Ginzburg (equivalently, mean-field) universality class.[10] This universality class is shared by a wide variety of physical systems, such as van der Waals fluids [12] and AdS black holes [1, 13, 14]. We again get a hint of holography here which we expand upon next.

4.2.6 Relation to Hawking-Page Transition in Black Holes*

In order to provide hints of holography, we first need to go to even further lower temperatures where we scale the temperature and the chemical potential as

$$\hat{T} \equiv \frac{T}{q^2}, \quad \hat{\mu} \equiv \frac{\mu}{q^2}. \tag{4.81}$$

We call this region as "very low" temperature regime. Contrast this with scaling relations in Eqs. (4.59) and (4.60), which we called the "low" temperature regime. Without going into the details, we present the results of what happens as very low temperature ranges. The details have been chalked out in Ref. [7] (in particular, see Appendix A.2).

At very low temperature, the system undergoes a symmetry-breaking transition from a symmetric Majorana state at charge density $\mathcal{Q} = 0$ to a finite charge density $\mathcal{Q} = \frac{1}{2}\sqrt{1 - e^{-4\hat{\beta}J_q}}$ where $\hat{\beta} = 1/\hat{T}$. This jump for the large-q limit with fixed $\hat{T}$, $\hat{\mu}$, has profound physical consequences:

[10] There are other critical exponents too associated with the critical point, however, we restrict ourselves to the four most important exponent as mentioned here. For further details, we refer the reader to Ref. [7].

- As has been shown in Appendix A.2 of Ref. [7], the charge density at very low temperatures scales as $\hat{\mathcal{Q}} = \frac{\mathcal{Q}}{q}$.
- The effective interaction strength $\mathcal{J}_q^2 \simeq e^{-2q\mathcal{Q}^2} J_q^2$ becomes suppressed, vanishing completely as $\mathcal{Q}$ approaches its finite value.
- This drives the system toward a free, integrable limit ($\nu \to 0$). Crucially, infinitesimal perturbations from the chemical potential $\hat{\mu}$ destabilize the $\mathcal{Q} = 0$ state, triggering a discontinuous transition to finite charge density. This spontaneously breaks the $U(1)$ symmetry and marks a transition from maximally chaotic to non-chaotic dynamics.

The phase diagram consisting of low and very-low temperature regimes is plotted in Fig. 4.3.

Such instabilities are characteristic of charged black holes too. We present a phase diagram of the charged black hole in Fig. 4.4. The figure is adapted from

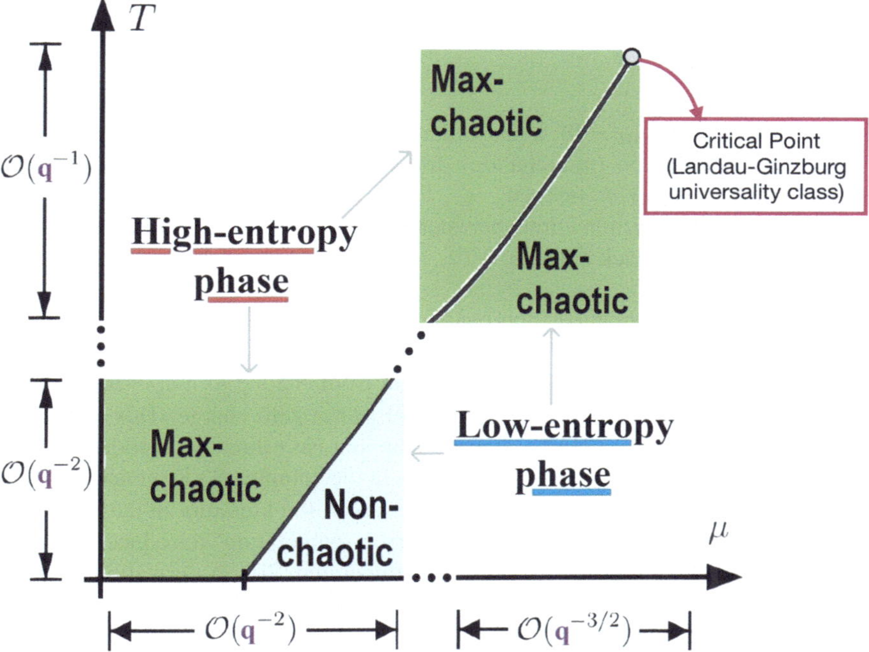

Fig. 4.3 Phase diagram of the complex SYK model, highlighting low- and very-low-temperature regimes (defined in the main text). A first-order transition line separates: (i) chaotic and non-chaotic (regular) phases at very low temperatures, (ii) two distinct chaotic phases at low temperatures. This line terminates at a critical point characterized by a continuous phase transition with Landau-Ginzburg critical exponents (Table 4.1), confirming its universality class

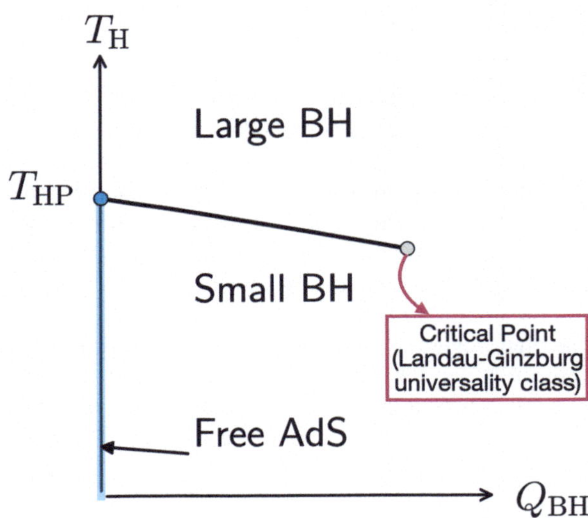

Fig. 4.4 Phase diagram of a charged Anti-de Sitter (AdS) black hole. A first-order transition line terminates at a critical point with Landau-Ginzburg universality (Table 4.1). At zero charge ($\mathcal{Q}_{BH}$=0), the Hawking-Page transition separates a chaotic large black hole phase from non-chaotic radiation at temperature T_{HP} (also known as "Zorro's temperature" in Ref. [1]). The similarity to the complex SYK phase diagram (Fig. 4.3) suggests a hint for holographic connection

Refs. [1, 15, 16], where

- At zero charge ($\mathcal{Q}_{BH} = 0$), a first-order Hawking-Page transition occurs: below T_{HP}, stable radiation (non-chaotic) dominates; above T_{HP}, large black holes (maximally chaotic) are favored.
- For $\mathcal{Q}_{BH} \neq 0$, only transitions between chaotic phases exist: large black holes (high T) to small black holes (low T).

Crucially, the chaos-to-regular (non-chaotic) transition occurs exclusively along the $\mathcal{Q}_{BH} = 0$ axis in black holes. This contrasts sharply with our SYK system at very low temperatures: here, a first-order transition from chaotic to non-chaotic phases persists for all charge densities (Fig. 4.3), not just at zero charge. This distinction explains why the standard Hawking-Page mapping isn't directly reproduced, but a generalized Hawking-Page transition appears in the complex SYK model.

We wish to clarify a terminology that appears in the literature in the context of charged black hole thermodynamics: the "Zorro's temperature" introduced in [1] for $\mathcal{Q}_{BH} = 0$ is identical to the Hawking-Page temperature T_{HP}. A derivation is shown in Appendix E of Ref. [8]. Both describe the same transition from chaotic black holes to non-chaotic radiation.

Finally, we note the remarkable robustness of the complex SYK phase transition. As demonstrated in Ref. [8], introducing an additional energy scale—which in the large-q limit behaves as a one-dimensional equilibrium chain—preserves the phase diagram topology shown in Fig. 4.3. This demonstrates that the observed universality transcends both the number of energy scales and system dimensionality. We will explore such SYK chain extensions later in this work.

4.3 Chaos*

The study of quantum chaos traces its origins to fundamental questions about thermalization in closed quantum systems. For a detailed review, we refer the reader to Ref. [17]. While motivated by classical chaos, it addresses a distinct challenge: How can signatures of classical chaotic behavior manifest in quantum systems? Given quantum mechanics' linear structure, true chaos in the classical sense cannot occur. Instead, quantum chaos examines what features—particularly in the classical limit when it exists—capture essential characteristics of classical chaotic dynamics.

Classically, chaos is defined by sensitivity to initial conditions, quantified by the classical Lyapunov exponent, which we denote as λ_{cl}. In quantum systems, an analogous quantum Lyapunov exponent (denoted as λ_L) characterizes chaos through the exponential growth of out-of-time-order correlators (OTOCs). Divergent OTOC behavior serves as a key diagnostic for many-body quantum chaos. Crucially, whereas classical Lyapunov exponents are unbounded, their quantum counterparts obey an upper limit (recall that we are using the natural units throughout, unless stated otherwise):

$$\lambda_L \leq 2\pi T, \tag{4.82}$$

where T is temperature. This is the celebrated Maldacena-Shenker-Stanford (MSS) bound [6]. Systems saturating this bound are termed "maximally chaotic".

Given this section's specialized focus (marked *), we provide a conceptual foundation rather than full technical derivations. After motivating quantum chaos through its classical roots, we outline how the SYK model achieves maximal chaos at low temperatures. Computational details are deferred to the literature; our goal is to highlight the essential mechanisms enabling SYK's saturation of the MSS bound.

4.3.1 Quantum Lyapunov Exponent

Classical chaos is fundamentally characterized by exponential sensitivity to initial conditions, quantified through the Lyapunov exponent. Consider two trajectories $\vec{x}\,(t, \vec{x}_0)$ and $\vec{x}\left(t, \vec{x}_0 + \vec{\delta}\right)$ in phase space, where $\vec{\delta}$ represents an infinitesimal displacement (Fig. 4.5). The classical Lyapunov exponent λ_{cl} is defined via the growth rate of phase space derivatives:

$$\left|\frac{\partial x^i(t)}{\partial x^j(0)}\right| = \left|\{x^i(t), p^j(0)\}_{\text{Poisson bracket}}\right| \sim e^{\lambda_{\text{cl}} t} \tag{4.83}$$

where $\{\cdot, \cdot\}_{\text{P}}$ denotes the Poisson bracket, $x^i(t)$ is the ith position component at time t, and $p^j(0)$ is the conjugate momentum component at $t = 0$. A positive λ_{cl} indicates exponential divergence, thereby implying that the classical system is chaotic in nature.

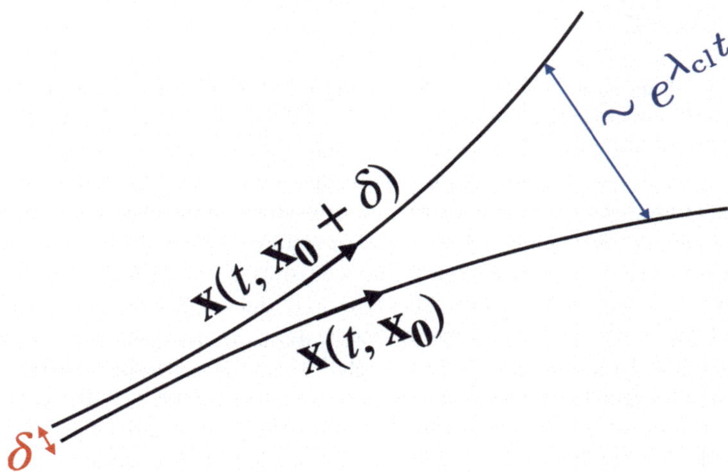

Fig. 4.5 Exponential trajectory divergence in classical chaos: Infinitesimally separated initial conditions ($\delta \lll 1$) evolve with asymptotic separation. The Lyapunov exponent $\lambda_{\rm cl}$ quantifies this instability rate

To construct the quantum analog, we apply the canonical quantization prescription: promote $\vec{x}$ and $\vec{p}$ to operators $\hat{X}$, $\hat{P}$, and replace Poisson brackets with commutators:

$$\left|\{x^i(t), p^j(0)\}_{\text{Poisson bracket}}\right|^2 \to \left\langle \left|\left[\hat{X}(t), \hat{P}(0)\right]\right|^2 \right\rangle = \text{TOC} - 2\text{Re}\,[\text{OTOC}] \tag{4.84}$$

where the correspondence from the Poisson bracket to commutator contains an additional factor of $\frac{1}{\imath\hbar}$. They don't show up because we are using the natural units where $\hbar = 1$ and $|1/\imath| = 1$. The squared commutator[11]—which quantifies operator growth—decomposes into time-ordered and out-of-time-ordered components, given by

$$\begin{aligned} \text{TOC} &= ||\rho^{1/2}\hat{X}(t)\hat{P}(0)||_F^2 + ||\hat{X}(t)\hat{P}(0)\rho^{1/2}||_F^2, \\ \text{OTOC} &= \text{Tr}\left[\left(\hat{X}(t)\hat{P}(0)\right)^2 \rho\right] = \left\langle \left(\hat{X}(t)\hat{P}(0)\right)^2 \right\rangle, \end{aligned} \tag{4.85}$$

[11] The commutator is anti-Hermitian: $[\hat{X}(t), \hat{P}(0)]^\dagger = -[\hat{X}(t), \hat{P}(0)]$, so its expectation value is imaginary. We need a real, positive quantity to compare with classical chaos. The square provides a gauge-invariant, real measure of operator spreading.

where $\langle\cdot\rangle = \mathrm{Tr}(\rho\cdot)$ denotes the thermal expectation value with respect to the density matrix $\rho = e^{-\beta\mathcal{H}}/\mathcal{Z}$ at temperature $T = 1/\beta$, and $||\ldots||$ denotes the Frobenius norm defined by $||A||_F \equiv \sqrt{\mathrm{Tr}[A^\dagger A]}$.[12]

The out-of-time-ordered correlator (OTOC) encodes quantum chaos. For many-body systems, motivated by the seminal work of Maldacena, Shenker and Stanford [6], we use the regularized OTOC formulation, denoted as $\mathcal{F}$, which we denote as $\mathcal{F}$

$$\mathcal{F}(t) = \mathrm{Tr}\left[\rho^{1/4}\hat{X}(t)\rho^{1/4}\hat{P}(0)\rho^{1/4}\hat{X}(t)\rho^{1/4}\hat{P}(0)\right]. \tag{4.86}$$

Accordingly, if there exists an exponential divergence in the system at the level of OTOCs, we capture the effect as OTOC $\sim e^{\lambda_L t}$ where λ_L is the quantum Lyapunov exponent, an analogue of the classical Lyapunov exponent λ_{cl}, where a non-zero value of λ_L is used as a proxy for the existence of quantum chaos. We also note that while expressed here in position and momentum operators, the discussion extends to any non-commuting operators.

The OTOC serves as a fundamental diagnostic for quantum chaos in many-body systems. Following Maldacena, Shenker, and Stanford [6], we employ the *regularized* OTOC at finite temperature (labeled as $\mathcal{F}(t)$)[13]

$$\mathcal{F}(t) = \mathrm{Tr}\left[\rho^{1/4}\hat{X}(t)\rho^{1/4}\hat{P}(0)\rho^{1/4}\hat{X}(t)\rho^{1/4}\hat{P}(0)\right], \tag{4.87}$$

where $\rho = e^{-\beta H}/\mathcal{Z}$ is the thermal density matrix. This regularization preserves unitarity and avoids artificial divergences. The unregularized OTOC can be denoted as $\mathcal{F}_{\mathrm{unreg}}(t)$.

When $\mathcal{F}(t)$ exhibits exponential decay of the form[14]

$$\mathcal{F}(t) \simeq 1 - f_0 e^{\lambda_L t} + \mathcal{O}(e^{2\lambda_L t}), \tag{4.89}$$

[12] Plugging the definition of Frobenius norm, an alternative expression for TOC is TOC $= \left\langle\hat{P}(0)\hat{X}^2(t)\hat{P}(0)\right\rangle + \left\langle\hat{X}(t)\hat{P}^2(0)\hat{X}(t)\right\rangle$.

[13] The unregularized OTOC (denoted by $\mathcal{F}_{\mathrm{unreg}}(t)$ as in Eq. (4.84)) and regularized OTOC (denoted by $\mathcal{F}(t)$) match only at zero temperature ($\beta \to \infty$).

[14] The unregularized OTOC$(t) = \mathcal{F}_{\mathrm{unreg}}(t) = C_1 - C_2 e^{\lambda_2 t} + \cdots$ (early times). Here, $C_1, C_2 > 0$, so OTOC decays exponentially. Therefore, we have

$$c(t) \equiv \left\langle \left|\left[\hat{X}(t), \hat{P}(0)\right]\right|^2 \right\rangle = \underbrace{\mathrm{TOC}}_{\text{slow}} - 2\mathrm{Re}[\underbrace{C_1 - C_2 e^{\lambda_L t}}_{\text{OTOC}}] = \underbrace{(\mathrm{TOC} - 2C_1)}_{\text{constant}} + \underbrace{2C_2 e^{\lambda_L t}}_{\text{exponential growth}}. \tag{4.88}$$

That's why, loosely speaking, quantum Lyapunov exponent is defined as $c(t) \sim e^{\lambda_L t}$. The constants C_1 and C_2 are not related to f_0 in Eq. (4.89). However, we are more careful here and follow Ref. [6] that the regularized OTOC, labeled as $\mathcal{F}(t)$, captures the real chaos exponent, not the unreguralized one.

the exponent $\lambda_L > 0$ defines the quantum Lyapunov exponent. This

- generalizes the classical Lyapunov exponent λ_{cl} to quantum systems,
- provides a proxy for many-body quantum chaos, and
- quantifies operator growth via $\|[\hat{X}(t), \hat{P}(0)]\|^2 \sim e^{\lambda_L t}$.

Here f_0 is the scrambling amplitude (a constant) that depends on the temperature, operators and system-specific details (such as $f_0 = \mathcal{O}(1/N)$ for large-N systems, such as the SYK model). Furthermore, f_0 sets the initial decay rate at $t \to 0$, $|\mathcal{F}(t) - 1| \sim f_0$, and is bounded by unitarity: $\mathcal{F}(t) \geq 0$ requires $f_0 e^{\lambda_L t} \leq 1$.

Crucially, while expressed here for conjugate variables $(\hat{X}, \hat{P})$, this framework extends to any noncommuting operators $(\hat{W}, \hat{V})$ through the universal form:

$$\mathrm{Tr}[\rho^{1/4}\hat{W}(t)\rho^{1/4}\hat{V}(0)\rho^{1/4}\hat{W}(t)\rho^{1/4}\hat{V}(0)]. \tag{4.90}$$

4.3.2 Four-Point Correlation Functions in the SYK Model

To compute the Lyapunov exponent, we begin with the four-point function of the coupled SYK system. In the large-N limit, the correlator for complex fermions c_j takes the form [4, 18] (where we have resorted to imaginary-time as the system is in equilibrium, but shortly afterward will return to the real-time picture)

$$\begin{aligned}&\frac{1}{N^2}\sum_{j,k}^{N}\langle T c_j(\tau_1)c_j^\dagger(\tau_2)c_k^\dagger(\tau_3)c_k(\tau_4)\rangle \\ &= \mathcal{G}(\tau_{12})\mathcal{G}(\tau_{34}) + \frac{1}{N}\mathcal{F}_{\mathrm{unreg}}(\tau_1, \tau_2, \tau_3, \tau_4) + \ldots\end{aligned} \tag{4.91}$$

where $\mathcal{F}_{\mathrm{unreg}} = \sum_n \mathcal{F}_n$ denotes the unregularized OTOC and T is the time-ordering operator. The leading disconnected term is supplemented by subleading $\mathcal{O}\left(N^{-1}\right)$ contributions that encode quantum chaos.

Each $\mathcal{F}_n$ corresponds to a unique ladder diagram [18] as shown in Fig. 4.6, obtained through resummation of dressed propagators [19]. The recursive structure emerges via the kernel operation:

$$\mathcal{F}_{n+1}(\tau_1, \tau_2, \tau_3, \tau_4) = \int d\tau d\tau'\, K(\tau_1, \tau_2, \tau, \tau')\mathcal{F}_n(\tau, \tau', \tau_3, \tau_4) \tag{4.92}$$

denoted compactly as $\mathcal{F}_{n+1} = K \odot \mathcal{F}_n$. At late times, the exponential growth dominates over zero-rung contributions, reducing the OTOC to an eigenvalue

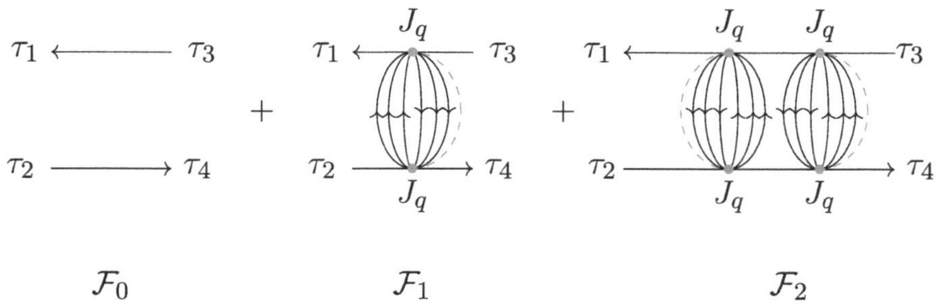

$\tau_1 \leftarrow \tau_3$ $\quad$ J_q $\quad$ $J_q \; J_q$

$\tau_2 \rightarrow \tau_4$ $\quad$ J_q $\quad$ $J_q \; J_q$

$\mathcal{F}_0 \qquad \mathcal{F}_1 \qquad \mathcal{F}_2$

Fig. 4.6 The diagrams containing J_q terms in the $\mathcal{O}\left(\frac{1}{N}\right)$ part of the correlator for the large-q complex SYK model. The diagram is drawn for $q = 8$, and also consists of a diagram with $\tau_3 \leftrightarrow \tau_4$, which have been omitted here. There are $q - 2 = 6$ lines connecting the two horizontal rails, out of which half of them ($q/2 - 1 = 3$ lines) run in one direction while the remaining half ($q/2 - 1 = 3$ lines) run in the opposite. The dotted line denotes disorder averaging

equation $\mathcal{F} = K \odot \mathcal{F}$. The equilibrium kernel for our coupled system is:

$$K(\tau_1, \tau_2; \tau, \tau') \equiv -\frac{2^{q-1} J_q^2}{q}(q-1)\, \mathcal{G}(\tau_1 - \tau)\mathcal{G}(\tau_2 - \tau')$$
$$\left[\mathcal{G}^>(\tau - \tau')\mathcal{G}^<(\tau' - \tau)\right]^{q/2-1} \tag{4.93}$$

As illustrated in Fig. 4.7, this kernel add single rungs.

To determine the exponential growth rate of the Out-of-Time-Order Correlator (OTOC) at late times, we employ the standard approach [2,6]. This involves defining

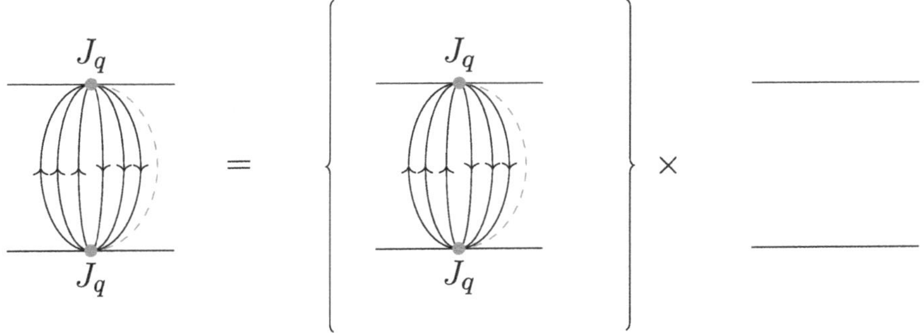

Fig. 4.7 In the large-N limit, the J_q contribution to the $\mathcal{O}\left(\frac{1}{N}\right)$ out-of-time-correlator (OTOC) diagrams for the large-q complex SYK system is constructed via the kernel K (shown in parentheses). The OTOC is given by $\mathcal{F} = \mathcal{F}_0 + K\mathcal{F}$, where $\mathcal{F} = \sum_{n=0}^{\infty} \mathcal{F}_n$ and $\mathcal{F}_0$ is defined in Fig. 4.6. This figure shows $\mathcal{F}_1 = K\mathcal{F}_0$ for $q = 8$, representing the J_q sector kernel action

the following regularized OTOC in real time:

$$\mathcal{F}(t_1, t_2, t_3, t_4) = \mathrm{Tr}\left[\rho(\beta)^{1/4}\, c_j(t_1)\, \rho(\beta)^{1/4}\, c_k^\dagger(t_3)\, \rho(\beta)^{1/4}\, c_j^\dagger(t_2)\, \rho(\beta)^{1/4}\, c_k(t_4)\right], \tag{4.94}$$

where $\rho(\beta)$ is the thermal density matrix. This regularization positions the fermions at quarter intervals on the thermal circle, with real-time separations between operator pairs.

Thermal Circle and OTOC in Keldysh Plane

The regularized OTOC in Eq. (4.94) with specific ordering of thermal factors $\rho(\beta)^{1/4}$ inserted between operators serves two key purposes: (1) It regularizes the correlator for analytic control, and (2) it preserves the Lyapunov exponent while simplifying computations [3, 6]. Figure 4.8 illustrates this configuration on the Keldysh contour. Crucially, the operators exhibit two distinct separations [4]:

- Thermal Separation: Operators are offset by $\frac{\beta}{4}$ along the imaginary-time axis (quarter thermal circle)
- Temporal Separation: $V(0)$ and $W(t)$ are separated by large real time t along the real-time contour

In the late-time limit ($t \to \infty$), non-interacting ("zero-rung") diagrams become negligible. The contour dynamics are governed by:

- Retarded propagators (Eq. (4.97)) for real-time segments
- Wightman functions (Eq. (4.97)) for the $\beta/2$-separated thermal segment

The real-time Wightman function emerges through analytic continuation: evaluating the Green's function (Eq. (4.47)) at $t \to it + \frac{\beta}{2}$ transforms imaginary-time dependence into real-time evolution.

The kernel K generates all interacting ladder diagrams (i.e., excluding the free propagator), satisfying $\sum_{n=1}^{\infty} \mathcal{F}_n = K \odot \sum_{n=0}^{\infty} \mathcal{F}_n$. Adding $\mathcal{F}_0$ (the zero-rung diagram) to both sides and assuming exponential OTOC growth at late times $(t_1, t_2 \to \infty)$, $\mathcal{F}_0$ becomes negligible. This yields the eigenvalue equation [6, 18]:

$$\mathcal{F}(t_1, t_2, t_3, t_4) = \int_{-\infty}^{\infty} \mathrm{d}t \int_{-\infty}^{\infty} \mathrm{d}t' K_R(t_1, t_2; t, t') \mathcal{F}(t, t', t_3, t_4). \tag{4.95}$$

The exponential growth is governed by the real-time component of $\mathcal{F}$, with diagrams generated by the retarded kernel K_R. Exploiting time-translation invariance in

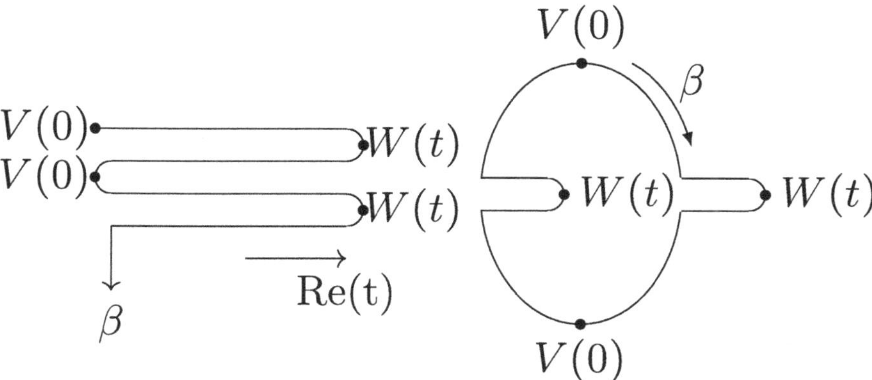

Fig. 4.8 The left panel depicts the Keldysh-Schwinger contour used to compute the OTOC in chaotic systems. As illustrated in the right panel, the operators $V(0)$ and $W(t)$ are separated by both: (i) a large real-time difference t, and (ii) a quarter-period along the thermal circle $\beta/4$

equilibrium where $t_{ij} \equiv t_i - t_j$, we express K_R as:

$$K_R(t_1, t_2; t_3, t_4) = \frac{2^{q-1} J_q^2}{q}(q-1)\mathcal{G}_R(t_{13})\mathcal{G}_R(-t_{24})\mathcal{G}_{lr}^{>}(t_{34})^{q/2-1}\mathcal{G}_{lr}^{<}(-t_{34})^{q/2-1}, \tag{4.96}$$

where the retarded propagator $\mathcal{G}_R$ and Wightman correlators $\mathcal{G}_{lr}$ are defined by:

$$\begin{aligned} \mathcal{G}_R(t) &\equiv [\mathcal{G}^>(t) - \mathcal{G}^<(t)]\,\theta(t) \overset{q\to\infty}{=} \theta(t), \\ \mathcal{G}_{lr}^{>}(t) &\equiv \mathcal{G}^>(\imath t + \beta/2), \\ \mathcal{G}_{lr}^{<}(t) &\equiv \mathcal{G}^<(\imath t - \beta/2). \end{aligned} \tag{4.97}$$

We have already solved for these Green's functions for the complex SYK model. Note that the retarded Green's function in frequency space relates to the Matsubara function via analytic continuation: $\mathcal{G}(\imath\omega_n \to \omega + \imath 0^+) = \mathcal{G}_R(\omega)$ [20]. Compared to the imaginary-time kernel (Eq. (4.93)), the overall sign change arises from the contour rotation $\tau \to \imath t, \tau' \to \imath t'$. The Jacobian $d\tau d\tau' = -dtdt'$ in Eq. (4.95) absorbs this sign into K_R. (See [21], Sec. 8, for details on the retarded kernel and chaos exponents).

Using the Green's functions in Eqs. (4.17) and (4.47) as well as the definition of the coupling in Eq. (4.20), we get

$$K_R(t_1, t_2; t_3, t_4) = \theta(t_{13})\theta(t_{24})\frac{2(\pi\nu T)^2}{\cosh(\pi\nu t_{34}T)^2} = \theta(t_{13})\theta(t_{24})W(t_{34}) \tag{4.98}$$

where ν is given by the closure relation in Eq. (4.48). Critically, while K_R generates all interacting diagrams, the total sum $\mathcal{F}$ remains invariant. Combined with the suppression of $\mathcal{F}_0$ at late times, $\mathcal{F}$ is an eigenfunction of the integral transform. To solve Eq. (4.95), we simplify by applying $\partial_{t_1}\partial_{t_2}$ to both sides to get

$$\partial_{t_1}\partial_{t_2}\mathcal{F}(t_1, t_2, t_3, t_4) = \int_{-\infty}^{\infty} dt \int_{-\infty}^{\infty} dt'\ \delta(t_1 - t)\delta(t_2 - t')W(t_{34})\ \mathcal{F}(t, t', t_3, t_4). \tag{4.99}$$

The delta functions enforce $t = t_1, t' = t_2$. For large t_1, t_2 (where exponential growth dominates), we may set $t_3 = t_4 = 0$ without loss of generality. Defining $\mathcal{F}(t_1, t_2, 0, 0) \equiv \mathcal{F}(t_1, t_2)$ simplifies notation.

We have the ansatz for $\mathcal{F}(t_1, t_2)$ in Eq. (4.89) where $t = \frac{t_1+t_2}{2}$ and $f_0 = f_0(t_{12}) = f_0(t_1 - t_2)$ which for late time and exponential growth becomes $\mathcal{F}(t_1, t_2) \sim -f_0(t_1 - t_2)e^{\lambda_L \frac{t_1+t_2}{2}}$. Accordingly, plugging this at $t_3 = t_4 = 0$ above, we get the following eigenvalue problem (see Ref. [4] for further details)

$$\left[\partial_t^2 + \frac{2(\pi\nu T)^2}{\cosh^2(\pi\nu T t)}\right] f_0(t) = \frac{\lambda_L^2}{4} f_0(t). \tag{4.100}$$

Re-defining $\tilde{x} = \pi\nu T t$, and using $\tilde{f}_0(\tilde{x}) = f_0(t)$, we are left with

$$\left[-\partial_{\tilde{x}}^2 - \frac{2}{\cosh^2 \tilde{x}}\right] \tilde{f}_0(\tilde{x}) = -\left(\frac{\lambda_L}{2\pi\nu T}\right)^2 \tilde{f}_0(\tilde{x}). \tag{4.101}$$

This is just a "Schrödinger" type equation with potential $V(x) = -\frac{2}{\cosh^2 \tilde{x}}$. We identify this potential with Pöschl-Teller potential [22].

Although dynamical systems typically possess a full spectrum of Lyapunov exponents λ_i—each quantifying exponential divergence along distinct phase-space directions—the dominant chaotic behavior is characterized by the maximum λ_L. This is because the largest exponent governs the fastest exponential growth rate, overwhelming slower instabilities at long timescales: $e^{\lambda_{\max} t} \gg e^{\lambda_i t}$ for $\lambda_i < \lambda_{\max}$ as $t \to \infty$. Furthermore, $\lambda_{\max}$ sets the fundamental timescale $\tau \sim 1/\lambda_L$ for information scrambling and loss of predictability in chaotic systems. The robustness matters: smaller exponents may reflect transient or constrained dynamics, while $\lambda_{\max}$ captures system-wide chaos. Thus, when we say 'the Lyapunov exponent', we implicitly refer to $\lambda_L \equiv \max_i \lambda_i$—the physically observable rate that defines chaoticity. The maximum value universally characterizes the fastest instability rate and the predictability horizon in chaotic systems.

Accordingly, we know the ground state value of the Pöschl-Teller potential which is $E_0 = -1$. Therefore, equating the eigenvalue above with E_0 gives

$$\lambda_L = 2\pi\nu T \tag{4.102}$$

where we know from the properties of ν (same for Majorana and complex cases) where ν can range between 0 and 1 with 0 denoting the free case while 1 signifies the infinitely strong coupling as the temperature $T \to 0$ (see the paragraph below

Eq. (3.88)). Therefore, the maximum value happens in the low-temperature limit where we have also seen before that the conformal symmetry arises. Accordingly, the maximum value that the quantum Lyapunov exponent can take for the SYK model is

$$\boxed{\lambda_L = 2\pi T} \quad (\nu \to 1). \tag{4.103}$$

We know that the quantum Lyapunov exponent is bounded from above as provided in Eq. (4.82), which turns out exactly to be $2\pi T$. Therefore, the SYK model saturates the MSS bound of quantum chaos. That's why, the SYK model is sometimes referred to as "maximally chaotic". This further provides a hint of holography as the black holes are known to be the fastest scramblers of Nature and they also saturated the MSS bound [23–26].

NOTE: The thermodynamic properties (including phase transitions and critical exponents) and chaotic properties appear robust in SYK-like systems. Specifically, introducing an additional scale via perturbations like $\mathcal{H} = J_q\mathcal{H}_q + J_{\frac{q}{2}}\mathcal{H}_{\frac{q}{2}}$ preserves the low-temperature universality class of critical exponents, and the saturation of the MSS bound for quantum chaos by the quantum Lyapunov exponent, as demonstrated in Ref. [8]. This robustness naturally raises the question of whether it extends to arbitrary combinations of large-q SYK models, $\mathcal{H} = \sum_\kappa J_{\kappa q}\mathcal{H}_{\kappa q}$. This remains an open problem.

4.4 Equilibrium SYK Chains

We now introduce one-dimensional chains composed of SYK dots connected by nearest-neighbor hopping. Our primary goals are to illustrate this construction and demonstrate that the chain inherits the solvability characteristics of the individual SYK dot. This solvability enables us to study quenched non-equilibrium dynamics, probing thermalization and charge transport, as explored in the next chapter.

Crucially, for SYK-type systems, a distinctive feature emerges: a uniform chain at equilibrium can be mapped exactly to its single-dot counterpart. This powerful mapping allows us to directly leverage the known solutions from the zero-dimensional case for analyzing the uniform equilibrium chain. This proves especially valuable for linear response calculations: when perturbing a uniform equilibrium chain, we can compute all non-equilibrium properties using the borrowed equilibrium solutions obtained via the single-dot mapping. We will apply this approach in the next chapter to investigate transport phenomena (Fig. 4.9).

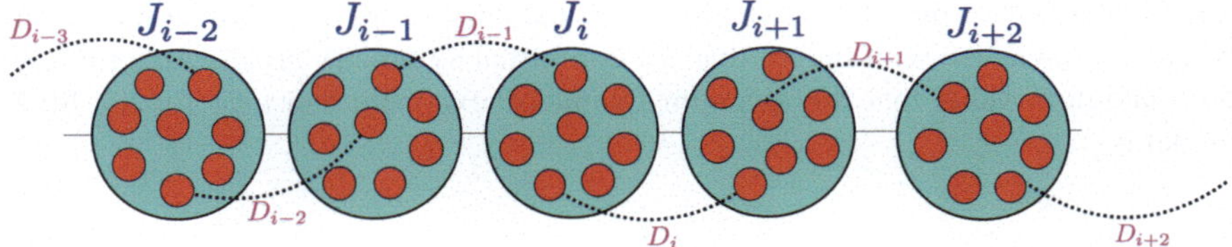

Fig. 4.9 Schematic representation of a SYK chain where each site is a complex SYK dot with nearest-neighbor hopping whose Hamiltonian is given in Eq. (4.104). The on-site strengths are given by $J_i(t)$ while the hopping strengths are governed by $D_i(t)$ where we have kept the most general, time-dependent case

4.4.1 General Framework for the SYK Chain Model

We consider a one-dimensional chain with L sites where each site is a large-q SYK dot ($q/2$-body interaction), with a general $r/2$-body nearest-neighbor hopping. We will deal with the real-time formalism in the most general bi-local formalism (no assumption of equilibrium), then later specialize to a uniform chain (where all on-site strengths are equal, so are the hopping strengths) at equilibrium (implying time-translational invariance: local in time).

The Hamiltonian is given by

$$\mathcal{H}(t) = \sum_{i=1}^{L} \left(\mathcal{H}_i(t) + \mathcal{H}_{i\to i+1}(t) + \mathcal{H}^{\dagger}_{i\to i+1}(t) \right) \tag{4.104}$$

where the on-site $q/2$-body interacting Hamiltonian is given by the large-q complex SYK dot (same as in Eq. (4.1) which we reproduce here with slight modification in notation)

$$\mathcal{H}_i = J_i(t) \sum_{\substack{\{\boldsymbol{\mu}_{1:\frac{q}{2}}\}_{\leq} \\ \{\boldsymbol{\nu}_{1:\frac{q}{2}}\}_{\leq}}} X(i)^{\boldsymbol{\mu_q}}_{\boldsymbol{\nu_q}} c^{\dagger}_{i;\mu_1} \cdots c^{\dagger}_{i;\mu_{q/2}} c_{i;\nu_{q/2}} \cdots c_{i;\nu_1} \tag{4.105}$$

where we adopt the same notation as before with slight modification of showing the indices they run over, namely $\{\boldsymbol{\mu}_{1:\frac{q}{2}}\}_{\leq} \equiv 1 \leq \mu_1 < \mu_2 < \cdots < \mu_{\frac{q}{2}-1} < \mu_{\frac{q}{2}} \leq N$ and $\boldsymbol{\mu_q} \equiv \{\mu_1, \mu_2, \mu_3, \ldots, \mu_{\frac{q}{2}}\}$ (where the label running from 1 to $q/2$ is assumed and has been consistent throughout this chapter). The fermionic creation and annihilation operators can be time-dependent and satisfy the usual equal-time anti-commutation algebra (Eq. (3.4)). Interactions are random and controlled by the random matrix $X(i)^{\boldsymbol{\mu_q}}_{\boldsymbol{\nu_q}}$, derived from a Gaussian ensemble with the following mean and variance (same as in Eq. (4.3)) which is the same for all lattice sites i:

$$\overline{X} = 0, \quad \overline{|X|^2} = \sigma_q^2 = \frac{4\left[\left(\frac{q}{2}\right)!\right]^2}{q^2\left(\frac{N}{2}\right)^{q-1}}. \tag{4.106}$$

The Gaussian probability distribution itself is given by

$$\mathcal{P}_q\left[X_{\boldsymbol{\nu}_q}^{\boldsymbol{\mu}_q}\right] = A_q \exp\Big(-\frac{1}{2\sigma_q^2}\sum_{\substack{\{\boldsymbol{\mu}_{1:\frac{q}{2}}\}_{\leq}\\ \{\boldsymbol{\nu}_{1:\frac{q}{2}}\}_{\leq}}}\left|X_{\boldsymbol{\nu}_q}^{\boldsymbol{\mu}_q}\right|^2\Big), \tag{4.107}$$

where $A_q = \sqrt{\frac{1}{2\pi\sigma_q^2}}$ is the normalization factor.

Next, we have $r/2$-body hopping from site i to $i+1$ is given by

$$\mathcal{H}_{i\to i+1}(t) = D_i(t)\sum_{\substack{\{\boldsymbol{\mu}_{1:\frac{r}{2}}\}_{\leq}\\ \{\boldsymbol{\nu}_{1:\frac{r}{2}}\}_{\leq}}} Y(i)_{\boldsymbol{\nu}_r}^{\boldsymbol{\mu}_r} c_{i+1;\mu_1}^{\dagger}\cdots c_{i+1;\mu_{\frac{r}{2}}}^{\dagger} c_{i;\nu_{\frac{r}{2}}}\cdots c_{i;\nu_1}. \tag{4.108}$$

where $\{\boldsymbol{\mu}_{1:\frac{r}{2}}\}_{\leq} \equiv 1 \leq \mu_1 < \mu_2 < \cdots < \mu_{\frac{r}{2}-1} < \mu_{\frac{r}{2}} \leq N$ and $\boldsymbol{\mu}_r \equiv \{\mu_1, \mu_2, \mu_3, \ldots, \mu_{\frac{r}{2}}\}$. The hopping is random, controlled by the random matrix $Y(i)_{\boldsymbol{\nu}_r}^{\boldsymbol{\mu}_r}$, derived from Gaussian ensemble with following mean and variance which is the same for all lattice sites i:

$$\overline{Y} = 0, \quad \overline{|Y|^2} = \sigma_r^2 = \frac{1}{q}\frac{(1/r)((r/2)!)^2}{(N/2)^{r-1}}. \tag{4.109}$$

The Gaussian probability distribution is

$$\mathcal{P}_q\left[Y_{\boldsymbol{\nu}_r}^{\boldsymbol{\mu}_r}\right] = A_q \exp\Big(-\frac{1}{2\sigma_r^2}\sum_{\substack{\{\boldsymbol{\mu}_{1:\frac{r}{2}}\}_{\leq}\\ \{\boldsymbol{\nu}_{1:\frac{r}{2}}\}_{\leq}}}\left|Y_{\boldsymbol{\nu}_r}^{\boldsymbol{\mu}_r}\right|^2\Big), \tag{4.110}$$

where $A_r = \sqrt{\frac{1}{2\pi\sigma_r^2}}$ is the normalization factor. When $r = 2$, this corresponds to the standard 1-body kinetic hopping, while for $r > 2$, there is a diffusive-type transport in the system. Our setup is general and the formalism developed here holds true for arbitrary r.

Both the on-site and hopping terms have time-dependent strengths $J_i(t)$ and $D_i(t)$, respectively. There is an associated $U(1)$ charge density for the chain which is given by

$$\mathcal{Q} = \sum_{i=1}^{L}\mathcal{Q}_i, \tag{4.111}$$

where $\mathcal{Q}_i$ is the local charge density at lattice site i:

$$\mathcal{Q}_i \equiv \frac{1}{N}\sum_{\alpha=1}^{N}[c_{i;\alpha}^{\dagger}c_{i;\alpha} - 1/2]. \tag{4.112}$$

NOTE: Latin indices are used to denote lattice dots (running from 1 to $2L$) where SYK dot sits, while Greek indices are used to denote the sites within a single SYK dot that constitute a single SYK model (running from 1 to N, where N is the number of sites in a SYK dot).

4.4.2 The Schwinger-Dyson Equations

We chose the convention for the Green's function as we have chosen for this chapter (Eq. (4.6)):

$$\mathcal{G}_{ij}(t_1, t_2) \equiv \frac{-1}{N}\sum_{\alpha=1}^{N}\langle T_{\mathcal{C}}c_{i;\alpha}(t_1)c_{j;\alpha}^{\dagger}(t_2)\rangle. \tag{4.113}$$

These obey Dyson's equation in matrix form: $\hat{\mathcal{G}}^{-1} = \hat{\mathcal{G}}_0^{-1} - \hat{\Sigma}$. Here $\hat{\Sigma}$ is the self-energy matrix which, for SYK-type systems, is diagonal in lattice indices ($\hat{\Sigma}_{i,j} \equiv \delta_{ij}\Sigma_i$), as justified by disorder averaging (see footnote 4). Consistent with this, we restrict solutions to lattice-diagonal Green's functions: $\hat{\mathcal{G}}_{i,j} = \delta_{ij}\mathcal{G}_i$, which reduces all physically relevant correlations to the on-site Green's functions $\mathcal{G}_i$.

Then we follow the same steps as in Sect. 4.1 to evaluate the disorder-averaged partition function, given by

$$\langle\mathcal{Z}\rangle = \int \mathcal{D}\mathcal{G}_i\mathcal{D}\Sigma_i e^{-S_J[\mathcal{G},\Sigma]-S_D[\mathcal{G},\Sigma]}, \tag{4.114}$$

where $S_J[\mathcal{G}, \Sigma]$ and $S_D[\mathcal{G}, \Sigma]$ are the effective actions corresponding to the on-site interaction and hopping term respectively, giving as the total effective action

$$\boxed{S_{\text{eff}} = S_J[\mathcal{G}, \Sigma] + S_D[\mathcal{G}, \Sigma]}, \tag{4.115}$$

where

$$\begin{aligned} \frac{S_J}{N} &= -\operatorname{Tr}\ln\left(\mathcal{G}_{o;i}^{-1} - \Sigma_i\right) \\ &\quad -\int dt_1 dt_2\left(\Sigma_i\left(t_1, t_2\right)\mathcal{G}_i\left(t_2, t_1\right) + \frac{J_i(t_1)J_i(t_2)}{q^2}\left(-4\mathcal{G}_i\left(t_1, t_2\right)\mathcal{G}_i\left(t_2, t_1\right)\right)^{\frac{q}{2}}\right) \\ \frac{S_D}{N} &= \int dt_1 dt_2\, \mathcal{L}_D[\mathcal{G}](t_1, t_2). \end{aligned} \tag{4.116}$$

Here $\mathcal{L}_D[\mathcal{G}](t_1,t_2)$ is defined as

$$\mathcal{L}_D[\mathcal{G}](t_1,t_2) = \frac{D^\star_{i-1}(t_1)D_{i-1}(t_2)}{qr}[-4\mathcal{G}_{i-1}(t_1,t_2)\mathcal{G}_i(t_2,t_1)]^{r/2} + \frac{D_i(t_1)D^\star_i(t_2)}{qr}[-4\mathcal{G}_{i+1}(t_1,t_2)\mathcal{G}_i(t_2,t_1)]^{r/2}. \tag{4.117}$$

Clearly, when $D_i = 0\,\forall\, i$, then we get isolated $2L$ complex SYK dots. Accordingly, we have the full effective action $S_{\text{eff}} = S_J[\mathcal{G},\Sigma] + S_D[\mathcal{G},\Sigma]$, which we extremize to get the Schwinger-Dyson equations:

$$\boxed{\mathcal{G}_i^{-1} = \mathcal{G}_{0;i}^{-1} - \Sigma_i, \qquad \Sigma_i(t_1,t_2) = \Sigma_{J;i}(t_1,t_2) + \Sigma_{D;i}(t_1,t_2)}. \tag{4.118}$$

where $\mathcal{G}_{0;i}^{-1}$ is the free inverse Green's function and $\Sigma_{J;i}(t_1,t_2)$, $\Sigma_{D;i}(t_1,t_2)$ are the self-energies coming from on-site interacting and hopping terms, respectively. They are given by

$$q\Sigma_{J;i}(t_1,t_2) = 2J_i(t_1)J_i(t_2)\mathcal{G}_i(t_1,t_2)\,[-4\mathcal{G}_i(t_1,t_2)\mathcal{G}_i(t_2,t_1)]^{\frac{q}{2}-1} \tag{4.119}$$

and

$$q\Sigma_{D;i}(t_1,t_2) = 2D^\star_{i-1}(t_1)D_{i-1}(t_2)[-4\mathcal{G}_i(t_2,t_1)\mathcal{G}_{i-1}(t_1,t_2)]^{\frac{r}{2}-1}\mathcal{G}_{i-1}(t_1,t_2) + 2D_i(t_1)D^\star_i(t_2)[-4\mathcal{G}_i(t_2,t_1)\mathcal{G}_{i+1}(t_1,t_2)]^{\frac{r}{2}-1}\mathcal{G}_{i+1}(t_1,t_2). \tag{4.120}$$

Then, just like the Green's function (Eq. (4.11)), one can read-off the conjugate relation for self-energies, namely $\Sigma_i(t_1,t_2)^\star = \Sigma_i(t_2,t_1)$.

4.4.3 The Kadanoff-Baym Equations

We have already discussed the Kadanoff-Baym equations at length in Sect. 4.1.3. We use the general form of the Kadanoff-Baym equations from Eq. (4.31) where $\mathcal{I}$ is simplified using the generalized Galitskii-Migdal sum rule in Eq. (4.36) (which is same as the equal-time Kadanoff-Baym equations). We have throughout ignored the imaginary contour in the Keldysh plane (Fig. 2.1) due to Bogoliubov's principle as explained many times throughout this work. Accordingly, we can write the Kadanoff-Baym equations as[15]

[15] Note that we are dealing with charged fermions, so we can always assign a "mass term" $\mathcal{H}_\mu(t) = -\sum_i \dot{\eta}_i(t)N\mathcal{Q}_i$ where η_i plays the role of the chemical potential. This is reflected in the free Green's function $\mathcal{G}_{0;i}$. Accordingly the Kadanoff-Baym equations has a left-hand side that is equal to $\left[\partial_{t_1} - \imath\dot{\eta}_i(t_1)\right]\mathcal{G}_i^{\gtrless}(t_1,t_2)$. We can always do a phase transformation of the full Green's function

$$\partial_{t_1}\mathcal{G}_i^{\gtrless}(t_1,t_2) = \int_{t_1}^{t_2} dt_3 \left(\Sigma_i^{\gtrless}(t_1,t_3)\mathcal{G}_i^{A}(t_3,t_2)\right) - \frac{\imath}{2q}\alpha_i(t_1,t_2), \tag{4.121}$$

where α_i is defined using Eq. (4.32) to get

$$\alpha_i(t_1,t_2) = 2\imath q \int_{-\infty}^{t_1} dt_3 \Big[\Sigma^{>}(t_1,t_3)\,\mathcal{G}^{<}(t_3,t_2) - \Sigma^{<}(t_1,t_3)\,\mathcal{G}^{>}(t_3,t_2)\Big]. \tag{4.122}$$

Here, the greater and the lesser self-energies have been defined along the same lines as discussed in Sect. 2.9. Using Eqs. (4.119) and (4.120), we get (ordering of time arguments are of essential importance)

$$\begin{aligned}
q\Sigma_i^{>}(t_1,t_2) &= 2J_i(t_1)J_i(t_2)[-4\mathcal{G}_i^{>}(t_1,t_2)\mathcal{G}_i^{<}(t_2,t_1)]^{q/2-1}\mathcal{G}_i^{>}(t_1,t_2)\\
&\quad + 2D_{i-1}(t_2)D_{i-1}^{\star}(t_1)[-4\mathcal{G}_{i-1}^{>}(t_1,t_2)\mathcal{G}_i^{<}(t_2,t_1)]^{r/2-1}\mathcal{G}_{i-1}^{>}(t_1,t_2)\\
&\quad + 2D_i(t_1)D_i^{\star}(t_2)[-4\mathcal{G}_{i+1}^{>}(t_1,t_2)\mathcal{G}_i^{<}(t_2,t_1)]^{r/2-1}\mathcal{G}_{i+1}^{>}(t_1,t_2)\\
q\Sigma_i^{<}(t_2,t_1) &= 2J_i(t_1)J_i(t_2)[-4\mathcal{G}_i^{>}(t_1,t_2)\mathcal{G}_i^{<}(t_2,t_1)]^{q/2-1}\mathcal{G}_i^{<}(t_2,t_1)\\
&\quad + 2D_{i-1}(t_1)D_{i-1}^{\star}(t_2)[-4\mathcal{G}_i^{>}(t_1,t_2)\mathcal{G}_{i-1}^{<}(t_2,t_1)]^{r/2-1}\mathcal{G}_{i-1}^{<}(t_2,t_1)\\
&\quad + 2D_i(t_2)D_i^{\star}(t_1)[-4\mathcal{G}_i^{>}(t_1,t_2)\mathcal{G}_{i+1}^{<}(t_2,t_1)]^{r/2-1}\mathcal{G}_{i+1}^{<}(t_2,t_1)
\end{aligned} \tag{4.123}$$

where $\Sigma_i^{\gtrless}(t_1,t_2)^{\star} = +\Sigma_i^{\gtrless}(t_2,t_1)$ clearly holds true (using the general conjugate relations for the Green's function in Eq. (4.11)).

4.4.4 The Large-q Limit

We hope that the reader must be familiar with the large-q ansatz until now. We take the same ansatz for the chain as before, namely

$$\mathcal{G}_i^{\gtrless}(t_1,t_2) = \left(\mathcal{Q}_i(t) \mp \frac{1}{2}\right) e^{\frac{g_i^{\gtrless}(t_1,t_2)}{q}} \tag{4.124}$$

where $\mathcal{Q}_i$ is defined in Eq. (4.112). The same equal-time boundary conditions are valid for $g_i(t,t) = 0$ and so is the general conjugate relation as in Eq. (4.18) for each g_i. As before, $g^{\gtrless} = \mathcal{O}(q^0)$.

where the chemical term creates a trivial shift and hide the chemical factor in the rotated basis as we have been doing all throughout. This is explained in detail in footnote 4 which we refer the reader to.

Energy Density and Equal-Time Kadanoff-Baym Equation
We realize that $\mathcal{I}$ (appearing in the Kadanoff-Baym equations) can be expressed in terms of energy density using the generalized Galitskii-Migdal sum rule. The expression is provided in the footnote a which we reproduce here for convenience: for a general SYK-like Hamiltonian $\mathcal{H} = \sum_q \mathcal{H}_q$, we have $\lim_{t'\to t} \partial_{t'}\mathcal{G}^<(t', t) = \mathcal{I}(t, t) = -\imath \sum_{q>0} \frac{q}{2}\mathcal{E}_q(t)$ where $\mathcal{E}_q \equiv \langle \mathcal{H}_q\rangle/N$. Accordingly, the equal-time Kadanoff-Baym equations from Eq. (4.121) can be written as (where we have assumed $t_2 >_\mathcal{C} t_1$ without loss of generality)

$$\boxed{2\imath q\partial_t \mathcal{G}_i^<(t, t) = \alpha_i(t, t)}. \tag{4.125}$$

Following the same steps as in the box below Eq. (4.32), we get for the left-hand side

$$2\imath q\partial_t \mathcal{G}_i^<(t, t) = \frac{2q}{N}\langle\sum_\mu c^\dagger_{i;\mu}\left[c_{i;\mu}, \mathcal{H}\right]\rangle(t). \tag{4.126}$$

Here $\mathcal{H}$ is the Hamiltonian of the SYK chain in Eq. (4.104). Accordingly, the sum rule gives us

$$2\sum_\mu c^\dagger_{i;\mu}[c_{i;\mu}, \mathcal{H}] = q\mathcal{H}_i + r\mathcal{H}^\dagger_{i\to i+1} + r\mathcal{H}_{i-1\to i}. \tag{4.127}$$

Accordingly, the left-hand side of the equal-time Kadanoff-Baym equation becomes

$$2\imath q\partial_{t_1}\mathcal{G}_i^<(t_1, t_1^+) = \frac{q^2}{N}\langle\mathcal{H}_i\rangle(t_1) + \frac{qr}{N}\langle\mathcal{H}^\dagger_{i\to i+1}\rangle(t_1) + \frac{qr}{N}\langle\mathcal{H}_{i-1\to i}\rangle(t_1). \tag{4.128}$$

Therefore, we identify for α_i

$$\boxed{\alpha_i(t_1, t_1) = \epsilon_i(t_1) + \frac{r}{q}\left[\epsilon^\star_{i\to i+1}(t_1) + \epsilon_{i-1\to i}(t_1)\right]}, \tag{4.129}$$

where

$$\epsilon_i(t_1) \equiv \frac{q^2}{N}\langle\mathcal{H}_i\rangle(t_1), \quad \epsilon_{i\to i+1}(t_1) \equiv \frac{q^2}{N}\langle\mathcal{H}_{i\to i+1}\rangle(t_1). \tag{4.130}$$

(continued)

Previously for a single dot complex SYK model, we have seen that $\mathcal{I}$ and α is related via the expression in Eq. (4.38). Recalling the scaling relation from Sect. 4.2.2, $\mathcal{Q} = \mathcal{O}(1/\sqrt{q})$. Accordingly, in the large-q limit, keeping everything to the leading order in $1/q$, we get using Eq. (4.39) $\alpha(t) = \epsilon_q(t)$ where $\epsilon_q(t) \equiv q^2\langle\mathcal{H}_q\rangle/N$. Indeed, if we switch off the hopping in the chain ($D_i = 0\ \forall\ i$), we boil down to single dots and the energy density $\alpha_i \to \alpha$ as it should. Note that we had to make use of the large-q limit to get this identification, otherwise a factor of $(1 - \mathcal{Q}^2)$ would also play a role (see Eq. (4.39)).

The core approach remains unchanged: we employ the large-q ansatz for the Green's functions to retain leading-order terms in $1/q$, enabling computation of self-energies via the Schwinger-Dyson equations (Eq. (4.118)) and subsequent simplification of the Kadanoff-Baym equations. While the resulting expressions are lengthy and lack intuitive transparency, the procedure itself is methodologically straightforward. That's why we avoid presenting the full resulting expressions due to their length and limited physical insight, but wish to highlight that the systematic procedure remains robust. We presented the large-q ansatz in this section to demonstrate that the chain construction preserves the diagonal solution structure for self-energies and Green's functions, mirroring the single-dot case—a feature that will be indispensable while studying transport in SYK chains in the next chapter. Further simplifications depend critically on the hopping type determined by r.[16]

Here we show the simplification at the leading order in $1/q$, while keeping the generality of the formalism. We start with the Kadanoff-Baym equations in Eq. (4.121) where we evaluate the left-hand side using the large-q ansatz in Eq. (4.124).

$$\text{Left-hand side} = \dot{\mathcal{Q}}_i(t)e^{g_i^{\gtrless}(t_1,t_2)/q} + \frac{1}{q}\mathcal{G}_i^{\gtrless}(t_1,t_2)\,\partial_{t_1} g_i^{\gtrless}(t_1,t_2)\,. \tag{4.131}$$

After plugging this for the left-hand side, the full Kadanoff-Baym equations become (with some re-arranging)

$$\begin{aligned}\partial_{t_1} g_i^{\gtrless}(t_1,t_2) = &\int_{t_1}^{t_2} dt_3 \frac{q\,\Sigma_i^{\gtrless}(t_1,t_3)\,\mathcal{G}_i^{A}(t_3,t_2)}{\mathcal{G}_i^{\gtrless}(t_1,t_2)} \\ &- \left(\frac{\imath\frac{\alpha_i(t_1,t_2)}{2} + q\,\dot{\mathcal{Q}}_i(t)e^{g_i^{\gtrless}(t_1,t_2)/q}}{\mathcal{G}_i^{\gtrless}(t_1,t_2)}\right).\end{aligned} \tag{4.132}$$

[16] As noted previously, the $r/2$-body hopping yields kinetic behavior for $r = 2$ and diffusive dynamics for $r > 2$. We consider both the scaling relation $r = \mathcal{O}(q^0)$ as well as the hyper-scaling $r = \mathcal{O}(q)$ in the large-q limit (e.g., $r = 2q$ or $r = q/2$) later.

Until now, this is exact in the large-N limit. Now we take the large-q limit. Using the large-q ansatz for the Green's function in Eq. (4.124), we have $\mathcal{G}_i^{\gtrless} \sim \mp\frac{1}{2}$. Recall from the expressions of self-energies that $\Sigma_i = \mathcal{O}(1/q)$. Using the definition of advanced Green's function (see Eq. (2.193), we get $\mathcal{G}^A(t_3, t_2) = \Theta(t_2 - t_3) + \mathcal{O}(1/q)$. Finally we need to evaluate the order for $\dot{\mathcal{Q}}_i$ where we already know from the scaling relations (Eq. (4.58)) that $\mathcal{Q} = \mathcal{O}(1/\sqrt{q})$.

Local Charge Density Transport

Using the boundary condition for $g^{\gtrless}(t_1, t_1) = 0$, we get from Eq. (4.124) (without loss of generality, we consider $t_1 < t_2$)

$$\mathcal{Q}_i(t_1) = \lim_{t_2 \to t_1} \mathcal{G}_i^<(t_1, t_2) - \frac{1}{2}. \tag{4.133}$$

The time derivative of the charge requires the total derivative of $\mathcal{G}_i^<(t_1, t_2)$ evaluated along the line $t_1 = t_2$:

$$\dot{\mathcal{Q}}_i(t_1) = \frac{d}{dt_1}\left[\lim_{t_2 \to t_1} \mathcal{G}_i^<(t_1, t_2)\right] = \lim_{\epsilon \to 0^+} \frac{d}{dt_1} \mathcal{G}_i^<(t_1, t_1 + \epsilon)\bigg|_{\epsilon=0}. \tag{4.134}$$

By the chain rule, this decomposes into:

$$\boxed{\dot{\mathcal{Q}}_i(t_1) = \left[\partial_{t_1} \mathcal{G}_i^<(t_1, t_2) + \partial_{t_2} \mathcal{G}_i^<(t_1, t_2)\right]_{t_2 = t_1}}. \tag{4.135}$$

This is how equal-time full derivatives are performed for bi-local functions.

The second term here can be re-written using $\mathcal{G}_i^<(t_1, t_2) = \mathcal{G}^<(t_2, t_1)^\star$ (see Eq. (4.11)) to get

$$\dot{\mathcal{Q}}_i(t_1) = \left[\partial_{t_1} \mathcal{G}_i^<(t_1, t_2) + \partial_{t_2} \mathcal{G}_i^<(t_2, t_1)^\star\right]_{t_2 = t_1}. \tag{4.136}$$

We know the equal-time Kadanoff-Baym equations from Eq. (4.125), namely (where derivative is performed on the first time argument before taking the equal-time limit)

$$\partial_t \mathcal{G}_i^<(t, t) = -\frac{\imath}{2q} \alpha_i(t, t), \quad \xrightarrow{\text{complex conjugation}} \quad \partial_t \mathcal{G}_i^<(t, t)^\star = +\frac{\imath}{2q} \alpha_i(t, t)^\star.$$

(continued)

Plugging back, we get the expression for the local charge density transport

$$\boxed{\dot{\mathcal{Q}}_i(t) = \frac{1}{q}\text{Im}[\alpha_i(t,t)]} \tag{4.137}$$

This relation will be used in the next chapter while studying non-equilibrium behavior to derive the expression for transport of charge dynamics in SYK chains. With the benefit of hindsight by evaluating the local charge density dynamics in Sect. 5.2.3 (see Eq. (5.54)) where we show $q\dot{\mathcal{Q}}_i = \mathcal{O}(\mathcal{Q}_i)$ for $r = \mathcal{O}(q^0)$ and $\dot{\mathcal{Q}}_i = \mathcal{O}(\mathcal{Q}_i)$ for $r = \mathcal{O}(q^1)$, we have

$$\boxed{\dot{\mathcal{Q}}_i = \begin{cases} \mathcal{O}\left(\frac{1}{q^{3/2}}\right) & (r = \mathcal{O}(q^0)) \\ \mathcal{O}\left(\frac{1}{q^{1/2}}\right) & (r = \mathcal{O}(q^1)) \end{cases}} \tag{4.138}$$

Returning to Eq. (4.132), we now have the missing ingredient for $\dot{\mathcal{Q}}_i$. We see that we must have $r = \mathcal{O}(q^0)$ for the large-q limit to make sense otherwise there will be a factor of $\sqrt{q}$ in the numerator if we wish to take $\mathcal{G}_i^{\gtrless} \sim \mp\frac{1}{2}$ as we are taking.

So, we specialize to the case where the hopping r is $\mathcal{O}(q^0)$ for further simplify the Kadanoff-Baym equations in Eq. (4.132) at leading order in $1/q$. We finally get

$$\boxed{\frac{1}{2}\partial_{t_1} g_i^{\gtrless}(t_1,t_2) = \mp\left(\int_{t_1}^{t_2} dt_3 q\Sigma_i^{\gtrless}(t_1,t_3)\right) \pm \frac{\imath\alpha_i(t_1)}{2}} \quad (r = \mathcal{O}(q^0)). \tag{4.139}$$

where the expressions for self-energies are provided in Eq. (4.123).

We can re-write the Kadanoff-Baym equations at leading order in $1/q$ in terms of $g_i^{\pm}(t_1,t_2)$ where we use the definition analogous to the dot in Eq. (4.20):

$$g_i^{\pm}(t_1,t_2) \equiv \frac{g_i^{>}(t_1,t_2) \pm g_i^{<}(t_2,t_1)}{2}. \tag{4.140}$$

The Kadanoff-Baym equations at leading order in $1/q$ with $r = \mathcal{O}(q^0)$ become

$$\boxed{\begin{aligned}\partial_{t_1} g_i^+(t_1,t_2) &= -\int_{t_1}^{t_2} dt_3 \left(q\Sigma_i^>(t_1,t_3) - q\Sigma_i^<(t_3,t_1)\right) + \imath\,\mathrm{Re}\left[\alpha_i(t_1)\right] \\ \partial_{t_1} g_i^-(t_1,t_2) &= -\int_{t_1}^{t_2} dt_3 \left(q\Sigma_i^>(t_1,t_3) + q\Sigma_i^<(t_3,t_1)\right) - \mathrm{Im}\left[\alpha_i(t_1)\right]\end{aligned}}$$
$$(r = \mathcal{O}(q^0)). \tag{4.141}$$

These equations will come in handy in the next chapter where we will deal with non-equilibrium behaviors. Note that we have never assumed time-translational invariance (equilibrium) conditions in deriving these relations.

4.4.5 Uniform Equilibrium Conditions

We now specialize to the case of a uniform chain (where all couplings are equal to each other, namely $J_i = J$ and $D_i = D\ \forall\ i$) which is at equilibrium (implying time-translational invariance). Since at equilibrium, there are no transport of charges or energy, each dot is at the same temperature and charge density. This simplifies the setup significantly because the Green's functions $\mathcal{G}_i$ becomes independent of the lattice site label i: $\mathcal{G}_i = \mathcal{G}\ \forall\ i$. Since the self-energies are determined completely by the Green's functions (see Eq. (4.123)), they also become independent of the site label i. All couplings are time-independent at equilibrium and the Schwinger-Dyson equations simplify significant (using Eq. (4.123) where we impose time-translational invariance $(t_1, t_2) \to (t_1 - t_2) = t$)

$$\begin{aligned} q\Sigma^>(t) &= 2J^2[-4\mathcal{G}^>(t)\mathcal{G}^<(-t)]^{q/2-1}\mathcal{G}^>(t) \\ &\quad + 4|D|^2[-4\mathcal{G}^>(t)\mathcal{G}^<(-t)]^{r/2-1}\mathcal{G}^>(t) \\ q\Sigma^<(-t) &= 2J^2[-4\mathcal{G}^>(t)\mathcal{G}^<(-t)]^{q/2-1}\mathcal{G}^<(-t) \\ &\quad + 4|D|^2[-4\mathcal{G}^>(t)\mathcal{G}^<(-t)]^{r/2-1}\mathcal{G}^<(-t), \end{aligned} \tag{4.142}$$

where we again note that $\Sigma^{\gtrless}(t)^\star = \Sigma^{\gtrless}(-t)$ (using $\mathcal{G}^{\gtrless}(t)^\star = +\mathcal{G}^{\gtrless}(-t)$ from Eq. (4.11)). Similarly, significant simplifications happen for the Kadanoff-Baym equations too.

The Schwinger-Dyson equations follow from the saddle point solutions of the effective action that we found above in Eq. (4.115). The effective action itself simplifies drastically and we get for uniform chain at equilibrium:

$$S_{\text{eff}} = S_J[\mathcal{G}, \Sigma] + S_D[\mathcal{G}, \Sigma], \tag{4.143}$$

where

$$\frac{S_J}{N} = -\operatorname{Tr}\ln\left(\mathcal{G}_{o;i}^{-1} - \Sigma\right) - \int dt\left(\Sigma(t)\,\mathcal{G}(-t) + \frac{J^2}{q^2}\left(-4\mathcal{G}(t)\,\mathcal{G}(-t)\right)^{\frac{q}{2}}\right)$$
$$\frac{S_D}{N} = \int dt_1 dt_2\, \mathcal{L}_D[\mathcal{G}](t), \tag{4.144}$$

and $\mathcal{L}_D[\mathcal{G}](t)$ is defined as

$$\mathcal{L}_D[\mathcal{G}](t) = \frac{2|D|^2}{qr}[-4\mathcal{G}(t)\mathcal{G}(-t)]^{r/2}. \tag{4.145}$$

Note that we have not yet used the large-q limit. We now see how these equations for equilibrium and uniform one-dimensional SYK chains get mapped onto a coupled SYK dot with proper identification of the couplings.

4.4.6 Mapping to a Coupled Dot

We consider a general coupled SYK (effectively zero-dimensional) dot where the presence of additional coupling introduces another energy scale in the system. The Hamiltonian is given by

$$\mathcal{H} = K_q\mathcal{H}_q + K_{\kappa q}\mathcal{H}_{\kappa q} \tag{4.146}$$

where

$$\mathcal{H}_{\kappa q} = \sum_{\substack{\{\mu_{1:\frac{\kappa q}{2}}\}_{\leq} \\ \{\nu_{1:\frac{\kappa q}{2}}\}_{\leq}}}^{N} Z_{\boldsymbol{\nu}_{\kappa q}}^{\boldsymbol{\mu}_{\kappa q}} c_{\mu_1}^{\dagger} \cdots c_{\mu_{\kappa q/2}}^{\dagger} c_{\nu_{\kappa q/2}} \cdots c_{\nu_1} \tag{4.147}$$

where the notation is consistent with what is followed for the Hamiltonian of the chain in Eq. (4.104), namely $\{\boldsymbol{\nu}_{1:\frac{\kappa q}{2}}\}_{\leq} \equiv 1 \leq \nu_1 < \mu_2 < \cdots < \nu_{\frac{\kappa q}{2}-1} < \nu_{\frac{\kappa q}{2}} \leq N$ and $\boldsymbol{\nu}_{\kappa q} \equiv \{\nu_1, \nu_2, \nu_3, \ldots, \nu_{\frac{\kappa q}{2}}\}$. Just like the SYK chain, $Z_{\boldsymbol{\nu}_{\kappa q}}^{\boldsymbol{\mu}_{\kappa q}}$ is a random matrix (modeling disorder), derived from a Gaussian ensemble with the following mean and variance

$$\overline{Z} = 0, \qquad \overline{|Z|^2} = \frac{4\left[\left(\frac{\kappa q}{2}\right)\right]^2}{(\kappa q)^2\left(\frac{N}{2}\right)^{\kappa q-1}}. \tag{4.148}$$

Note that $\kappa = 1$ recovers $\mathcal{H}_q$ in Eq. (4.146).

As stated multiple times, we follow a consistent convention for the Green's function throughout each chapter. For completely, we repeat our convention: $\mathcal{G}(t_1, t_2) \equiv \frac{-1}{N} \sum_{j=1}^{N} \left\langle T_{\mathcal{C}} c_j(t_1) c_j^{\dagger}(t_2) \right\rangle$.

Then we repeat the same procedure for the SYK model as throughout this work to get the disorder-averaged partition function for this coupled (effectively zero-dimensional dot)

$$\langle \mathcal{Z} \rangle = \int \mathcal{D}\mathcal{G}\mathcal{D}\Sigma e^{-S_{K_q}[\mathcal{G},\Sigma]-S_{K_{\kappa q}}[\mathcal{G},\Sigma]}, \tag{4.149}$$

where the effective action is

$$S_{\text{eff}} = S_{K_q}[\mathcal{G}, \Sigma] + S_{K_{\kappa q}}[\mathcal{G}, \Sigma]. \tag{4.150}$$

Here, $S_{K_q}[\mathcal{G}, \Sigma]$ and $S_{K_{\kappa q}}[\mathcal{G}, \Sigma]$ are the contributions corresponding to $\mathcal{H}_q$ and $\mathcal{H}_{\kappa q}$ in the Hamiltonian (4.146). They are given by

$$\begin{aligned} \frac{S_{K_q}}{N} &= -\operatorname{Tr}\ln\left(\mathcal{G}_{o;i}^{-1} - \Sigma\right) - \int dt \left(\Sigma(t)\,\mathcal{G}(-t) + \frac{K_q^2}{q^2}\left(-4\mathcal{G}(t)\,\mathcal{G}(-t)\right)^{\frac{q}{2}}\right) \\ \frac{S_{K_{\kappa q}}}{N} &= \int dt_1 dt_2\, \mathcal{L}_{K_{\kappa q}}[\mathcal{G}](t), \end{aligned} \tag{4.151}$$

and $\mathcal{L}_{K_{\kappa q}}[\mathcal{G}](t)$ is defined as

$$\mathcal{L}_{K_{\kappa q}}[\mathcal{G}](t) = \frac{K_{\kappa q}^2}{qr}[-4\mathcal{G}(t)\mathcal{G}(-t)]^{r/2}. \tag{4.152}$$

Again, note that we have only used the large-N limit (and not the large-q limit).

By comparing the effective action of a uniform 1D SYK chain at equilibrium (Eq. (4.143), with Hamiltonian Eq. (4.104)) to the effective action of the coupled SYK dot (Eq. (4.150), with Hamiltonian Eq. (4.146)), we find that their large-N effective actions map *exactly* under the coupling identifications:

$$\boxed{J^2 \longleftrightarrow K_q^2, \quad 2|D|^2 \longleftrightarrow K_{\kappa q}^2} \qquad \text{(chain} \longleftrightarrow \text{dot).} \tag{4.153}$$

This equivalence implies identical Schwinger-Dyson (and thus Kadanoff-Baym) equations for both systems. Crucially, the *uniformity* and *equilibrium conditions* of the chain are essential for this mapping, as they enforce translational invariance and equal treatment of couplings/lattice sites. This correspondence will be pivotal in the next chapter for studying SYK chain transport via linear response theory, where equilibrium results quantify non-equilibrium behavior to linear order.

References

1. Kubizňák, D., Mann, R.B.: P - V criticality of charged AdS black holes. J. High Energy Phys. **2012**(7), 033 (2012). Publisher: Springer-Verlag
2. Kitaev, A.: A simple model of quantum holography. Talks given at "Entanglement in Strongly-Correlated QuantumMatter". (Part1, Part2), KITP (2015)
3. Stanford, D.: Many-body chaos at weak coupling. J. High Energy Phys. **2016**(10), 1–18 (2016)
4. Maldacena, J., Stanford, D.: Remarks on the Sachdev-Ye-Kitaev model. Phys. Rev. D **94**(10), 106002 (2016)
5. Louw, J.C., Kehrein, S.: Thermalization of many many-body interacting Sachdev-Ye-Kitaev models. Phys. Rev. B **105**(7), 075117 (2022)
6. Maldacena, J., Shenker, S.H., Stanford, D.: A bound on chaos. J. High Energy Phys. **2016**(8), 1–17 (2016)
7. Louw, J.C., Kehrein, S.: Shared universality of charged black holes and the complex large-q Sachdev-Ye-Kitaev model. Phys. Rev. B **107**(7), 075132 (2023)
8. Louw, J.C., van Manen, L.M., Jha, R.: Thermodynamics and dynamics of coupled complex SYK models. J. Phys.: Condens. Matter **36**(49), 495601 (2024)
9. Jha, R., Louw, J.C.: Dynamics and charge fluctuations in large-q Sachdev-Ye-Kitaev lattices. Phys. Rev. B **107**(23), 235114 (2023)
10. Sorokhaibam, N.: Phase transition and chaos in charged SYK model. J. High Energy Phys. **2020**(7), 1–32 (2020)
11. Wang, J., Anisimov, M.A.: Nature of vapor-liquid asymmetry in fluid criticality. Phys. Rev. E **75**(5), 051107 (2007)
12. Dehyadegari, A., Majhi, B.R., Sheykhi, A., Montakhab, A.: Universality class of alternative phase space and Van der Waals criticality. Phys. Lett. B **791**, 30 (2019). Publisher: North-Holland.
13. Mandal, A., Samanta, S., Majhi, B.R.: Phase transition and critical phenomena of black holes: a general approach. Phys. Rev. D **94**(6), 064069 (2016). Publisher: American Physical Society
14. Majhi, B.R., Samanta, S.: P-V criticality of AdS black holes in a general framework. Phys. Lett. B **773**, 203–207 (2017)
15. Chamblin, A., Emparan, R., Johnson, C.V., Myers, R.C.: Charged AdS black holes and catastrophic holography. Phys. Rev. D **60**(6), 064018 (1999)
16. Chamblin, A., Emparan, R., Johnson, C.V., Myers, R.C.: Holography, thermodynamics, and fluctuations of charged AdS black holes. Phys. Rev. D **60**(10), 104026 (1999)
17. D'Alessio, L., Kafri, Y., Polkovnikov, A., Rigol, M.: From quantum chaos and eigenstate thermalization to statistical mechanics and thermodynamics. Adv. Phys. **65**, 239–362 (2016)
18. Bhattacharya, R., Chakrabarti, S., Jatkar, D.P., Kundu, A.: SYK model, chaos and conserved charge. J. High Energy Phys. **2017**(11), 1–16 (2017)
19. Klebanov, I.R., Tarnopolsky, G.: Uncolored random tensors, melon diagrams, and the Sachdev-Ye-Kitaev models. Phys. Rev. D **95**(4), 046004 (2017)
20. Jishi, R.A.: Feynman Diagram Techniques in Condensed Matter Physics. Cambridge University Press, Cambridge (2013)
21. Murugan, J., Stanford, D., Witten, E.: More on supersymmetric and 2d analogs of the SYK model. J. High Energy Phys. **2017**(8), 1–99 (2017)
22. Pöschl, G., Teller, E.: Bemerkungen zur Quantenmechanik des anharmonischen Oszillators. Z. Phys. **83**(3), 143–151 (1933)
23. Sekino, Y., Susskind, L.: Fast scramblers. J. High Energy Phys. **2008**(10), 065 (2008). Publisher: Springer Science and Business Media LLC
24. Shenker, S.H., Stanford, D.: Black holes and the butterfly effect. J. High Energy Phys. **03**(3), 067 (2014). Publisher: Springer Berlin Heidelberg

25. Cotler, J.S., Gur-Ari, G., Hanada, M., Polchinski, J., Saad, P., Shenker, S.H., Stanford, D., Streicher, A., Tezuka, M.: Black holes and random matrices. J. High Energy Phys. **2017**(5), 1–54 (2017)
26. Sachdev, S.: Statistical mechanics of strange metals and black holes (2022). _eprint: 2205.02285

Non-equilibrium Properties and Transport 5

Abstract

This chapter develops a real time, large-q framework for non equilibrium dynamics and transport in Sachdev-Ye-Kitaev (SYK) systems. Starting from a time dependent Hamiltonian that mixes kinetic SYK_2 and higher-body $\mathrm{SYK}_{\kappa q}$ interactions, the chapter formulates Keldysh Schwinger-Dyson and Kadanoff-Baym equations for quenches in a single complex SYK dot and shows that, in the large-q limit, the dot exhibits instantaneous thermalization at the level of Green's functions, with a final temperature fixed by an emergent Liouville solution and its KMS relation. Extending to one dimensional SYK chains, the chapter derives on-site and transport energy densities via Keldysh contour deformations, proves that chains cannot thermalize instantaneously, and obtains charge density evolution governed by local continuity equations. The second part of the chapter applies contour deformations and linear response theory to uniform SYK chains, mapping them to coupled SYK dots and solving three classes of chains that differ in the competition between on-site and inter-site interactions. Using currents defined microscopically from the continuity equation, Kubo formula, and the fluctuation-dissipation theorem, the chapter expresses conductivity entirely in terms of current-current correlators and extracts DC resistivity scalings, including linear-in-temperature strange-metal behavior and insulating T^{-2} regimes suggestive of holographic insulators.

In this chapter, we will delve into non-equilibrium properties of SYK systems, ranging from thermalization to transport. We use quench protocols to introduce non-equilibrium behavior. Surprisingly, the large-q SYK model admits instantaneous thermalization with respect to the Green's functions. However, there exists a finite thermalization rate for SYK chains where charge transport happens which will be studied analytically. In an effectively zero-dimensional dot, there is no transport, but as we will see, we can perform a linear response to an external field to study

R. Jha, *Theory of Strongly Correlated Systems*, Lecture Notes in Physics 1051,
https://doi.org/10.1007/978-3-032-20910-8_5

transport properties such as resistivity and current-current correlations in SYK chains. We will see the power of Keldysh formalism that allows us to deform the Keldysh contours in Fig. 2.1 which allows us to evaluate current-current correlations that govern resistivity.

5.1 Quench to a Single Dot: Instantaneous Thermalization

5.1.1 Model

We first define the setup (which is mostly a summary of the topics we have already seen above) and later proceed to introduce time-dependence that will allow us to perform a quench protocol. The Hamiltonian is given by

$$\mathcal{H}(t) = J_2\mathcal{H}_2(t) + \sum_{\kappa>0} J_{\kappa q}(t)\mathcal{H}_{\kappa q}(t). \tag{5.1}$$

Here $\mathcal{H}_{\kappa q}$ is defined in Eq. (4.147) and the associated mean and variance of the Gaussian ensemble from which the random couplings of $\mathcal{H}_{\kappa q}$ are derived is given in Eq. (4.148). The kinetic SYK$_2$ Hamiltonian is given by

$$\mathcal{H}_2 = \sum_{i=1}^{N}\sum_{j=1}^{N} Y_j^i c_i^\dagger c_j, \quad \text{where} \quad \overline{Y} = 0, \quad \overline{\left|Y_j^i\right|^2} = \frac{2}{Nq}. \tag{5.2}$$

The additional $1/q$ scaling in the kinetic term is required to enable the competition between the kinetic and the interaction terms and that the kinetic term does not dominate solely at large-q. This is a similar situation for the quench we considered and solved in the Majorana case in Sect. 2.9.5. Note that various terms in the Hamiltonian are non-commuting and show chaotic properties on their own, that's why Eq. (5.1) is far from a trivial construct that we are considering here.

Our convention for the Green's function is

$$\mathcal{G}(t_1, t_2) \equiv \frac{-1}{N}\sum_{k=1}^{N}\left\langle \mathcal{T}_\mathcal{C} c_k(t_1)\, c_k^\dagger(t_2)\right\rangle, \tag{5.3}$$

which satisfies the general conjugate relation $\mathcal{G}(t_1, t_2)^\star = \mathcal{G}(t_2, t_1)$. This Green's function can be written as $\Theta(t_1 - t_2)\mathcal{G}^>(t_1, t_2) + \Theta(t_2 - t_1)\mathcal{G}^<(t_1, t_2)$ where the greater and the lesser Green's function are given by

$$\mathcal{G}^>(t_1, t_2) = -\frac{1}{N}\sum_{k=1}^{N}\left\langle c_k(t_1)\, c_k^\dagger(t_2)\right\rangle \quad \mathcal{G}^<(t_1, t_2) = \frac{1}{N}\sum_{k=1}^{N}\left\langle c_k^\dagger(t_2)\, c_k(t_1)\right\rangle. \tag{5.4}$$

We repeat the same procedure as in Sects. 4.1.2 and 4.1.3 to get the for the Dyson equation in matrix form for the Green's function and the self-energy $\hat{\mathcal{G}}^{-1} = \hat{\mathcal{G}}_0^{-1} - \hat{\Sigma}$

(also see below Eq. (4.113)). The Schwinger-Dyson equation for the self-energy is given by

$$\boxed{\Sigma(t_1,t_2) = \frac{1}{q}\mathcal{L}^{\gtrless}(t_1,t_2)\mathcal{G}^{\lessgtr}(t_1,t_2)}, \tag{5.5}$$

where $\mathcal{L}^{\gtrless}$ is given by

$$\mathcal{L}^{\gtrless}(t_1,t_2) = 2J_2(t_1)\,J_2(t_2) + \sum_{\kappa>0}\mathcal{L}^{\gtrless}_{\kappa q}(t_1,t_2)\,. \tag{5.6}$$

Here $\mathcal{L}_{\kappa q}$ is the same as in Eq. (4.22) for each of the SYK $\mathcal{H}_{\kappa q}$ Hamiltonian. Explicitly,

$$\mathcal{L}^{\gtrless}_{\kappa q}(t_1,t_2) \equiv 2J_{\kappa q}(t_1)\,J_{\kappa q}(t_2)\left[-4\mathcal{G}^{\gtrless}(t_1,t_2)\,\mathcal{G}^{\lessgtr}(t_2,t_1)\right]^{\frac{\kappa q}{2}-1}. \tag{5.7}$$

We now take the large-q ansatz as before

$$\mathcal{G}^{\gtrless}(t_1,t_2) = \left[\mathcal{Q} \mp \frac{1}{2}\right]e^{g^{\gtrless}(t_1,t_2)/q}, \tag{5.8}$$

where $\mathcal{Q}$ is the conserved $U(1)$ charge of the system as before (Eq. (3.9)):

$$\mathcal{Q} \equiv \frac{1}{N}\sum_{i=1}^{N}\left\langle c_i^\dagger c_i\right\rangle - \frac{1}{2}. \tag{5.9}$$

Following the steps of Sect. 4.1.4, we get in the large-q limit the same Kadanoff-Baym equations for $g^{\gtrless}$ as in Eq. (4.42) with the exception that the definition of $\mathcal{L}^{\gtrless}$ is given in Eq. (5.7). We reproduce the equation for convenience:

$$\begin{aligned}
\partial_{t_1} g^{>}(t_1,t_2) &= \int_{t_1}^{t_2} dt_3 \mathcal{L}^{>}(t_t,t_3) + 2\imath\left(\mathcal{Q}+\frac{1}{2}\right)\alpha(t_1)\\
\partial_{t_1} g^{<}(t_1,t_2) &= \int_{t_1}^{t_2} dt_3 \mathcal{L}^{<}(t_t,t_3) + 2\imath\left(\mathcal{Q}-\frac{1}{2}\right)\alpha(t_1),
\end{aligned} \tag{5.10}$$

where α is already evaluated in footnote 5, namely:[1]

$$\left(1-4\mathcal{Q}^2\right)\alpha(t) = q\mathcal{E}_2(t) + q^2\sum_{\kappa>0}\kappa\mathcal{E}_{\kappa q}(t). \tag{5.11}$$

[1] We have already seen in the box below Eq. (4.124) that at leading order in $1/q$, the factor $(1-\mathcal{Q}^2)$ does not play a role.

Defining the rescaled coupling $\mathcal{J}_{\kappa q}$ and $g^{\pm}$ as in Eq. (4.20):

$$\mathcal{J}_{\kappa q}^2 \equiv J_q^2(1-4\mathcal{Q}^2)^{\frac{\kappa q}{2}-1}, \quad g^{\pm}(t_1,t_2) \equiv \frac{g^{>}(t_1,t_2) \pm g^{<}(t_2,t_1)}{2}, \tag{5.12}$$

we get for the Kadanoff-Baym equations as (similar to Eq. (4.45) with the definition of $\mathcal{L}^{\gtrless}$ in Eq. (5.7))

$$\partial_{t_1}\partial_{t_2} g^{+}(t_1,t_2) = \mathcal{L}^{>}(t_1,t_2), \qquad \partial_{t_1} g^{-}(t_1,t_2) = 2\imath\mathcal{Q}\alpha(t_1). \tag{5.13}$$

We have summarized the model and the associated equations and are ready to perform a quench protocol to introduce non-equilibrium behavior.

5.1.2 Quench Protocol

We follow the Refs. [3, 11] where the results were derived for the Majorana and the complex SYK models. We will focus on the complex SYK model as the Majorana case becomes a subset by taking $\mathcal{Q} = 0$ (half-filling) which implies the Green's function $g^{-} = 0$. We perform the following quench

$$\mathcal{H}(t) = \begin{cases} \mathcal{H}_2(t) + \sum_{\kappa>0} \mathcal{H}_{\kappa q}(t) & t < 0 \\ \mathcal{H}_q & t \geq 0 \end{cases}, \tag{5.14}$$

where for pre-quench regime $t < 0$, we have the Hamiltonian as in Eq. (5.1) whose equations are derived in above. Then we switch off all couplings J_2 and $J_{\kappa q}$ except for $\kappa = 1$ case. We can also write this quench in terms of the time-dependent couplings:

$$\begin{aligned} J_2(t) = J_2\Theta(-t), \quad J_{\kappa q}(t) = J_{\kappa q}\Theta(-t) \ \ (\text{except}\ \kappa = 1), \\ J_{\kappa q}\big|_{\kappa=1}(t) = J_q(t) = J_q \forall t. \end{aligned} \tag{5.15}$$

All couplings on the right-hand side of the equations are time-independent and time-dependence is captured by the Heaviside function. This introduces non-equilibrium dynamics in the system where the post-quench protocol is exactly what we studied in Sect. 4.1.1. We strongly recommend the reader to revise Sect. 4.1.1 before proceeding as we are going to use the results from there.[2]

The post-quench Kadanoff-Baym equations take the form

$$\partial_{t_1}\partial_{t_2} g^{+}(t_1,t_2) = \mathcal{L}_q^{>}(t_1,t_2), \qquad \partial_{t_1} g^{-}(t_1,t_2) = 2\imath\mathcal{Q}\alpha(t_1). \tag{5.16}$$

[2] The convention for the Green's function and the Gaussian ensemble for the random couplings are the same in Sect. 4.1.1 as here, so we can safely carry over the results and use them here.

where $\mathcal{L}_q^>(t_1, t_2)$ is the same as in Eq. (4.23), namely

$$\mathcal{L}^>(t_1, t_2) \equiv 2\mathcal{J}_q^2 e^{g_+(t_1,t_2)}. \tag{5.17}$$

Recall $\mathcal{L}^<(t_1, t_2) = \mathcal{L}^>(t_1, t_2)^\star$.

We now have to solve these equations post-quench to study the thermalization dynamics. Post-quench, we are left with a single $\mathcal{H}_q$ SYK Hamiltonian whose couplings are time-independent. This implies that α will be a time-independent constant post-quench. This allows us to solve for g^- which we simply get as $g^-(t) = 2\imath\mathcal{Q}\alpha t$, immediately post-quench ($t \geq 0^+$). Thus Green's function admits a time-translational invariance right after the quench.

Next we focus on g^+. The equation $\partial_{t_1}\partial_{t_2} g^+(t_1, t_2) = 2\mathcal{J}_q^2 e^{g_+(t_1,t_2)}$ is a Liouville equation whose most general form of solution can be written as [13]:

$$e^{g^+(t_1,t_2)} = \frac{-\dot{u}(t_1)\,\dot{u}^\star(t_2)}{\mathcal{J}_q^2\,[u(t_1) - u^\star(t_2)]^2}, \tag{5.18}$$

where we use $g^+(t_1, t_2)^\star = g^+(t_2, t_1)$. We can verify this using pen and paper but we recommend using Mathematica to do so. We have provided Mathematica codes in Appendix H which the reader is encouraged to play with. We will present the results here. Since the most general form of Liouville's solution is given in Eq. (5.18), we now take an ansatz for u as follows:

$$u(t) = \frac{a e^{\imath\pi\nu/2} e^{\sigma t} + \imath b}{c e^{2\pi\nu/2} e^{\sigma t} + \imath d}, \quad \nu \in [-1, 1] \tag{5.19}$$

where ν is the same parameter that was introduced while evaluating the solutions of the Green's function for both the Majorana and the complex SYK models (for instance, see Eqs. (2.149) or (3.88)). Here $a, b, c, d, \sigma \in \mathbb{R}$. When we plug this ansatz in Eq. (5.18), we get a perfect cancellation of a, b, c and d, and we are left with (see Appendix H)

$$e^{g^+(t_1,t_2)} = \left(\frac{\sigma}{2\mathcal{J}_q}\right)^2 \frac{1}{\cos^2\left(\pi\nu/2 - \sigma\imath\,(t_1 - t_2)/2\right)}, \tag{5.20}$$

where $\sigma \geq 0$. We see that g^+ immediately admits a time-translational invariance post-quench ($t_1, t_2 \geq 0+$). Accordingly, $e^{g(t_1,t_2)} \to e^{g(t_1-t_2)}$ right after quench. As discussed in Appendix H, the KMS relation $g(t) = g(-t - \imath 2\pi\nu/\sigma)$ is inherited to $g^{\gtrless}$ and g^+ (see Eq. (4.49)) but $g^-(t)$ does not inherit this property. Accordingly, we expect the solution in Eq. (5.20) to satisfy the KMS relation which we explicitly verify in Appendix H. The KMS relation allows us to associate a temperature to the system, namely $\beta_f = \frac{2\pi\nu}{\sigma}$, where the subscript f denotes the final temperature of the system post-quench.[3] Thus, we conclude that the quench performed here (that leads to non-equilibrium dynamics) admits an *instantaneous thermalization* with respect to the Green's functions.[4] Since we derived for the large-q complex SYK model, the Majorana limit naturally admits an instantaneous thermalization.

5.2 Quench to a Chain: Charge Dynamics and Thermalization

The emergence of instantaneous thermalization in the large-q complex SYK dot following a quench is particularly striking given the generic initial Hamiltonian (Eq. (5.1)), which contains non-commuting, chaotic, and ergodic interactions that typically induce complex thermalization dynamics. This result—where Green's functions equilibrate immediately despite the absence of spatial structure—motivates our investigation of thermalization in spatially extended systems. We now extend this analysis to a one-dimensional SYK chain, where spatial correlations and charge transport are expected to introduce *finite* thermalization timescales. Using proof by contradiction, we first demonstrate the absence of instantaneous thermalization (i.e., a finite thermalization time) and then compute the post-quench dynamics of local charge density, following Ref. [9]. We start by evaluating the energy densities for both the on-site energy and the transport energy.

5.2.1 On-Site and Transport Energy Densities

We consider the same model as considered in Sect. 4.4. All conventions for the Green's function and the Gaussian distribution for the random variables are the same as there. We have already derived the equations there, here will simply cite the equations and use them to show via proof by contradiction that the chain does not admit instantaneous thermalization.

We start with the Kadanoff-Baym equations derived in Eq. (4.141) which at equal time becomes

$$\partial_t g_i^+(t_1, t_1) = \imath \mathrm{Re}\left[\alpha_i(t_1)\right], \quad \partial_t g_i^-(t_1, t_1) = -\mathrm{Im}\left[\alpha_i(t_1)\right]. \tag{5.21}$$

[3] Note that Eq. (5.20) must match with the solution we have already calculated for the complex SYK model in Eq. (4.47). Comparing the two also leads to the same expression for the final temperature as via the KMS relation, as it should because both methods have the same origin.

[4] It is essential to specify the physical object with respect to which thermalization occurs, as different quantities in non-equilibrium systems relax over distinct timescales. The slowest of these—governed by system-wide propagation of correlations—defines the *Thouless time*. The thermalization of closed quantum systems remains an intensely studied and debated topic, particularly regarding the hierarchy of relaxation scales and their dependence on initial conditions. We direct readers to Ref. [4] for a detailed review of these phenomena.

where, as before, equal-time limit is taken after taking the derivative. The expression for α_i is derived in Eq. (4.129), which we reproduce here for convenience

$$\alpha_i(t_1, t_1) = \epsilon_i(t_1) + \frac{r}{q}\left[\epsilon^{\star}_{i\to i+1}(t_1) + \epsilon_{i-1\to i}(t_1)\right], \tag{5.22}$$

where $\epsilon_i(t_1) \equiv \frac{q^2}{N}\langle \mathcal{H}_i\rangle(t_1)$ is the *on-site energy density* and $\epsilon_{i\to i+1}(t_1) \equiv \frac{q^2}{N}\langle \mathcal{H}_{i\to i+1}\rangle(t_1)$ is the *transport energy density*.

We now need to evaluate the expressions for $\epsilon_i(t_1)$ and $\epsilon_{i-1\to i}(t_1)$[5] This brings us the topic of Keldysh contour deformation, first encountered in Sect. 2.9.4.2. Here, we will provide a fresh perspective by deforming the Keldysh contour at the level of coupling strengths that will lead to the expectation values of the Hamiltonian, such as $\mathcal{H}_i(t_1)$. Later on, this method will be developed at a rather generalized level, suited for calculations in any SYK-like systems.

Keldysh Contour Deformation for Energy Density

Without loss of generality, we consider $t_2 < t_1$ in the Keldysh plane where we reserve the sign $+$ and $-$ to denote earlier time (t_2) and later time (t_1), respectively. Then we deform the couplings $J_i^-(t_1)$ and $D_{i-1}^-(t_1)$ on the forward Keldysh contour (earlier time t_1) as (each term must have the same dimension)

$$\begin{aligned} J_i^-(t_1) &= J_i(t_1) + \eta J_i(\tau)\delta(\tau - t_1), \\ D_{i-1}^-(t_1) &= D_{i-1}(t_1) + \lambda D_{i-1}(\tau)\delta(\tau - t_1) \end{aligned} \tag{5.23}$$

where the limits η, λ (units of time) $\to 0$ recovers the original coupling and we keep the backward Keldysh couplings (later times) unperturbed, namely $J_i^+(t_2) = J_i(t_2)$ and $D_{i-1}^+(t_2) = D_{i-1}(t_2)$. We have already evaluated the Keldysh partition function and associated effective action in Eqs. (4.114) (we simply denote as $\mathcal{Z}[\eta, \lambda]$ here where η and λ are the deformation parameters), (4.115), (4.116) and (4.117). We recall that the un-deformed Keldysh contour has normalized partition function $\mathcal{Z}[0, 0] = 1$. Then we have

$$\langle \mathcal{H}_i\rangle = \imath \frac{\delta \mathcal{Z}[\eta, \lambda]}{\delta \eta}\bigg|_{\eta,\lambda=0}, \qquad \langle \mathcal{H}_{i-1\to i}\rangle = \imath \frac{\delta \mathcal{Z}[\eta, \lambda]}{\delta \lambda}\bigg|_{\eta,\lambda=0}. \tag{5.24}$$

(continued)

[5] Note that $i \to i+1$ in $\epsilon_{i-1\to i}(t_1)$ gives $\epsilon_{i\to i+1}(t_1)$.

Evaluating the on-site interaction energy first, we get

$$\epsilon_i(t_1) \equiv \frac{q^2}{N}\langle\mathcal{H}_i\rangle(t_1) = -\imath\int_{\mathcal{C}} dt_2 J_i(t_1)J_i(t_2)\left(-4\mathcal{G}_i(t_1,t_2)\,\mathcal{G}_i(t_2,t_1)\right)^{\frac{q}{2}} \tag{5.25}$$

where we unfold the Keldysh integration as (keeping in mind Keldysh time ordering can have $t_2 >_{\mathcal{C}} t_1$ when, for instance, t_2 is on $\mathcal{C}_-$ while t_1 is on $\mathcal{C}_+$ in Fig. 2.1 while retaining real-time ordering $t_2 < t_1$ as assumed here without loss of generality): $\int_{\mathcal{C}} dt_2 = \int_{-\infty}^{t_1} dt_2 + \int_{t_1}^{-\infty} dt_2 = \int_{-\infty}^{t_1} dt_2 - \int_{-\infty}^{t_1} dt_2$. Thus, we get (the on-site SYK interaction strengths J_i are always real)

$$\epsilon_i(t_1) = \mathrm{Im}\int_{-\infty}^{t_1} dt_2\, 2J_i(t_1)J_i(t_2)\left(-4\mathcal{G}_i^{>}(t_1,t_2)\,\mathcal{G}_i^{<}(t_2,t_1)\right)^{\frac{q}{2}}. \tag{5.26}$$

Clearly $\mathrm{Im}[\epsilon_i(t_1)] = 0$. Therefore, $\epsilon_i(t_1)$ is an always real quantity.

Similarly, we get for $\epsilon_{i-1\to i}(t_1) \equiv \frac{q^2}{N}\langle\mathcal{H}_{i-1\to i}\rangle(t_1)$ the following:

$$\begin{aligned}\epsilon_{i-1\to i}(t_1) = \imath\frac{q}{r}\int_{-\infty}^{t_1} dt_2\, D_{i-1}^{\star}(t_1)D_{i-1}(t_2)\Big([-4\mathcal{G}_{i-1}^{>}(t_1,t_2)\mathcal{G}_i^{<}(t_2,t_1)]^{\frac{r}{2}}\\ -[-4\mathcal{G}_{i-1}^{<}(t_1,t_2)\mathcal{G}_i^{>}(t_2,t_1)]^{\frac{r}{2}}\Big)\end{aligned} \tag{5.27}$$

which implies $(i \to i+1)$

$$\begin{aligned}\epsilon_{i\to i+1}(t_1) = \imath\frac{q}{r}\int_{-\infty}^{t_1} dt_2\, D_i^{\star}(t_1)D_i(t_2)\Big([-4\mathcal{G}_i^{>}(t_1,t_2)\mathcal{G}_{i+1}^{<}(t_2,t_1)]^{\frac{r}{2}}\\ -[-4\mathcal{G}_i^{<}(t_1,t_2)\mathcal{G}_{i+1}^{>}(t_2,t_1)]^{\frac{r}{2}}\Big)\end{aligned} \tag{5.28}$$

and its complex conjugate (using general conjugate relation in Eq. (4.11))

$$\begin{aligned}\epsilon_{i\to i+1}(t_1)^{\star} = -\imath\frac{q}{r}\int_{-\infty}^{t_1} dt_2\, D_i(t_1)D_i(t_2)^{\star}\Big([-4\mathcal{G}_i^{>}(t_2,t_1)\mathcal{G}_{i+1}^{<}(t_1,t_2)]^{\frac{r}{2}}\\ -[-4\mathcal{G}_i^{<}(t_2,t_1)\mathcal{G}_{i+1}^{>}(t_1,t_2)]^{\frac{r}{2}}\Big).\end{aligned} \tag{5.29}$$

We have all the expressions for the energy density that constitute $\alpha_i(t_1)$ in Eq. (5.22). We now go to the large-q limit where the Green's function ansatz is given in Eq. (4.124), namely $\mathcal{G}_i^{\gtrless}(t_1,t_2) = \left(\mathcal{Q}_i(t) \mp \frac{1}{2}\right)e^{\frac{g_i^{\gtrless}(t_1,t_2)}{q}}$. Keeping to leading order in $\mathcal{O}(\mathcal{Q})$ where $\mathcal{Q} = \mathcal{O}(1/\sqrt{q})$ (Eq. (4.58)), we get for the Green's functions

$$-2\mathcal{G}_i^>(t_1,t_2) = e^{-2[\mathcal{Q}_i(t)+\mathcal{Q}_i(t)^2]+\frac{g_i^>(t_1,t_2)}{q}},$$
$$2\mathcal{G}_i^<(t_1,t_2) = e^{-2[-\mathcal{Q}_i(t)+\mathcal{Q}_i(t)^2]+\frac{g_i^<(t_1,t_2)}{q}}, \tag{5.30}$$

where we took $t = \frac{t_1+t_2}{2}$. One can expand the exponential and keep the terms to leading order in $\mathcal{Q}$ and verify. Accordingly, we have

$$[-4\mathcal{G}_{i-1}^>(t_1,t_2)\mathcal{G}_i^<(t_2,t_1)]^{\frac{r}{2}}$$
$$= e^{-r[\mathcal{Q}_{i-1}(t)+\mathcal{Q}_{i-1}(t)^2-\mathcal{Q}_i(t)+\mathcal{Q}_i(t)^2]+\frac{r}{2q}(g_i^>(t_1,t_2)+g_i^<(t_2,t_1))} \tag{5.31}$$

and

$$[-4\mathcal{G}_{i-1}^<(t_1,t_2)\mathcal{G}_i^>(t_2,t_1)]^{\frac{r}{2}}$$
$$= e^{-r[-\mathcal{Q}_{i-1}(t)+\mathcal{Q}_{i-1}(t)^2+\mathcal{Q}_i(t)+\mathcal{Q}_i(t)^2]+\frac{r}{2q}(g_i^<(t_1,t_2)+g_i^>(t_2,t_1))}. \tag{5.32}$$

Therefore, we get for $\epsilon_{i-1\to i}(t_1)$ in Eq. (5.27) the following (at leading order in $1/q$):

$$\epsilon_{i-1\to i}(t_1) = \imath\frac{q}{r}\int_{-\infty}^{t_1} dt_2\, D_{i-1}^\star(t_1)D_{i-1}(t_2)$$
$$\Big[1 - r(\mathcal{Q}_{i-1}(t) + \mathcal{Q}_{i-1}(t)^2 - \mathcal{Q}_i(t) + \mathcal{Q}_i(t)^2) \tag{5.33}$$
$$- 1 + r(-\mathcal{Q}_{i-1}(t) + \mathcal{Q}_{i-1}(t)^2 + \mathcal{Q}_i(t) + \mathcal{Q}_i(t)^2)\Big]$$

where there is a perfect cancellation of $\mathcal{Q}^2$ terms and we are left with

$$\epsilon_{i-1\to i}(t_1) = 2\imath q\int_{-\infty}^{t_1} dt_2\, D_{i-1}^\star(t_1)D_{i-1}(t_2)\Big(\mathcal{Q}_i(t) - \mathcal{Q}_{i-1}(t)\Big). \tag{5.34}$$

which immediate gives (Eq. (5.28))

$$\epsilon_{i\to i+1}(t_1) = 2\imath q\int_{-\infty}^{t_1} dt_2\, D_i^\star(t_1)D_i(t_2)\Big(\mathcal{Q}_{i+1}(t) - \mathcal{Q}_i(t)\Big). \tag{5.35}$$

Similarly for $\epsilon_{i\to i+1}(t_1)^\star$ in Eq. (5.29), we get

$$\epsilon_{i\to i+1}(t_1)^\star = -2\imath q\int_{-\infty}^{t_1} dt_2\, D_i(t_1)D_i^\star(t_2)\Big(\mathcal{Q}_{i+1}(t) - \mathcal{Q}_i(t)\Big). \tag{5.36}$$

Finally, the on-site energy density becomes (using Eq. (5.26))

$$\epsilon_i(t_1) = \mathrm{Im} \int_{-\infty}^{t_1} dt_2\; 2J_i(t_1)J_i(t_2)e^{-2q\mathcal{Q}^2}e^{g_i^+(t_1,t_2)} \tag{5.37}$$

where we used the definition of $g^+(t_1, t_2)$ from Eq. (4.20).

Now, we assume that all couplings are real, namely $J_i, D_i \in \mathbb{R} \forall\, i$ to get from Eqs. (5.35) and (5.36) the following:

$$\epsilon_{i\to i+1}(t_1)^\star = -\epsilon_{i\to i+1}(t_1) \qquad (J_i, D_i \in \mathbb{R}\;\forall\; i) \tag{5.38}$$

which implies that the transport energy density $\epsilon_{i\to i+1}(t_1)$ is *purely imaginary*. We have already seen that the on-site energy density $\epsilon_i(t_1)$ is *purely real*. Therefore, we have in the large-q limit for real couplings

$$\mathrm{Im}[\epsilon_i(t_1)] = 0, \quad \mathrm{Re}[\epsilon_{i\to i+1}(t_1)] = 0 \qquad (J_i, D_i \in \mathbb{R}\;\forall\; i). \tag{5.39}$$

Using the expression for $\alpha_i(t_1)$ from Eq. (5.22), we get

$$\begin{aligned} \mathrm{Re}[\alpha_i(t_1)] = \epsilon_i(t_1), \quad \mathrm{Im}[\alpha_i(t_1)] &= \frac{r}{q}\mathrm{Im}\left[\epsilon^\star_{i\to i+1}(t_1) + \epsilon_{i-1\to i}(t_1)\right] \\ &= \frac{r}{q}\mathrm{Im}\left[-\epsilon_{i\to i+1}(t_1) + \epsilon_{i-1\to i}(t_1)\right]. \end{aligned} \tag{5.40}$$

We now have all the ingredients to establish lack of instantaneous thermalization via proof by contradiction as well as evaluate the local charge density dynamics post-quench. The quench protocol we are following for the remaining two subsections is

$$D_i(t) = R_i\Theta(t) \qquad \text{(quench protocol)}, \tag{5.41}$$

where $R_i \in \mathbb{R}$ are time-independent coupling strengths (see Fig. 4.9 for a visualization of D_i).

5.2.2 Absence of Instantaneous Thermalization

We begin by visualizing with the help of Fig. 4.9 the quench protocol: for $t < 0$, the system consists of disconnected large-q SYK dots. Each dot independently thermalizes instantaneously (as proved in Sect. 5.1 above), establishing a pre-quench equilibrium where the full system is in a product state of thermal ensembles. At $t = 0$, we introduce an $r/2$-particle nearest-neighbor hopping term, coupling the dots. This quench disrupts equilibrium, triggering non-equilibrium dynamics for $t > 0$.

We assume that the chain instantaneously thermalizes with respect to the Green's functions. If we find a contradiction, then this assumption must be wrong and we will establish a finite thermalization rate (lack of instantaneous thermalization). Mathematically, the rate of change of transport energy density with respect to time must vanish, namely $\ddot{\epsilon}_{i\to i+1}(0^+) = 0$ if the system thermalizes instantaneously after the quench at $t = 0^+$. We will use this criteria to find a contradiction.

A word of caution: We know in the general case that $\mathrm{Re}[\epsilon_{i\to i+1}(t_1)] = 0$ (Eq. (5.39)) from which it seems that this requirement must hold for all times t, including the post-quench time $t = 0^+$. However, recall that Eq. (5.39) is derived in the large-q limit where we ignored the little Green's functions $g^{\gtrless}(t_1, t_2)$. As we will see below, here we consider the expansion of the Green's function keeping the little Green's functions that does not guarantee Eq. (5.39). Fortunately, as we will see, even if we include the corrections at orders containing little Green's functions, we find Eqs. (5.39) and (5.40) to hold post-quench at $t = 0^+$ where $g_i^+(t = 0^+)$ is purely real (see Eq. (4.47)).

So we start by re-expressing the real part of the full transport energy density in Eq. (5.28) where we have used the quench protocol (5.41):

$$\begin{aligned}\mathrm{Re}[\epsilon_{i\to i+1}(t_1)] = -\frac{q}{r}\int_0^{t_1} dt_2\, R_i^2\, \mathrm{Im}\Big(& [-4\mathcal{G}_i^>(t_1, t_2)\mathcal{G}_{i+1}^<(t_2, t_1)]^{\frac{r}{2}} \\ & -[-4\mathcal{G}_i^<(t_2, t_1)^\star \mathcal{G}_{i+1}^>(t_1, t_2)^\star]^{\frac{r}{2}}\Big)\end{aligned} \tag{5.42}$$

where we used the general conjugate relations for the Green's function in Eq. (4.11).

We focus on the argument inside Im[...], namely

$$A \equiv [-4\mathcal{G}_i^>(t_1, t_2)\mathcal{G}_{i+1}^<(t_2, t_1)]^{\frac{r}{2}} - [-4\mathcal{G}_i^<(t_2, t_1)^\star \mathcal{G}_{i+1}^>(t_1, t_2)^\star]^{\frac{r}{2}}. \tag{5.43}$$

Then we plug the large-q ansatz from Eq. (5.30) which when kept to the leading order in q gives us Eq. (5.35). However, we go to the next order where the information of the dynamics gets captured at the level of the little Green's functions. We get

$$\begin{aligned} A &= -2r(\mathcal{Q}_i(t) - \mathcal{Q}_{i+1}(t)) + \frac{r}{2q}\left(g_i^>(t_1, t_2) - g_i^<(t_2, t_1)^\star\right) \\ &\quad + \frac{r}{2q}\left(g_{i+1}^<(t_2, t_1) - g_{i+1}^>(t_1, t_2)^\star\right) \\ &= -2r(\mathcal{Q}_i(t) - \mathcal{Q}_{i+1}(t)) + \frac{r}{2q}\left(g_i^+(t_1, t_2) - 2\mathrm{Re}[g_i^<(t_2, t_1)]\right) \\ &\quad + \frac{r}{2q}\left(g_{i+1}^+(t_1, t_2) - 2\mathrm{Re}[g_{i+1}^>(t_1, t_2)]\right) \end{aligned} \tag{5.44}$$

where we used the definition in Eq. (4.20).

Now we impose the assumption of instantaneous thermalization of the Green's function post-quench which implies time-translational invariance $(t_1, t_2) \to (t_1 - t_2)$. We immediately see that if we plug this back in Eq. (5.28) and use the expression for $g_i^+(0^+)$ from Eq. (4.47) (since we have assumed instantaneous thermalization) to find that it's always real, accordingly Eqs. (5.39) and (5.40) follow at $t = 0^+$, namely $\mathrm{Re}[\epsilon_{i\to i+1}(0^+)] = 0$ which implies $\mathrm{Re}[\alpha_i(0^+)] = \epsilon_i(0^+)$. This relation will be used below.

Then we take the imaginary part of A (see Eq. (5.42)) to get

$$\mathrm{Im}[A] = \frac{r}{2q}\mathrm{Im}[g_i^+(t_1, t_2) + g_{i+1}^+(t_1, t_2)]. \tag{5.45}$$

Plugging back in Eq. (5.42), we get

$$\mathrm{Re}[\epsilon_{i\to i+1}(t_1)] = -\frac{1}{2}\int_0^{t_1} dt_2\; R_i^2\; \mathrm{Im}[g_i^+(t_1 - t_2) + g_{i+1}^+(t_1 - t_2)]. \tag{5.46}$$

where we imposed the time-translational invariance based on our assumption of instantaneous thermalization of the Green's function post-quench. Applying the Leibniz rule:

$$\frac{d}{dt_1}\mathrm{Re}\left[\epsilon_{i\to i+1}(t_1)\right] = -\frac{1}{2}\left[f(t_1, t_1) + \int_0^{t_1} \frac{\partial}{\partial t_1} f(t_1, t_2)\, dt_2\right]$$

where $f(t_1, t_2) = R_i^2 \mathrm{Im}\left[g_i^+(t_1 - t_2) + g_{i+1}^+(t_1 - t_2)\right]$. At $t_2 = t_1, t_1 - t_2 = 0$, so $f(t_1, t_1) = R_i^2 \mathrm{Im}\left[g_i^+(0) + g_{i+1}^+(0)\right]$. The partial derivative with respect to t_1 (holding t_2 fixed) is:

$$\frac{\partial}{\partial t_1} f(t_1, t_2) = R_i^2 \mathrm{Im}\left[\dot{g}_i^+(t_1 - t_2) + \dot{g}_{i+1}^+(t_1 - t_2)\right],$$

where $\frac{\partial}{\partial t_1} g_i^+(t_1 - t_2) = \dot{g}_i^+(t_1 - t_2)$.

The second derivative is the derivative of the first derivative. The first derivative can be written as:

$$\frac{d}{dt_1}\mathrm{Re}\left[\epsilon_{i\to i+1}(t_1)\right] = A + B(t_1),$$

where $A = -\frac{1}{2} R_i^2 \mathrm{Im}\left[g_i^+(0) + g_{i+1}^+(0)\right]$ (constant) and

$$B(t_1) = -\frac{1}{2} R_i^2 \int_0^{t_1} dt_2 \mathrm{Im}\left[\dot{g}_i^+(t_1 - t_2) + \dot{g}_{i+1}^+(t_1 - t_2)\right].$$

Since A is constant, $\frac{dA}{dt_1} = 0$. Applying Leibniz rule to $B(t_1)$:

$$\frac{d}{dt_1} B(t_1) = -\frac{1}{2} R_i^2 \left[k(t_1, t_1) + \int_0^{t_1} \frac{\partial}{\partial t_1} k(t_1, t_2)\, dt_2 \right]$$

where $k(t_1, t_2) = \text{Im}\left[\dot{g}_i^+(t_1 - t_2) + \dot{g}_{i+1}^+(t_1 - t_2)\right]$. At $t_2 = t_1, t_1 - t_2 = 0$, so $k(t_1, t_1) = \text{Im}\left[\dot{g}_i^+(0) + \dot{g}_{i+1}^+(0)\right]$. The partial derivative with respect to t_1 is:

$$\frac{\partial}{\partial t_1} k(t_1, t_2) = \text{Im}\left[\ddot{g}_i^+(t_1 - t_2) + \ddot{g}_{i+1}^+(t_1 - t_2)\right]$$

where $\frac{\partial}{\partial t_1} \dot{g}_i^+(t_1 - t_2) = \ddot{g}_i^+(t_1 - t_2)$. Thus, the second derivative is:

$$\frac{d^2}{dt_1^2} \text{Re}\left[\epsilon_{i \to i+1}(t_1)\right] = -\frac{1}{2} R_i^2 \text{Im}\left[\dot{g}_i^+(0) + \dot{g}_{i+1}^+(0)\right] - \frac{1}{2} R_i^2 \int_0^{t_1} dt_2 \text{Im}\left[\ddot{g}_i^+(t_1 - t_2) + \ddot{g}_{i+1}^+(t_1 - t_2)\right]. \tag{5.47}$$

Then we go to the post-quench $t_1 \to 0$ limit where the second integral vanishes and we are left with

$$\left. \frac{d^2}{dt_1^2} \text{Re}\left[\epsilon_{i \to i+1}(t_1)\right] \right|_{t_1 = 0} = -\frac{1}{2} R_i^2 \text{Im}\left[\dot{g}_i^+(0) + \dot{g}_{i+1}^+(0)\right]. \tag{5.48}$$

Using the equal-time Kadanoff-Baym equations in Eq. (5.21), we have post-quench with the assumption of instantaneous thermalization with respect to the Green's function: $\dot{g}_i^+(0) = \imath \text{Re}\left[\alpha_i(0)\right]$. Also, $\text{Re}[\alpha_i(0)] = \epsilon_i(0)$. Therefore, we have

$$\text{Re}\left[\ddot{\epsilon}_{i \to i+1}(0^+)\right] = -\frac{1}{2} R_i^2 \left(\text{Re}\left[\epsilon_{i+1}(0^+)\right] + \text{Re}\left[\epsilon_i(0^+)\right]\right). \tag{5.49}$$

However, as established, $\text{Re}\left[\ddot{\epsilon}_{i \to i+1}\left(0^+\right)\right] = 0$ would hold if the system thermalized instantaneously after the quench at $t = 0^+$. This necessitates that for non-zero coupling ($R_i \neq 0$), the energy densities $\epsilon_{i+1}\left(0^+\right)$and $\epsilon_i\left(0^+\right)$must have opposite signs. Yet, at any finite temperature, both energy densities are inherently negative because any single SYK dot at any given temperature have negative energy density. This contradiction proves that instantaneous thermalization does not occur for the coupled chain, despite individual blobs thermalizing instantaneously in isolation. Notice that this relation in Eq. (5.49) is trivially satisfied for individual dots (that do thermalize instantaneously) where $R_i = 0$ and transport energy densities are zero (since we switched off the transport).

Another solution satisfying Eq. (5.49) is $J_i = 0 \ \forall \ i$. This corresponds to a pure transport chain where exactly $r/2$ particles hop to nearest neighbors, which *can* support instantaneous thermalization. However, as we will show in the next subsection, the result for the local charge density shows that for any finite q, persistent current flow prevents equilibrium in the chain, thereby removing the possibility of instantaneous thermalization. Only in the limit $q \to \infty$—where the local charge density becomes effectively constant (as shown in the next subsection)—does Eq. (5.49) permit the possibility of instantaneous thermalization for this pure transport case.

NOTE: Computing the thermalization rate (inverse thermalization time) as a function of final equilibrium temperature is analytically intractable for generic initial temperatures. This arises from the non-Markovian (memory-dependent) and nonlinear structure of the Kadanoff-Baym equations—which for SYK-like systems emerge in the large-N limit and represent thermodynamic-limit results—typically requiring computationally intensive numerical methods. For the special case of $r = 2$ kinetic hopping in the Majorana limit ($\mathcal{Q}_i = 0 \ \forall i$), such a calculation has been achieved in Ref. [8].

5.2.3 Dynamics of Local Charge Density for $r = \mathcal{O}(q^0)$

We now proceed to calculate the local charge transport dynamics in the system post-quench, where the quench protocol is provided in Eq. (5.41). We first proceed with the general situation and then impose the quench to get the final result.

The charge transport has already been considered in Eq. (4.137), which we reproduce here for convenience:

$$\dot{\mathcal{Q}}_i(t_1) = \frac{1}{q}\mathrm{Im}[\alpha_i(t_1)]. \tag{5.50}$$

We have already evaluated in the large-q limit in Eq. (5.40) that $\mathrm{Im}[\alpha_i(t)] = \frac{r}{q}\mathrm{Im}\left[-\epsilon_{i\to i+1}(t_1) + \epsilon_{i-1\to i}(t_1)\right]$. We have also evaluated the explicit expressions for $\epsilon_{i-1\to i}(t_1)$ and $\epsilon_{i\to i+1}(t_1)$ in Eqs. (5.34) and (5.35), respectively. Thus, we are left with the following local charge density transport equation (where $D_i \in \mathbb{R} \ \forall \ i$)

$$\begin{aligned}\dot{\mathcal{Q}}_i(t_1) = \frac{2r}{q}\int_{-\infty}^{t_1} dt_2 \Big[& D_{i-1}(t_1)D_{i-1}(t_2)\Big(\mathcal{Q}_i(t) - \mathcal{Q}_{i-1}(t)\Big) \\ & - D_i(t_1)D_i(t_2)\Big(\mathcal{Q}_{i+1}(t) - \mathcal{Q}_i(t)\Big)\Big].\end{aligned} \tag{5.51}$$

We can simplify this by re-writing

$$\begin{aligned}\dot{\mathcal{Q}}_i(t_1) = \frac{2r}{q}\int_{-\infty}^{t_1} dt_2 \Big[& - D_{i-1}(t_1)D_{i-1}(t_2)\mathcal{Q}_{i-1}(t) + D_{i-1}(t_1)D_{i-1}(t_2)\mathcal{Q}_i(t) \\ & + D_i(t_1)D_i(t_2)\mathcal{Q}_i(t) - D_i(t_1)D_i(t_2)\mathcal{Q}_{i+1}(t)\Big]\end{aligned} \tag{5.52}$$

where we introduce the matrix of hopping strengths H_{ij}

$$H_{ij} \equiv 2\Big[- D_{i-1}(t_1)D_{i-1}(t_2)\delta_{j,i-1} - D_i(t_1)D_i(t_2)\delta_{j,i+1} + (D_{i-1}(t_1)D_{i-1}(t_2) + D_i(t_1)D_i(t_2))\delta_{j,i}\Big] \tag{5.53}$$

and we get the matrix equation for the local charge density transport

$$\boxed{\dot{\mathcal{Q}}_i(t_1) = \frac{r}{q}\int_{-\infty}^{t_1} dt_2 \sum_j H_{ij}(t_1,t_2)\mathcal{Q}_j(t)}. \tag{5.54}$$

The term with $j = i - 1$ contributes $-2\left[D_{i-1}(t_1)\, D_{i-1}(t_2)\right] \mathcal{Q}_{i-1}(t_2)$. The term with $j = i + 1$ contributes $-2\left[D_i(t_1)\, D_i(t_2)\right] \mathcal{Q}_{i+1}(t_2)$. The term with $j = i$ contributes $2\left[D_{i-1}(t_1)\, D_{i-1}(t_2) + D_i(t_1)\, D_i(t_2)\right] \mathcal{Q}_i(t_2)$.

Before proceeding to the quench protocol, we mention that Eq. (5.54) is responsible for the origin of scaling of $\dot{\mathcal{Q}}_i(t)$ in Eq. (4.138). Depending on whether $r = \mathcal{O}(q^0)$ or $r = \mathcal{O}(q^1)$, we get the scaling for $\mathcal{O}(\dot{\mathcal{Q}}_i) = \mathcal{O}\left(\frac{r}{q}\mathcal{Q}_i\right)$ where $\mathcal{Q}_i$ scales as in Eq. (4.58).

Now, we focus on the quench protocol in Eq. (5.41), namely $D_i(t) = R_i\Theta(t)$. Then taking the second derivative gives us (recall that we are using the Kadanoff-Baym equations in Eq. (4.141) which is evaluated in the large-q for $r = \mathcal{O}(q^0)$)

$$\boxed{\ddot{\boldsymbol{\mathcal{Q}}} = \frac{r}{q}\boldsymbol{H}\boldsymbol{\mathcal{Q}}} \quad (r = \mathcal{O}(q^0)) \tag{5.55}$$

where

$$H_{ij} = -|R_i|^2\delta_{j,i+1} - |R_{i-1}|^2\delta_{j,i-1} + \left[|R_i|^2 + |R_{i-1}|^2\right]\delta_{ij}. \tag{5.56}$$

Thus, we obtain a discrete wave equation that depends solely on local charge densities and transport coupling strengths – *independent of on-site interactions* – under the boundary conditions $R_{L=0} = R_{L+1} = 0$. This resolves the question of instantaneous thermalization where there are no on-site interactions and Eq. (5.49) is satisfied, implying there is a possibility of instantaneous thermalization. However, we see that there exists a local charge density dynamics post-quench (albeit for a small time) even if there are no on-site interactions, ruling out the possibility of instantaneous thermalization for any finite but large-q. Thus we have successfully calculated the local charge transport dynamical equation post-quench in a closed form, independent of J_i. We refer the reader to Ref. [9] for further analyses of the transport equation along with generalizing the results for absence of instantaneous thermalization and local charge transport dynamics to arbitrary d-dimensional lattices.

NOTE: The local charge density can evolve on timescales $t = \mathcal{O}(q^0)$, but exhibits only sub-leading fluctuations of order $\mathcal{O}\left(\mathcal{Q}q^{-1}\right)$. Thus, to leading order in $1/q$, the charge density remains effectively constant. This large-q construction ensures competition between transport and onsite interactions: "Smal" transport terms $r = \mathcal{O}(q^0)$ (suppressing charge fluctuations while retaining their influence on Green's functions) balance onsite effects. The case of $r = \mathcal{O}(q)$ scaling for full competition remains an open problem. The Kadanoff-Baym structure will have to consider a different scaling for $\dot{\mathcal{Q}}_i$ in Eq. (4.132) and a proper time scaling will have to be introduced to allow for a true competition, otherwise without time rescaling ($t \neq q^{3/2}\tau$), no charge flow occurs for any finite $t = \mathcal{O}\left(q^0\right)$.

5.3 Keldysh Contour Deformations: Theory

Following Ref. [10], we now develop the theory of Keldysh contour deformation introduced previously in Sect. 2.9.4.2 and the box below Eq. (5.22), which allowed us to calculate the expectation values of various physical quantities. We systematize this theory into a general framework for SYK-like systems, covering both the general case and the equilibrium specialization. The equilibrium case is of central importance, as it forms the basis for applying linear response theory to evaluate quantities like the DC resistivity—a key application we explore in the next section.

NOTE: Comprehensive pedagogical derivations are presented in Ref. [10]. To avoid duplication and enhance readability, we cite key results (referencing specific sections and equations) from this work. For complete step-by-step derivations, we strongly recommend consulting Ref. [10] where indicated. Accordingly, we follow the same notation and convention as Ref. [10] (consistent with the rest of this work), except for the notation for forward and backward Keldysh contours (Fig. 2.1). Here, we denote the forward contour by $\mathcal{C}_+$ and the backward contour by $\mathcal{C}_-$, while Ref. [10] uses the reverse notation ($\mathcal{C}_-$ for forward, $\mathcal{C}_+$ for backward). We have still tried to be as self-complete as possible, filling in conceptual and mathematical gaps that exist in the literature.

5.3.1 Time Evolution in the Deformed Keldysh Plane

We establish how time-dependent expectation values derive from the Keldysh generating functional $\mathcal{Z}$, defined as the trace of the contour-ordered evolution operator:

$$\mathcal{Z} = \mathrm{Tr}\left[U_{\mathcal{C}}\right] \quad \text{with} \quad U_{\mathcal{C}} = \mathcal{T}\exp\left(-\imath \int_{\mathcal{C}} dt \lambda(t) \mathcal{H}(t)\right). \tag{5.57}$$

Here $\mathcal{C}$ denotes the full Keldysh contour (Fig. 2.1), traversing $t_0 \to \infty$ ($\mathcal{C}_+$) then back to t_0 ($\mathcal{C}_-$). The Hamiltonian

$$\mathcal{H}(\tau) = U(t_0, \tau)\mathcal{H}(t_0)U^\dagger(t_0, \tau) \tag{5.58}$$

contains both implicit time dependence via $\lambda(t)$ and explicit dependence through U. Crucially, only the implicit component $\lambda(t)\mathcal{H}(t_0)$ governs unitary evolution.

Decomposing $U_\mathcal{C}$ into forward ($\mathcal{C}_+$) and backward ($\mathcal{C}_-$) segments (see Fig. 2.1) yields

$$U_{\mathcal{C}_+} = \lim_{\delta t\to 0}\prod_k (1 - \imath\delta t\mathcal{H}(t_k)), \tag{5.59a}$$

$$U_{\mathcal{C}_-} = \lim_{\delta t\to 0}\prod_k (1 + \imath\delta t\mathcal{H}(t_k)), \tag{5.59b}$$

where t_k partitions $[t_0, t_f]$. The closed contour yields $U_\mathcal{C} = \mathbb{1}$ identically. To extract observables, we deform $\mathcal{C}_+$ by introducing contour-dependent couplings:

$$\lambda_+(t) = \lambda(t) + \Delta(t), \quad \lambda_-(t) = \lambda(t), \tag{5.60}$$

with a source term localized on $\mathcal{C}_+$ (where we mark the source terms in color for more visibility, following Ref. [10]):

$$\Delta(t) = \sum_n \lambda_n D(\tau_n)\delta(\tau_n - t). \tag{5.61}$$

(General deformations of both contours will be developed subsequently.)

The functional derivative of $U_{\mathcal{C}_+}$ with respect to $\vec{\lambda}(\tau)$ (denoting the set of all λ_n as $\vec{\lambda}$) at $\vec{\lambda} = 0$ is

$$\begin{aligned}\left.\frac{\delta U_{\mathcal{C}_+}}{\delta\lambda(\tau)}\right|_{\vec{\lambda}=0} &= \lim_{\vec{\lambda}\to 0}\frac{U_{\mathcal{C}_+}[D(t) + \vec{\lambda}D(\tau)\delta(\tau - t)] - U_{\mathcal{C}_+}[D(t)]}{\vec{\lambda}} \\ &= U(t_0, \tau)\,(-\imath\mathcal{H}(t_0))\,U(\tau, t_f) \\ &= U(t_0, \tau)(-\imath U(t_0, \tau)^\dagger\mathcal{H}(\tau)U(t_0, \tau))U(\tau, t_f) \\ &= -\imath\mathcal{H}(\tau)U(t_0, t_f).\end{aligned} \tag{5.62}$$

Since $U_{\mathcal{C}_-}$ is $\vec{\lambda}$-independent, the full $U_\mathcal{C}$ derivative at $\lambda_n = 0$ is

$$\partial_{\lambda_n} U_\mathcal{C}|_{\vec{\lambda}=0} = U(t_0, \tau_n)(-\imath\mathcal{H}(t_0))U(\tau_n, t_f)U(t_f, t_0) = -\imath\hat{\mathcal{H}}(\tau_n) \tag{5.63}$$

Thus, the energy expectation at τ_n follows as

$$\langle H(\tau_n)\rangle = \imath\partial_{\lambda_n}\mathcal{Z}|_{\vec{\lambda}=0}, \tag{5.64}$$

which matches with Eq. (2.227) used earlier as well as coincides with the generalized Galitskii-Migdal relation when expressed via Green's functions. (We note this predicts negative energy (verified above for the Majorana case already) and positive conductivity, to be validated later.)

5.3.2 Effective Interacting Action under Keldysh Deformations

We have seen that the effective interacting action of the dot (for example, see Eq. (4.8)) or for the chain (see Eq. (4.116)) share the structure and in general, the effective interacting action of the SYK-type systems can be written as

$$S_I[\mathcal{G}] \equiv \int_C dt_1 \int_C dt_2 \frac{D_C(t_1)D_C(t_2)}{2} F[\mathcal{G}(t_1, t_2)\mathcal{G}(t_2, t_1)], \tag{5.65}$$

where C denotes the Keldysh contour (Fig. 2.1). We have used D as generic real-valued coupling strengths (for a single dot, this is the same as J_q for instance). This formulation handles general scenarios without time-translation invariance. Results derived here will later be adapted to equilibrium. The effective interacting action $S_I[\mathcal{G}]$ is the *only* term dependent on the real-valued coupling constants in the SYK-type systems (see, for example, Eq. (4.8) or (4.116)). The functional F *exclusively* depends on the composite variable $\mathcal{G}(t_1, t_2)\mathcal{G}(t_2, t_1)$.

The subscript in D, namely D_C denotes the contour-dependent coupling. Explicitly,

$$D(t)|_{C_+} \equiv D_+(t) \qquad D(t)|_{C_-} \equiv D_-(t), \tag{5.66}$$

leading to a contour discontinuity due to contour deformation

$$D(t)|_{C_+} - D(t)|_{C_-} = D_+(t) - D_-(t) \equiv \Delta(t). \tag{5.67}$$

This discontinuity $\Delta(t)$ will play a critical role in dynamics.

We now proceed to simplify F by decomposing it in terms of greater and lesser Green's function. We begin by decomposing

$$\mathcal{G}(t_1, t_2) = \Theta_C(t_1 - t_2)\mathcal{G}^>(t_1, t_2) + \Theta_C(t_2 - t_1)\mathcal{G}^<(t_1, t_2) \tag{5.68}$$

where Θ_C is the Heaviside step function in the Keldysh plane (defined in Eq. (2.203)). Therefore, we have

$$\mathcal{G}(t_1, t_2)\mathcal{G}(t_2, t_1) = \Theta_C(t_1 - t_2)\mathcal{G}^>(t_1, t_2)\mathcal{G}^<(t_2, t_1) + \Theta_C(t_2 - t_1) \left[\mathcal{G}^>(t_1, t_2)\mathcal{G}^<(t_2, t_1)\right]^\star \tag{5.69}$$

where we used the general conjugate relation (Eq. (4.11))

$$\left[\mathcal{G}^>(t_1, t_2)\mathcal{G}^<(t_2, t_1)\right]^\star = \mathcal{G}^>(t_2, t_1)\mathcal{G}^<(t_1, t_2). \tag{5.70}$$

Substituting this in F gives

$$\begin{aligned} F[\mathcal{G}(t_1,t_2)\mathcal{G}(t_2,t_1)] &= \Theta_{\mathcal{C}}(t_1-t_2)F[\mathcal{G}^{>}(t_1,t_2)\mathcal{G}^{<}(t_2,t_1)] \\ &\quad + \Theta_{\mathcal{C}}(t_2-t_1)F^{\star}[\mathcal{G}^{>}(t_1,t_2)\mathcal{G}^{<}(t_2,t_1)]. \end{aligned} \tag{5.71}$$

Now, we can separate $F[\mathcal{G}^{>}\mathcal{G}^{<}]$ into real and imaginary parts:

$$F[\mathcal{G}^{>}(t_1,t_2)\mathcal{G}^{<}(t_2,t_1)] \equiv X(t_1,t_2) + \imath Y(t_1,t_2) = F(t_1,t_2), \tag{5.72}$$

where X, Y are real-valued. Complex conjugation yields:

$$F^{\star}[\mathcal{G}^{>}(t_1,t_2)\mathcal{G}^{<}(t_2,t_1)] = F[\mathcal{G}^{>}(t_2,t_1)\mathcal{G}^{<}(t_1,t_2)] \tag{5.73a}$$

$$[X(t_1,t_2) + \imath Y(t_1,t_2)]^{\star} = X(t_1,t_2) - \imath Y(t_1,t_2). \tag{5.73b}$$

We can read-off the symmetry by comparing the expressions for $F[\mathcal{G}^{>}(t_1,t_2)$ $\mathcal{G}^{<}(t_2,t_1)]$ and $F^{\star} = F[\mathcal{G}^{>}(t_2,t_1)\mathcal{G}^{<}(t_1,t_2)]$, we get

$$X(t_1,t_2) = X(t_2,t_1) \qquad Y(t_1,t_2) = -Y(t_2,t_1) \tag{5.74}$$

Thus, X is symmetric and Y antisymmetric under $t_1 \leftrightarrow t_2$.

Inserting Eq. (5.74) into (5.71):

$$\begin{aligned} F[\mathcal{G}(t_1,t_2)\mathcal{G}(t_2,t_1)] &= [\Theta_{\mathcal{C}}(t_1-t_2) + \Theta_{\mathcal{C}}(t_2-t_1)]\, X(t_1,t_2) \\ &\quad + \imath\, [\Theta_{\mathcal{C}}(t_1-t_2) - \Theta_{\mathcal{C}}(t_2-t_1)]\, Y(t_1,t_2) \\ &= X(t_1,t_2) + \mathrm{sgn}_{\mathcal{C}}(t_1-t_2)\imath Y(t_1,t_2) \end{aligned} \tag{5.75}$$

where $\mathrm{sgn}_{\mathcal{C}}$ function is defined already in Eq. (2.204).

The coupling dependence of the action resides entirely in S_I (Eq. (5.65)). We have reduced its functional $F\,[\mathcal{G}\,(t_1,t_2)\,\mathcal{G}\,(t_2,t_1)]$ to a sum of (in Eq. (5.75))

- a time-symmetric component $X\,(t_1,t_2)$, and
- a time-antisymmetric component $Y\,(t_1,t_2)$,

where X and Y are real-functions defined in Eq. (5.72) and obey the symmetries in Eq. (5.74). This decomposition is crucial for subsequent non-equilibrium analysis.

5.3.3 Interacting Action Under Forward and Backward Contour Deformations

We now address the most general scenario where

- both forward ($\mathcal{C}_+$) and backward ($\mathcal{C}_-$) Keldysh contours (Fig. 2.1) are deformed,

- coupling constants are complex-valued, and
- the system is in full non-equilibrium (no time-translation invariance).

Results derived here will later be applied to transport analyses with real couplings. As the next section employs linear response theory to study transport—demonstrating an application of Keldysh contour deformations—we develop the full non-equilibrium picture here, followed by the simplified equilibrium situation as a corollary.

The interacting action (Eq. (5.65)) generalizes to:

$$\begin{aligned} S_I[\mathcal{G}] &= \int_{\mathcal{C}} \mathrm{d}t_1 \int_{\mathcal{C}} \mathrm{d}t_2 \frac{D_{\mathcal{C}}(t_1)\bar{D}_{\mathcal{C}}(t_2)}{2} F(\mathcal{G}(t_1,t_2)\mathcal{G}(t_2,t_1)) \\ &= \int_{\mathcal{C}} \mathrm{d}t_1 \int_{\mathcal{C}} \mathrm{d}t_2 \frac{D_{\mathcal{C}}(t_1)\bar{D}_{\mathcal{C}}(t_2)}{2} \big(X(t_1,t_2) + \mathrm{sgn}_{\mathcal{C}}(t_1-t_2)\imath Y(t_1,t_2)\big) \\ &= \mathrm{Re}S_I + \imath \mathrm{Im}S_I \qquad (X(t_1,t_2) = X(t_2,t_1) \text{ and } Y(t_1,t_2) = -Y(t_2,t_1))\,, \end{aligned} \tag{5.76}$$

where

- $D_{\mathcal{C}}(t) = D_+(t) + D_-(t)$ and $\bar{D}_{\mathcal{C}}(t) = \bar{D}_+(t) + \bar{D}_-(t)$ are complex-valued couplings
- $X(t_1, t_2) = X(t_2, t_1)$ and $Y(t_1, t_2) = -Y(t_2, t_1)$ retain their symmetry (Eq. (5.74))

We deform both contours through the generic delta-function perturbations:

$$D_+(t) = D(t) + \Delta_+(t) = D(t) + \sum_n \lambda_n D(\tau_n)\delta(\tau_n - t), \tag{5.77a}$$

$$\bar{D}_+(t) = D^\star(t) + \bar{\Delta}_+(t) = D^\star(t) + \sum_n \bar{\lambda}_n D^\star(\tau_n)\delta(\tau_n - t), \tag{5.77b}$$

$$D_-(t) = D(t) - \Delta_-(t) = D(t) - \sum_n \eta_n D(\tau_n)\delta(\tau_n - t), \tag{5.77c}$$

$$\bar{D}_-(t) = D^\star(t) - \bar{\Delta}_-(t) = D^\star(t) - \sum_n \bar{\eta}_n D^\star(\tau_n)\delta(\tau_n - t), \tag{5.77d}$$

where λ_n, $\bar{\lambda}_n$, η_n and $\bar{\eta}_n$ are the deformation parameters. Collectively, we denote $\vec{\lambda} = (\{\lambda_n\}, \{\bar{\lambda}_m\})$ and $\vec{\eta} = (\{\eta_n\}, \{\bar{\eta}_m\})$.

Accounting for time-directionality, we have

$$\int_{\mathcal{C}_+} \mathrm{d}t(\cdots) = \int_{t_0}^{t_f} \mathrm{d}t(\cdots) \tag{5.78a}$$

$$\int_{\mathcal{C}_-} \mathrm{d}t(\cdots) = \int_{t_f}^{t_0} \mathrm{d}t(\cdots) = -\int_{t_0}^{t_f} \mathrm{d}t(\cdots). \tag{5.78b}$$

This ensures correct time-ordering on the reverse-oriented backward contour. Now we have all the ingredients to start deforming the contour and evaluate the deformed effective interacting action in Eq. (5.76). We can do the deformation separately for the real and the imaginary parts of S_I. Starting from the real part, we have

$$
\begin{aligned}
\mathrm{Re} S_I &= \int_C \mathrm{d}t_1 \int_C \mathrm{d}t_2 \frac{D_C(t_1)\bar{D}_C(t_2)}{2} X(t_1, t_2) \\
&= \int_{C_-} \mathrm{d}t_1 \int_{C_+} \mathrm{d}t_2 \frac{D_-(t_1)\bar{D}_+(t_2)}{2} X(t_1, t_2) \\
&\quad + \int_{C_+} \mathrm{d}t_1 \int_{C_-} \mathrm{d}t_2 \frac{D_+(t_1)\bar{D}_-(t_2)}{2} X(t_1, t_2) \\
&\int_{C_-} \mathrm{d}t_1 \int_{C_-} \mathrm{d}t_2 \frac{D_-(t_1)\bar{D}_-(t_2)}{2} X(t_1, t_2) \\
&\quad + \int_{C_+} \mathrm{d}t_1 \int_{C_+} \mathrm{d}t_2 \frac{D_+(t_1)\bar{D}_+(t_2)}{2} X(t_1, t_2) \\
&= \int \mathrm{d}t_1 \int \mathrm{d}t_2 \frac{\begin{bmatrix} -D_-(t_1)\bar{D}_+(t_2) - D_+(t_1)\bar{D}_-(t_2) \\ +D_-(t_1)\bar{D}_-(t_2) + D_+(t_1)\bar{D}_+(t_2) \end{bmatrix}}{2} X(t_1, t_2) \\
&= \frac{1}{2} \int \mathrm{d}t_1 \int \mathrm{d}t_2 X(t_1, t_2) \left[D_+(t_1)\bar{D}_+(t_2) + D_-(t_1)\bar{D}_-(t_2) \right. \\
&\quad \left. - D_+(t_1)\bar{D}_-(t_2) - D_-(t_1)\bar{D}_+(t_2) \right]
\end{aligned}
\tag{5.79}
$$

where we plug in Eq. (5.77) to get

$$
\begin{aligned}
\mathrm{Re} S_I &= \mathrm{Re} S_I|_{\vec{\lambda},\vec{\eta}=0} \\
&\quad + \frac{1}{2} \int \mathrm{d}t_1 \int \mathrm{d}t_2 X(t_1, t_2) \left[\Delta_+(t_1)\bar{\Delta}_+(t_2) + \Delta_-(t_1)\bar{\Delta}_-(t_2) \right. \\
&\quad \left. + \Delta_+(t_1)\bar{\Delta}_-(t_2) + \Delta_-(t_1)\bar{\Delta}_+(t_2) \right].
\end{aligned}
\tag{5.80}
$$

Then finally using the definition of $\Delta_\pm$ and $\bar{\Delta}_\pm$ from Eq. (5.77), we get

$$
\boxed{
\begin{aligned}
\mathrm{Re} S_I =& \mathrm{Re} S_I|_{\vec{\lambda},\vec{\eta}=0} \\
&+ \frac{1}{2} \sum_{n=m} X(\tau_n, \tau_m) \left\{ D(\tau_n) D^\star(\tau_m) \left[\lambda_n \bar{\lambda}_m + \eta_n \bar{\eta}_m + \lambda_n \bar{\eta}_m + \eta_n \bar{\lambda}_m \right] \right. \\
&\qquad \left. + D(\tau_m) D^\star(\tau_n) \left[\lambda_m \bar{\lambda}_n + \eta_m \bar{\eta}_n + \lambda_m \bar{\eta}_n + \eta_m \bar{\lambda}_n \right] \right\} \\
&+ \frac{1}{2} \sum_n |D(\tau_n)|^2 \left[|\lambda_n|^2 + |\eta_n|^2 + \left(\lambda_n \bar{\eta}_n + \eta_n \bar{\lambda}_n \right) \right] X(\tau_n, \tau_n).
\end{aligned}
}
\tag{5.81}
$$

These calculations are provided in great details in Appendix A of Ref. [10]. We strongly recommend the reader to refer that. Similarly, we solve for the imaginary part of S_I. The derivation is provided in Appendix A of [10] (Eq. (A13)) which we quote here

$$\boxed{\begin{aligned}\mathrm{Im}S_I =&\mathrm{Im}S_I|_{\vec{\lambda},\vec{\eta}=0}\\ &+\frac{1}{2}\sum_{n>m} Y(\tau_n,\tau_m)\left\{D(\tau_n)D^\star(\tau_m)\left[\lambda_n\bar{\lambda}_m-\eta_n\bar{\eta}_m-\lambda_n\bar{\eta}_m+\eta_n\bar{\lambda}_m\right]\right.\\ &\left.\qquad+D(\tau_m)D^\star(\tau_n)\left[\lambda_m\bar{\lambda}_n-\eta_m\bar{\eta}_n+\lambda_m\bar{\eta}_n-\eta_m\bar{\lambda}_n\right]\right\}\\ &-\sum_m\int_{t_0}^{\tau_m}\mathrm{d}t_1 Y(t_1,\tau_m)\left[D(t_1)D^\star(\tau_m)\bar{\lambda}_m\right.\\ &\qquad\left.+D^\star(t_1)D(\tau_m)\lambda_m+D^\star(t_1)D(\tau_m)\eta_m+D(t_1)D^\star(\tau_m)\bar{\eta}_m\right].\end{aligned}} \tag{5.82}$$

Then, we get for the total action S_I as $\mathrm{Re}S_I + \imath\mathrm{Im}S_I$ where we simplify (see Appendix A, Eq. (A15) of Ref. [10]) and use the definition of $F = X(t_1,t_2) + \mathrm{sgn}_{\mathcal{C}}(t_1-t_2)\imath Y(t_1,t_2)$ to get

$$\boxed{\begin{aligned}S_I =&S_I|_{\vec{\lambda},\vec{\eta}=0}\\ &+\frac{1}{2}\sum_{n>m} D(\tau_n)D^\star(\tau_m)\left(\lambda_n\bar{\lambda}_m+\eta_n\bar{\lambda}_m\right)F(\tau_n,\tau_m)\\ &+\frac{1}{2}\sum_{n>m} D(\tau_n)D^\star(\tau_m)\left(\eta_n\bar{\eta}_m+\lambda_n\bar{\eta}_m\right)F^\star(\tau_n,\tau_m)\\ &+\frac{1}{2}\sum_{n>m} D(\tau_m)D^\star(\tau_n)\left(\lambda_m\bar{\lambda}_n+\lambda_m\bar{\eta}_n\right)F(\tau_n,\tau_m)\\ &+\frac{1}{2}\sum_{n>m} D(\tau_m)D^\star(\tau_n)\left(\eta_m\bar{\eta}_n+\eta_m\bar{\lambda}_n\right)F^\star(\tau_n,\tau_m)\\ &-\imath\sum_m\int_{t_0}^{\tau_m}\mathrm{d}t_1 Y(t_1,\tau_m)\left[D(t_1)D^\star(\tau_m)\bar{\lambda}_m+D^\star(t_1)D(\tau_m)\lambda_m\right.\\ &\qquad\left.+D^\star(t_1)D(\tau_m)\eta_m+D(t_1)D^\star(\tau_m)\bar{\eta}_m\right]\\ &+\frac{1}{2}\sum_n |D(\tau_n)|^2\left[|\lambda_n|^2+|\eta_n|^2+\left(\lambda_n\bar{\eta}_n+\eta_n\bar{\lambda}_n\right)\right]X(\tau_n,\tau_n),\end{aligned}} \tag{5.83}$$

where $S_I|_{\vec{\lambda},\vec{\eta}=0} = \mathrm{Re}S_I|_{\vec{\lambda},\vec{\eta}=0} + \imath\mathrm{Im}S_I|_{\vec{\lambda},\vec{\eta}=0}$. This is the most general result for non-equilibrium situation.

We now specialize to the equilibrium situation which will be used in the next section to study transport using the linear response theory. For equilibrium, we make the following simplifications

- Time-independent couplings ($D(t) \to D$)
- Initial time $t_0 \to -\infty$ (recall that we have throughout ignored the imaginary part of the Keldysh contour in Fig. 2.1, as explained above several times due to the Bogoliubov's principle of weakening correlations)
- Translation invariance implies $X(t_1, t_2) = X(t_1 - t_2)$, $Y(t_1, t_2) = Y(t_1 - t_2)$

We refer the reader to Appendix A of Ref. [10] for step-by-step calculations to get the following form for the equilibrium condition

$$\boxed{\begin{aligned}\frac{S_I}{|D|^2} =& S_I|_{\vec{\lambda},\vec{\eta}=0}\\ &+\frac{1}{2}\sum_{n>m}\left[(\lambda_n+\eta_n)(\bar{\lambda}_m F(\tau_n-\tau_m)+\bar{\eta}_m F^\star(\tau_n-\tau_m))\right.\\ &\qquad\left.+(\bar{\lambda}_n+\bar{\eta}_n)(\lambda_m F(\tau_n-\tau_m)+\eta_m F^\star(\tau_n-\tau_m))\right]\\ &-\imath\sum_m\left[\bar{\lambda}_m+\lambda_m+\eta_m+\bar{\eta}_m\right]\left(\int_{-\infty}^0 \mathrm{d}t\,Y(t)\right)\\ &+\frac{1}{2}\sum_n(\lambda_n+\eta_n)(\bar{\lambda}_n+\bar{\eta}_n)X(0). \qquad \text{(Equilibrium)}\end{aligned}} \tag{5.84}$$

We now specialize to the large-q limit, employing the Green's function ansatz (Eq. (4.17)). Since we are going to study transport via linear response theory in the next section, we will deal with uniform equilibrium chains which has already been developed in Sect. 4.4.5. The key functional $F(t)$—which determines the interaction structure—was derived in Eq. (4.145) (excluding the coupling prefactor $2|D|^2$). Using our large-q ansatz, we simplify this to:

$$\begin{aligned}F(t) =&\frac{1}{\kappa q^2}\left[-4\mathcal{G}^>(t)\mathcal{G}^<(-t)\right]^{\kappa q/2} = \frac{1}{\kappa q^2}\left[\left(1-4\mathcal{Q}^2\right)e^{2g^+(t)/q}\right]^{\kappa q/2}\\ =&\frac{1}{\kappa q^2}\left[e^{-4\mathcal{Q}^2}e^{2g^+(t)/q}\right]^{\kappa q/2} = \frac{1}{\kappa q^2}e^{-2\kappa q\mathcal{Q}^2}e^{\kappa g^+(t)}\\ =&X(t)+\imath Y(t).\end{aligned} \tag{5.85}$$

5.4 Keldysh Contour Deformations: Applications to DC Resistivity

We now apply our Keldysh contour deformation framework to calculate DC resistivity—a fundamental transport property—for SYK chains in the thermodynamic limit ($N \to \infty$). Using linear response theory, we compute non-equilibrium responses by perturbing equilibrium solutions to leading order. Our approach systematically addresses:

1. Model Classification: Three distinct SYK chain architectures are analyzed, each with different scaling competition between the on-site and the hopping terms.
2. Current Formulation: We derive microscopically consistent current operators from the continuity equation, highlighting their relationship to charge transport.
3. Kubo Framework: DC conductivity σ_{DC} (which provides the DC resistivity $\rho_{DC} = 1/\sigma_{DC}$) is expressed through retarded current-current correlation functions:

$$\sigma_{DC} = \lim_{\omega \to 0} -\frac{\mathrm{Im}\chi^R(\omega)}{\omega}$$

   where χ^R is the susceptibility which we will evaluate using our contour deformation techniques.
4. Fluctuation-Dissipation Theorem: We establish this cornerstone result (relating correlation functions to response functions) within our formalism.
5. Holographic Signatures: For one chain variant, the temperature-scaling of ρ_{DC} reveals a striking correspondence with holographic insulators, providing us with a hint of holography.

This methodology demonstrates how Keldysh techniques resolve transport in strongly correlated non-Fermi liquids beyond conventional approaches.

NOTE: As mentioned earlier at the start of Sect. 5.3, while step-by-step derivations are thoroughly documented in Ref. [10], we prioritize conceptual clarity by citing essential results (with specific section/equation pointers). Our notation aligns with Ref. [10] except for one deliberate choice: The forward contour is denoted $\mathcal{C}_+$ (backward: $\mathcal{C}_-$) in this work consistently throughout, whereas Ref. [10] uses $\mathcal{C}_-$ for forward and $\mathcal{C}_+$ for backward contours in Fig. 2.1. Readers seeking mathematical details will find them comprehensively treated in Ref. [10] and are encouraged to refer wherever indicated. We have tried to be as self-complete as possible, highlighting conceptual and mathematical steps required for a smooth read.

5.4.1 Models for Three Chains

We consider the chain as discussed in Sect. 4.4. Since we have already performed the analysis out-of-equilibrium as well as in equilibrium (Sect. 4.4.5), we will simply reproduce the results from these sections and use here. We start by reproducing the Hamiltonian for the three chains we will be considering in this section.

We consider a uniformly coupled chain of L SYK dots

$$\mathcal{H} = \sum_{i=1}^{L} \left(\mathcal{H}_i + \mathcal{H}_{i\to i+1} + \mathcal{H}^{\dagger}_{i\to i+1} \right) \tag{5.86}$$

where the on-site q-body interaction is given by

$$\mathcal{H}_i = J \sum_{\substack{\{\boldsymbol{\mu}_{1:\frac{q}{2}}\}_{\le} \\ \{\boldsymbol{\nu}_{1:\frac{q}{2}}\}_{\le}}} X(i)^{\boldsymbol{\mu}_q}_{\boldsymbol{\nu}_q} c^{\dagger}_{i;\mu_1} \cdots c^{\dagger}_{i;\mu_{q/2}} c_{i;\nu_{q/2}} \cdots c_{i;\nu_1} \tag{5.87}$$

where the same notation as in Sect. 4.4 is followed: $\{\boldsymbol{\mu}_{1:\frac{q}{2}}\}_{\le} \equiv 1 \le \mu_1 < \mu_2 < \cdots < \mu_{\frac{q}{2}-1} < \mu_{\frac{q}{2}} \le N$ and $\boldsymbol{\mu}_q \equiv \{\mu_1, \mu_2, \mu_3, \ldots, \mu_{\frac{q}{2}}\}$ Transport between nearest neighbors is mediated by $r/2$-body SYK-like Hamiltonian given by

$$\mathcal{H}_{i\to i+1} \equiv D \sum_{\substack{\{\boldsymbol{\mu}_{1:\frac{r}{2}}\}_{\le} \\ \{\boldsymbol{\nu}_{1:\frac{r}{2}}\}_{\le}}} Y(i)^{\boldsymbol{\mu}}_{\boldsymbol{\nu}} c^{\dagger}_{i+1;\mu_1} \cdots c^{\dagger}_{i+1;\mu_{\frac{r}{2}}} c_{i;\nu_{\frac{r}{2}}} \cdots c_{i;\nu_1}. \tag{5.88}$$

Both $X(i)^{\boldsymbol{\mu}}_{\boldsymbol{\nu}}$ and $Y(i)^{\boldsymbol{\mu}}_{\boldsymbol{\nu}}$ are random variables derived from Gaussian distribution given in Eqs. (4.107) and (4.110), respectively. The model has an associated $U(1)$ conserved charge given in Eq. (4.111) where the local charge density is defined in Eq. (4.112).

We can re-express the Hamiltonian by defining

$$\mathcal{H}_{\to} \equiv \sum_{i=1}^{L} \mathcal{H}_{i\to i+1}, \tag{5.89}$$

and

$$\mathcal{H}_{\text{dot}} \equiv \sum_{i=1}^{L} \mathcal{H}_i, \qquad \mathcal{H}_{\text{trans}} \equiv \mathcal{H}_{\to} + \mathcal{H}^{\dagger}_{\to}, \tag{5.90}$$

enabling to to re-write Eq. (5.86) as

$$\mathcal{H} = \mathcal{H}_{\text{dot}} + \mathcal{H}_{\text{trans}}. \tag{5.91}$$

We focus on three different types of transport chains, namely

$$r = \kappa q \qquad \left(\kappa = \left\{ \frac{1}{2}, 1, 2 \right\} \right). \tag{5.92}$$

5.4.2 Solving the Three Chains

As discussed already in Sect. 4.4.6, the three chains can be mapped onto the following coupled SYK dot (Eq. (4.146)):

$$\mathcal{H} = K_q \mathcal{H}_q + K_{\kappa q} \mathcal{H}_{\kappa q} \tag{5.93}$$

whose effective action is evaluated in Eq. (4.150). We know that using the mapping in Eq. (4.153), namely

$$\boxed{J^2 \longleftrightarrow K_q^2, \quad 2|D|^2 \longleftrightarrow K_{\kappa q}^2}, \qquad \text{(chain} \leftrightarrow \text{coupled dot)}, \tag{5.94}$$

we successfully map the effective action, consequently the Schwinger-Dyson equations and the Kadanoff-Baym equations of the two theory. Recall that we will perform linear response shortly which requires us to treat our (uniform) chains at equilibrium.

So we focus on the solutions of Eq. (5.93) and keep the identification in Eq. (5.94) to map back to the chain. The associated Schwinger-Dyson equations using the effective action in Eq. (4.150) are given by

$$\mathcal{G}_0^{-1} - \mathcal{G}^{-1} = \Sigma, \quad \Sigma = \Sigma_q + \Sigma_{\kappa q}. \tag{5.95}$$

Here the self-energies are given by (using the single dot results for q and κq dots from Eq. (4.22))

$$\begin{aligned} \Sigma_q^{\gtrless}(t_1, t_2) &= \frac{1}{q} \tilde{\mathcal{L}}_q^{\gtrless}(t_1, t_2)\, \mathcal{G}^{\lessgtr}(t_1, t_2) \\ \Sigma_{\kappa q}^{\gtrless}(t_1, t_2) &= \frac{1}{q} \mathcal{L}_{\kappa q}^{\gtrless}(t_1, t_2)\, \mathcal{G}^{\lessgtr}(t_1, t_2) \end{aligned} \tag{5.96}$$

where the large-q ansatz for the Green's function from Eq. (4.17), namely $\mathcal{G}^{\gtrless}(t_1, t_2) = \left(\mathcal{Q} \mp \frac{1}{2}\right) e^{g^{\gtrless}(t_1,t_2)/q}$ is used

$$\begin{aligned} \tilde{\mathcal{L}}_q^{>}(t_1, t_2) &\equiv 2\mathcal{J}_q^2 e^{g_+(t_1,t_2)}, \ \tilde{\mathcal{L}}^{<}(t_1, t_2) = \tilde{\mathcal{L}}^{>}(t_1, t_2)^\star \\ \mathcal{L}_{\kappa q}^{>}(t_1, t_2) &\equiv 2\mathcal{K}_{\kappa q}^2 e^{\kappa g_+(t_1,t_2)}, \ \mathcal{L}^{<}(t_1, t_2) = \mathcal{L}^{>}(t_1, t_2)^\star \end{aligned} \tag{5.97}$$

and the couplings are redefined as before[6]

[6] Note that while $\mathcal{K}_{\kappa q}$ reduces to $\mathcal{J}_q$ when $\kappa = 1$, we maintain distinct notation $\left(\mathcal{J}_q \text{ vs. } \mathcal{K}_{\kappa q}\right)$ to emphasize their different physical origins in the coupled SYK system (Eq. (5.93)): $\mathcal{J}_q$ governs the intra-dot q-body interactions, and $\mathcal{K}_{\kappa q}$ controls the inter-dot κq-body couplings that translates to hopping terms for the chain (see Eq. (5.94)). This distinction highlights how identical functional

$$\mathcal{J}_q^2 \equiv \left(1 - 4\mathcal{Q}^2\right)^{q/2-1} K_q^2,$$
$$\mathcal{K}_{\kappa q}^2 \equiv \left(1 - 4\mathcal{Q}^2\right)^{\kappa q/2-1} K_{\kappa q}^2 \quad \text{(coupled dot couplings).} \tag{5.98}$$

At the level of chain where the on-site interaction strength is governed by J and hopping strength is given by D (see Eq. (5.86)), we can re-define the couplings as[7]

$$\mathcal{J}^2 \equiv (1 - 4\mathcal{Q}^2)^{q/2-1} J^2, \quad |\mathcal{D}|^2 \equiv (1 - 4\mathcal{Q}^2)^{\kappa q/2-1} 2|D|^2 \quad \text{(chain couplings).} \tag{5.99}$$

Thus the chain to dot mapping in Eq. (5.94) translates to

$$\boxed{\mathcal{J}^2 \longleftrightarrow \mathcal{J}_q^2, \quad |\mathcal{D}|^2 \longleftrightarrow \mathcal{K}_{\kappa q}^2} \quad \text{(chain} \leftrightarrow \text{coupled dot).} \tag{5.100}$$

The associated Kadanoff-Baym equations have the form similar to Eq. (4.45) where α for the coupled dot is given by (see footnote 5)

$$\left(1 - 4\mathcal{Q}^2\right) \alpha(t) = \epsilon_q(t) + \kappa \epsilon_{\kappa q}(t) \tag{5.101}$$

where $\mathcal{Q}$ is the total charge density of associated with the coupled dot and $\epsilon_{\kappa q}$ is given by

$$\epsilon_q(t) \equiv q^2 \frac{K_q \langle \mathcal{H}_q \rangle}{N}, \quad \epsilon_{\kappa q}(t) \equiv q^2 \frac{K_{\kappa q} \langle \mathcal{H}_{\kappa q} \rangle}{N}. \tag{5.102}$$

Accordingly, the relation in Eq. (4.50) also serves us here

$$\dot{g}^+(0) = \imath \alpha(0) \tag{5.103}$$

where α is a constant in equilibrium as it should (due to energy conservation).

The form of the Kadanoff-Baym equations are the same as derived in Sect. 4.1.4 (in particular, Eq. (4.45)) with α given by Eq. (5.101). At equilibrium, we can solve for $g^-(t)$ for all three cases of $\kappa = \{1/2, 1, 2\}$

$$\boxed{g^-(t) = 2\imath \mathcal{Q} \alpha t} \tag{5.104}$$

where recall that $\mathcal{Q} = \mathcal{O}(q^{-1/2})$ (Eq. (4.58)). Now, if we solve for $g^+(t)$ for all three cases of κ, we have successfully solved the model (the three uniform chains) at equilibrium. The differential equation for $g^+(t)$ is obtained by taking

forms ($\mathcal{K}_q \equiv \mathcal{J}_q$ at $\kappa = 1$) arise from separate terms in the Hamiltonian. The notation preserves conceptual clarity when analyzing their individual contributions to dynamics.

[7] Since $\mathcal{Q} = \mathcal{O}(q^{-1/2})$, to leading order in $1/q$, these become $\mathcal{J}^2 = e^{-2q\mathcal{Q}^2} J^2$ and $|\mathcal{D}|^2 = 2|D|^2 e^{-2\kappa q \mathcal{Q}^2}$.

a second derivative of the Kadanoff-Baym equation that results in Eq. (4.45) which we reproduce here for convenience (at equilibrium)[8]

$$\ddot{g}^+(t) = -\mathcal{L}^>(t) \tag{5.105}$$

where $\ddot{g}^+(t) = \frac{\partial^2 g(t)}{\partial t^2}$ and $\mathcal{L}^>(t) = \tilde{\mathcal{L}}^>_q(t) + \mathcal{L}^>_{\kappa q}(t)$ (see Eq. (5.97)). So we solve this differential equation for all three values of κ.

- $\kappa = \frac{1}{2}$: The time evolution of the two-point correlation function $g^+(t)$ is governed by the nonlinear differential equation:

$$\ddot{g}^+(t) = -2\mathcal{J}_q^2 e^{g^+(t)} - 2\mathcal{K}_{q/2}^2 e^{g^+(t)/2}. \tag{5.106}$$

  This system admits an exact solution (can be directly verified by substituting in the above differential equation):

$$e^{g^+(t)/2} = \frac{1}{\left(\beta\mathcal{K}_{q/2}\right)^2} \frac{(\pi v)^2}{1 + \sqrt{A^2+1}\cos(\pi v(1/2 - \imath t/\beta))} \tag{5.107}$$

  with the dimensionless parameter A characterizing the coupling ratio:

$$A \equiv \frac{\pi v \beta \mathcal{J}_q}{\left(\beta\mathcal{K}_{q/2}\right)^2}. \tag{5.108}$$

  The boundary condition $g^+(0) = 0$ yields the self-consistency relation:

$$\pi v = \sqrt{\left(\beta\mathcal{J}_q\right)^2 + \left(\frac{\left(\beta\mathcal{K}_{q/2}\right)^2}{\pi v}\right)^2}\cos(\pi v/2) + \frac{\left(\beta\mathcal{K}_{q/2}\right)^2}{\pi v} \tag{5.109}$$

- $\kappa = 1$: Similarly, the differential equation becomes

$$\ddot{g}^+(t) = -\left(2\mathcal{J}_q^2 + 2\mathcal{K}_q^2\right) e^{g^+(t)}, \tag{5.110}$$

  which can be solved exactly for $g^+(t)$ to get

$$e^{g^+(t)} = \frac{(\pi v)^2}{\beta^2(\mathcal{J}_q^2 + \mathcal{K}_q^2)\cos^2(\pi v(1/2 - \imath t/\beta))} \tag{5.111}$$

[8] The minus sign originates from the chain rule when differentiating with respect to t_2 under time-translation invariance. Since the system depends only on $t \equiv t_1 - t_2$, we have $\partial_{t_2} = -\partial_t$, yielding: $\frac{\partial}{\partial t_2} f(t_1 - t_2) = \underbrace{\frac{df}{dt}}_{\text{Function derivative}} \cdot \underbrace{\frac{\partial}{\partial t_2}(t_1 - t_2)}_{=-1} = -f'(t).$

with following closure relation

$$\pi\nu = \beta\sqrt{\mathcal{J}_q^2 + \mathcal{K}_q^2}\cos(\pi\nu/2). \tag{5.112}$$

- $\kappa = 2$: The differential equation is given by

$$\ddot{g}^+(t) = -2\mathcal{K}_{2q}^2 e^{2g^+(t)} - 2\mathcal{J}_q^2 e^{g^+(t)}, \tag{5.113}$$

whose solution is given by

$$e^{g^+(t)} = \frac{1}{2\left(\beta\mathcal{J}_q\right)^2}\frac{(\pi\nu)^2}{1+\sqrt{B^2+1}\cos(\pi\nu(1/2-\imath t/\beta))} \tag{5.114}$$

where we define

$$B \equiv \frac{\pi\nu\beta\mathcal{K}_{2q}}{\sqrt{2}\left(\beta\mathcal{J}_q\right)^2}, \tag{5.115}$$

and the closure relation is given by

$$\pi\nu = \sqrt{2\left(\beta\mathcal{K}_{2q}\right)^2 + \left(\frac{2\left(\beta\mathcal{J}_q\right)^2}{\pi\nu}\right)^2}\cos(\pi\nu/2) + \frac{2\left(\beta\mathcal{J}_q\right)^2}{\pi\nu}. \tag{5.116}$$

Therefore, we have successfully solved all three of the uniform chains at equilibrium.

NOTE: For a comprehensive discussion of scaling arguments—including the standard expectation that Hamiltonian terms with fewer fermion operators dominate low-energy physics—we refer readers to Sections III.D and VII.F of Ref. [10]. With this context established, we now define the current operator for our SYK chain chains.

5.4.3 Current

In this section, we establish fundamental relations for electrical currents in correlated quantum systems such as our SYK chains. Our derivation proceeds through three interlinked stages: (1) Microscopic Current Definition: We begin with the continuity equation, which defines the current operator via charge conservation. (2) Linear Response Framework: Applying stability analysis to equilibrium states, we compute current response to weak perturbations using time-dependent Hamiltonian deformations and Kubo's formula. (3) Conductivity: This allows us to evaluate the dynamical conductivity as frequency-dependent (ω) current response which is expressed via current-current correlation functions. The DC conductivity is obtained by taking the limit $\omega \to 0$. Accordingly, the DC resistivity is obtained by the inverse of DC conductivity.

5.4.3.1 Continuity Equation and the Definition of Current

Charge conservation in quantum many-body systems is encoded in the continuity equation. For a lattice with sites indexed by i, the time evolution of the local charge density $\mathcal{Q}_i$ (intensive in N) is governed by

$$\partial_t \mathcal{Q}_i(t) = -\nabla j^{\mathcal{Q}} \approx -\frac{j^{\mathcal{Q}}(x_i + \Delta x) - j^{\mathcal{Q}}(x_i)}{\Delta x} \qquad \text{(in } 1D) \tag{5.117}$$

where $j^{\mathcal{Q}}$ is the current density operator, $x_i = i/L$, $\Delta x = 1/L$ (lattice spacing a which we have set to unity without loss of generality) and L is the total number of sites (see Fig. 4.9). Note that $\mathcal{Q}_i$ and $j^{\mathcal{Q}}$ are intensive in N.

The time derivative of $\mathcal{Q}_i$ gives the current in the continuity equation and is governed by the Heisenberg's equation of motion (see footnote 2) where $\partial_t \mathcal{Q}_i(t) = \imath\,[\mathcal{H}, \mathcal{Q}_i(t)]$ where the local charge density transport remains intensive. However, note that for a single dot SYK $\mathcal{H}_q$ commutes with $\mathcal{Q}_i$ since it's a conserved quantity. Therefore the on-site interaction in Eq. (5.86) does not contribute to this commutator. However, the transport Hamiltonian connecting lattice sites $i-1$ with i as well as i with $i+1$ will contribute. We define this bond Hamiltonian connecting different lattice sites as

$$\mathcal{H}^{i,i+1} \equiv \mathcal{H}_{i\to i+1} + \mathcal{H}^{\dagger}_{i\to i+1}. \tag{5.118}$$

Thus, using physical picture that incoming current increases the local charge density while the outgoing current reduces it, the Heisenberg equation of motion simplifies to $\partial_t \mathcal{Q}_i(t) = \imath\left[\mathcal{H}^{i-1,i}, \mathcal{Q}_i\right] - \imath\left[\mathcal{H}^{i,i+1}, \mathcal{Q}_i\right]$ for the SYK chain Hamiltonian in Eq. (5.86). Thus the continuity equation centered at site i becomes

$$\frac{j^{\mathcal{Q}}(x_i + \Delta x) - j^{\mathcal{Q}}(x_i)}{\Delta x} = -\imath\left[\mathcal{Q}_i, \mathcal{H}^{i,i+1}\right] + \imath\left[\mathcal{Q}_i, \mathcal{H}^{i-1,i}\right] \tag{5.119}$$

Therefore, we have a natural identification for the current, namely $j^{\mathcal{Q}}(x_i)/\Delta x = -\imath\left[\mathcal{Q}_i, \mathcal{H}^{i-1,i}\right]$ where we use the definition for $\mathcal{H}^{i-1,i}$ from Eq. (5.118) ($i \to i-1$). Using the generalized Galtiskii-Migdal relation (see above Eq. (4.36)), we get[9]

$$-\imath\left[\mathcal{Q}_i, \mathcal{H}^{i-1,i}\right] = -\imath\,[\mathcal{Q}_i, \mathcal{H}_{i-1\to i}] - \imath\left[\mathcal{Q}_i, \mathcal{H}^{\dagger}_{i-1\to i}\right]$$

[9] Note that we can take derivative of $\mathcal{Q}_i$ in terms of lesser Green's function as in Eq. (4.16). Then the generalized Galitskii-Migdal relation follows because of the way derivative is performed, as explained properly below Eq. (4.33). Further, only the Hamiltonian terms consisting of sites $i-1$, i and $i+1$ contribute to the commutator, making the local charge density transport intensive.

$$= \frac{\iota r}{2N}\mathcal{H}_{i-1\to i} - \frac{\iota r}{2N}\mathcal{H}^{\dagger}_{i-1\to i} = \frac{I_i}{N} \tag{5.120}$$

where $r = \kappa q$ and we define the local current operator (extensive in N) as

$$\boxed{I_i \equiv \frac{\iota r}{2}\left(\mathcal{H}_{i-1\to i} - \mathcal{H}^{\dagger}_{i-1\to i}\right)}. \tag{5.121}$$

The total current is obtained by summing over lattice sites (making it extensive in L as well) to get

$$\boxed{I \equiv \sum_{i=1}^{L} I_i = \frac{\iota r}{2}\left(\mathcal{H}_{\to} - \mathcal{H}^{\dagger}_{\to}\right)} \tag{5.122}$$

where we borrowed the notation from Eq. (5.89), namely $\mathcal{H}_{\to} \equiv \sum_{i=1}^{L} \mathcal{H}_{i\to i+1}$. Recall that we have considered an open chain where the hopping term D connecting sites 0 with 1 and sites L with $L+1$ are zero. Accordingly, the continuity equation gets simplified to

$$\boxed{\frac{dQ_i(t)}{dt} = \frac{1}{N}(I_i - I_{i+1})} \tag{5.123}$$

where current coming to site i increases the local charge density at site i while the current going away from the site i reduces the local charge density at site i. We will use this relation later.

5.4.3.2 Linear Response and Dynamical Conductivity

We begin by explaining the linear response where we measure equilibrium response to weak perturbations. Consider a quantum system initially in thermal equilibrium. When weakly perturbed by a uniform electric field $\vec{E}(t)$ in a translationally invariant lattice, the induced current obeys linear response theory. For a d-dimensional system with periodic boundary conditions, the current in direction α at position $\mathbf{x}$ and time t is

$$I_\alpha(\mathbf{x}, t) = \int dt' \sum_{\mathbf{x}'} \sum_{\beta} \sigma^{\alpha\beta}\left(\mathbf{x} - \mathbf{x}', t - t'\right) E_\beta\left(\mathbf{x}', t'\right), \tag{5.124}$$

where $\sigma^{\alpha\beta}$ is the conductivity tensor, and spatial and temporal translation invariance restrict $\sigma^{\alpha\beta}$ to depend only on $\mathbf{x} - \mathbf{x}'$ and $t - t'$.

We are interested in studying direct current (DC) responses where we impose a uniform field which simplifies the equations. When the electric field is uniform and aligned along $x_1(\vec{E} = (E(t), 0, \ldots, 0))$, the response simplifies to

$$I_x(t) = \int_{-\infty}^{t} dt' \sigma(t - t') E(t'), \tag{5.125}$$

where $\sigma(t - t')$ is the longitudinal conductivity along the direction of the applied external field. Fourier transforming yields the frequency-dependent generalization of Ohm's law

$$I_x(\omega) = \sigma(\omega) E(\omega). \tag{5.126}$$

This defines the frequency-dependent conductivity $\sigma(\omega)$.

Kubo formula provides the microscopic foundation for studying phenomenological (mesoscopic) properties such as conductivity. The conductivity emerges from the linear response to a perturbation

$$\mathcal{H}_{\text{pert}}(t) = -E(t) X, \tag{5.127}$$

where X is the extensive (in N and L) polarization operator:

$$X(t) \equiv N \sum_{j=1}^{L} j \mathcal{Q}_j(t). \tag{5.128}$$

Physical interpretation of $\mathcal{Q}_j =$ is the local charge density at site j (intensive in N and L) and X represents the system's total dipole moment (extensive in N and L).

The total current I is fundamentally linked to the polarization dynamics

$$\boxed{\partial_t X(t) = \imath[\mathcal{H}, X] = I} \tag{5.129}$$

where $\mathcal{H}$ is the Hamiltonian of the system and I is the total current (as defined for SYK chains in Eq. (5.122)).

Proof We expand the commutator $\imath[\mathcal{H}, X] = \imath N \sum_{j=1}^{L} j \left[\mathcal{H}, \mathcal{Q}_j\right]$. But $\imath[\mathcal{H}, \mathcal{Q}_j] = \partial_t \mathcal{Q}_j(t)$. Then we use the continuity equation in Eq. (5.123) to get

$$\imath[\mathcal{H}, X] = \sum_{j=1}^{L} j \left(I_j - I_{j-1}\right). \tag{5.130}$$

By expanding the summation, we find a cancellation of many terms that leave us with the total current $I \equiv \sum_{j=1}^{L} I_j$. This gives us Eq. (5.129). Thus, current arises from dipole moment evolution.

We now use the Kubo's formula to calculate the current response to the applied electric field of the equilibrium chains. The general Kubo formula for the linear response of an operator $\boldsymbol{A}$ to a perturbation coupling to operator B is:

$$\langle A(t)\rangle = \langle A\rangle_0 - \imath \int_{-\infty}^{t} dt' \left\langle \left[A(t), B\left(t'\right)\right]\right\rangle_0 \lambda\left(t'\right), \tag{5.131}$$

where $\lambda\left(t'\right)$ is the time-dependent coupling and $\langle\ldots\rangle_0$ denotes expectation values with respect to the equilibrium states when the perturbation is off. Here, $A = I$ (total current) and $B = X$ (polarization), with $\lambda\left(t'\right) = -E$. Thus

$$\langle I(t)\rangle = \langle I\rangle_0 + \imath E \int_{-\infty}^{t} dt' \left\langle \left[I(t), X\left(t'\right)\right]\right\rangle_0 . \tag{5.132}$$

In equilibrium, $\langle I\rangle_0 = 0$, so

$$\langle I(t)\rangle = \imath E \int_{-\infty}^{t} dt' \left\langle \left[I(t), X\left(t'\right)\right]\right\rangle_0 . \tag{5.133}$$

Time-translational invariance implies the commutator depends only on $t-t'$. Setting $\tau = t - t'$, the kernel is

$$\tilde{\sigma}(\tau) = \imath\langle[I(\tau), X(0)]\rangle_0 \Theta(\tau), \tag{5.134}$$

where $\Theta(\tau)$ is the step function ensuring causality. Using time-translation invariance

$$\langle[I(\tau), X(0)]\rangle_0 = \langle[I(0), X(-\tau)]\rangle_0 = -\langle[X(-\tau), I(0)]\rangle_0. \tag{5.135}$$

Substituting, we get

$$\tilde{\sigma}(\tau) = \imath\left(-\langle[X(-\tau), I(0)]\rangle_0\right)\Theta(\tau) = i\langle[X(-\tau), I(0)]\rangle_0 \Theta(\tau). \tag{5.136}$$

Relabeling τ as t

$$\tilde{\sigma}(t) = -\imath\langle[X(-t), I(0)]\rangle_0 \Theta(t). \tag{5.137}$$

The conductivity kernel $\sigma(t)$ governing current response to a uniform electric field E is then defined via the intensive retarded commutator (where we drop the subscript label $\langle\ldots\rangle_0$ for brevity)

$$\sigma(t) \equiv -\Theta(t)\frac{\imath}{L}\langle[X(-t), I(0)]\rangle \tag{5.138}$$

where $I = \sum_x I_x$ (since we have applied the field along x-axis and are interested in the longitudinal response) is the total current operator (extensive in N and L) and X is the polarization operator as defined above (extensive in N and L). Time-translational invariance in equilibrium permits $\langle I_x(t)X(t-\tau)\rangle = \langle I_x(0)X(-\tau)\rangle$.

The current can be written as a convolution integral. The induced current follows:

$$I_i(t) = E\int_0^t dt'\sigma\left(t'\right) \tag{5.139}$$

with initial condition derived from energy density:

$$\sigma(0) = \frac{r^2}{4L}\langle\mathcal{H}_{\text{trans}}\rangle = -\frac{2}{L}\text{Im}\int_{-\infty}^{0} d\tau\langle I(0)I(\tau)\rangle \tag{5.140}$$

where $\mathcal{H}_{\text{trans}}$ is the translational coupling (see Eq. (5.90)). The derivation is shown in Appendix I.

Now we can obtain the differential equation for conductivity by taking time derivative of Eq. (5.138) where we can take $t > 0$ without loss of generality.

$$\begin{aligned}\dot{\sigma}(t) =& \frac{\imath\langle[I(-t), I(0)]\rangle}{L} = \frac{\imath\langle[I(0), I(t)]\rangle}{L}\\ =& -\frac{2}{L}\text{Im}\langle I(0)I(t)\rangle,\end{aligned} \tag{5.141}$$

where we used

$$\langle I(0)I(t)\rangle = \text{Tr}\{I(0)I(t)\rho\} = \text{Tr}\{(\rho I(t)I(0))^\dagger\} = \langle I(t)I(0)\rangle^\star \tag{5.142}$$

to get the last line. Integrating the equation yields

$$\boxed{\sigma(t) = -\frac{2}{L}\text{Im}\int_{-\infty}^{t} d\tau\langle I(0)I(\tau)\rangle}. \tag{5.143}$$

This exact expression holds for $t > 0$ and underpins frequency-domain conductivity $\sigma(\omega)$. Therefore, the universal form for $\sigma(t)$ depends solely on current-current correlations (derived in the next Sect. 5.4.4) where the initial condition for this differential equation is given by Eq. (5.140).

5.4.3.3 Derivation of DC Conductivity via Linear Response Theory

The dynamical conductivity $\sigma(\omega)$ is obtained through Fourier transformation of the kernel $\sigma(t)$, namely $\sigma(\omega) = \mathcal{F}[\sigma](\omega)$

$$\mathcal{F}[\sigma](\omega) = \sigma(\omega) = \int_{-\infty}^{\infty} d\tau e^{\imath\omega\tau}\sigma(\tau). \tag{5.144}$$

Integration by parts (with the assumption of boundary term vanishing) relates this to the susceptibility response function[10] $\boxed{\chi^R(\tau) \equiv \dot{\sigma}(\tau)}$[11]

$$\sigma(\omega) = -\frac{1}{i\omega}\int_{-\infty}^{\infty} d\tau e^{\imath\omega\tau}\chi^R(\tau). \tag{5.145}$$

The dissipative (real) part follows as[12]

$$\mathrm{Re}[\sigma(\omega)] = -\frac{1}{\omega}\mathrm{Im}\left[\mathcal{F}[\chi^R](\omega)\right] = -\frac{1}{\omega}\mathrm{Im}\left[\chi^R(\omega)\right], \tag{5.146}$$

where $\mathcal{F}[\chi^R](\omega) = \chi^R(\omega)$ for brevity (just like we did for $\sigma(\omega)$ above).

As derived below, for quantum systems in equilibrium, the fluctuation-dissipation theorem connects the retarded response $\chi^R(\omega)$ to current fluctuations:

$$\boxed{\frac{\mathrm{Im}\chi^R(\omega)}{\omega} = \underbrace{\frac{1-e^{-\beta\omega}}{\omega}}_{\text{quantum correction}} \mathrm{Im}\mathcal{F}\left[\Pi^R\right](\omega)}, \tag{5.147}$$

where $\Pi^R(t) \equiv \Theta(t)\frac{\langle I(t)I(0)\rangle}{\imath L}$ is the retarded current-current correlation. Then the DC conductivity is given by

$$\sigma_{\mathrm{DC}} = \lim_{\omega\to 0}\mathrm{Re}[\sigma(\omega)] = \lim_{\omega\to 0} -\frac{1}{\omega}\mathrm{Im}\left[\chi^R(\omega)\right] \tag{5.148}$$

where for $\omega \to 0$, we get from the fluctuation-dissipation theorem in Eq. (5.147) that $-\frac{\mathrm{Im}\chi^R(\omega)}{\omega} \approx -\beta\mathrm{Im}\mathcal{F}\left[\Pi^R\right](\omega) = -\frac{\beta}{L}\mathrm{Im}\int_0^\infty dt\frac{\langle I(t)I(0)\rangle}{\imath}$ (since the Fourier transform term $e^{\imath\omega t} \approx 1$). We use footnote 12 to get $-\frac{\mathrm{Im}\chi^R(\omega)}{\omega} = \frac{\beta}{L}\mathrm{Re}\int_0^\infty dt\langle I(t)I(0)\rangle$. Therefore, we get for the DC conductivity

$$\boxed{\sigma_{\mathrm{DC}} = \lim_{\omega\to 0}\mathrm{Re}[\sigma(\omega)] = \frac{\beta}{L}\mathrm{Re}\int_0^\infty \mathrm{d}t\langle I(t)I(0)\rangle = \frac{\beta}{L}\mathrm{Re}\int_{-\infty}^0 \mathrm{d}t\langle I(0)I(t)\rangle}. \tag{5.149}$$

This expresses DC conductivity as the time-integrated equilibrium current autocorrelation.

[10] Using the expression in Eq. (5.141) where we restore the Heaviside step function to remind that $t > 0$, the functional form of the response function is $\chi^R(\tau) \equiv \dot{\sigma}(\tau) = -\frac{2}{L}\Theta(t)\mathrm{Im}\langle I(0)I(t)\rangle$.

[11] $\mathcal{F}[\sigma](\omega) = \int \mathrm{d}\tau e^{\imath\omega\tau}\sigma(\tau) = \int \mathrm{d}\tau\left(\partial_\tau\frac{e^{\imath\omega\tau}}{\imath\omega}\right)\sigma(\tau) = -\int \mathrm{d}\tau\frac{e^{\imath\omega\tau}}{\imath\omega}\dot{\sigma}(\tau) = -\frac{1}{\imath\omega}\mathcal{F}[\dot{\sigma}](\omega)$

[12] If $z = a + \imath b$ and $t = \imath z$, then $\mathrm{Re}[t] = -\mathrm{Im}[z]$ and $\mathrm{Im}[t] = \mathrm{Re}[z]$. If $w = \frac{z}{\imath} = -\imath z$, then $\mathrm{Re}[w] = \mathrm{Im}[z]$ and $\mathrm{Im}[w] = -\mathrm{Re}[z]$.

Thus, we have successfully evaluated the dynamical conductivity in Eq. (5.143) and DC conductivity in Eq. (5.149). We define a master integral encapsulating current correlations for general coupling strength κ

$$\boxed{\mathcal{W}_\kappa(t) \equiv \int_{-\infty}^{t} d\tau \frac{2\langle I(0)I(\tau)\rangle}{\kappa N L}}. \tag{5.150}$$

This enables compact expressions for conductivities across all cases ($\kappa = \{1/2, 1, 2\}$):

$$\boxed{\sigma^{(\kappa)}(t) = -\kappa N \mathrm{Im}\mathcal{W}_\kappa(t), \quad \sigma_{\mathrm{DC}}^{(\kappa)} = \frac{\beta\kappa N}{2}\mathrm{Re}\mathcal{W}_\kappa(0)}. \tag{5.151}$$

So we focus on evaluating the main integral in Eq. (5.150) where we already know the solutions coresponding to the three chains that we evaluated above in Sect. 5.4.2.

Fluctuation-Dissipation Theorem

We derive the fluctuation-dissipation theorem used above in Eq. (5.147). We start with the definition of susceptibility response function

$$\chi^R(t) \equiv -\frac{2}{L}\Theta(t)\mathrm{Im}\langle I(0)I(t)\rangle. \tag{5.152}$$

The complex conjugate property also holds the same as in Eq. (5.142), which allows us to rewrite

$$\chi^R(t) = -\frac{2}{L}\Theta(t)\mathrm{Im}\langle I(0)I(t)\rangle = +\frac{2}{L}\Theta(t)\mathrm{Im}\langle I(t)I(0)\rangle. \tag{5.153}$$

Then we employ the definition of retarded current-current correlation

$$\Pi^R(t) \equiv \Theta(t)\frac{\langle I(t)I(0)\rangle}{\imath L}, \tag{5.154}$$

to get

$$\boxed{\chi^R(t) = 2\,\mathrm{Re}\Pi^R(t)}. \tag{5.155}$$

Recall that the superscript R denotes the retarded function, characterized by the inclusion of Θ-functions (Heaviside step functions) in its definition. The

(continued)

relation given in Eq. (5.155) holds generally, even without the restriction to retarded functions. So we continue without assuming the retarded condition where we have

$$\boxed{\chi(t) = 2\,\mathrm{Re}\Pi(t)}. \tag{5.156}$$

Then the central identity to establish fluctuation dissipation theorem is

$$\begin{aligned}\langle I(0)I(t)\rangle &= \mathrm{Tr}\,\{e^{-\beta\mathcal{H}}I(0)I(t)\} = \mathrm{Tr}\,\{I(t)e^{-\beta\mathcal{H}}I(0)\}\\ &= \mathrm{Tr}\,\{e^{-\beta\mathcal{H}}\underbrace{e^{\beta\mathcal{H}}I(t)e^{-\beta\mathcal{H}}}_{I(t-\imath\beta)}I(0)\}\\ &= \langle I(t-\imath\beta)I(0)\rangle\end{aligned} \tag{5.157}$$

which implies

$$\begin{aligned}\chi(t) =& \frac{2\,\mathrm{Im}\langle I(t)I(0)\rangle}{L} = \frac{\langle I(t)I(0)\rangle}{\imath L} - \frac{\langle I(0)I(t)\rangle}{\imath L}\\ &= \Pi(t) - \frac{\langle I(t-\imath\beta)I(0)\rangle}{\imath L}.\end{aligned} \tag{5.158}$$

Fourier transform implies

$$\begin{aligned}\chi(\omega) &= \int_{-\infty}^{\infty} \mathrm{d}t e^{\imath\omega t}\chi(t)\\ &= \mathcal{F}[\Pi](\omega) - \int_{-\infty}^{\infty} \mathrm{d}t e^{\imath\omega t}\frac{\langle I(t-\imath\beta)I(0)\rangle}{\imath L}\\ &= \mathcal{F}[\Pi](\omega) - e^{-\beta\omega}\int_{-\infty}^{\infty} \mathrm{d}t e^{\imath\omega(t-\imath\beta)}\frac{\langle I(t-\imath\beta)I(0)\rangle}{\imath L}\\ &= \mathcal{F}[\Pi](\omega) - e^{-\beta\omega}\int_{-\infty}^{\infty} \mathrm{d}\tau e^{\imath\omega\tau}\frac{\langle I(\tau)I(0)\rangle}{\imath L}\end{aligned} \tag{5.159}$$

which gives the fluctuation-dissipation theorem for the full (not just retarded) functions

$$\boxed{\frac{\mathrm{Im}\chi(\omega)}{\omega} = \frac{[1-e^{-\beta\omega}]}{\omega}\mathrm{Im}\mathcal{F}[\Pi](\omega)}. \tag{5.160}$$

(continued)

Now, we specialize to the case of retarded functions where we start by noticing the odd symmetry of the unrestricted $\chi(t)$ (where Eq. (5.142) is used)

$$\chi(t) = -\chi(-t). \tag{5.161}$$

Then using the definition of $\chi^R(t)$ from Eq. (5.153) to take the imaginary component of it Fourier transform

$$\mathrm{Im}\chi^R(\omega) = \int_0^\infty dt \sin(\omega t)\chi(t) = \frac{1}{2}\int_{-\infty}^\infty dt \sin(\omega t)\chi(t) = \frac{1}{2}\mathrm{Im}\chi(\omega), \tag{5.162}$$

where the $\Theta(t)$ in $\chi^R(t)$ restricts integration to $t \geq 0$, and the odd symmetry of $\chi(t)$ simplifies the full Fourier integral.

Then using the unrestricted definition of $\Pi(t)$ from Eq. (5.154) (i.e., without the $\Theta(t)$), we get its symmetry property as

$$\Pi(t)^\star = -\Pi(-t), \tag{5.163}$$

which implies (1) Real part: $\mathrm{Re}\Pi(t) = \frac{\Pi(t)-\Pi(-t)}{2}$, and (2) imaginary part: $\mathrm{Im}\Pi(t) = \frac{\Pi(t)+\Pi(-t)}{2\imath}$. Therefore, $\mathrm{Re}\Pi(-t) = -\mathrm{Re}\Pi(t)$ is an odd function while $\mathrm{Im}\Pi(-t) = \mathrm{Im}\Pi(t)$ is an even function in time.The Fourier transform of the retarded correlator $\Pi^R(t) = \Theta(t)\Pi(t)$ is:

$$\begin{aligned}
\mathrm{Im}\mathcal{F}\left[\Pi^R\right](\omega) &= \mathrm{Im}\int_0^\infty dt e^{\imath\omega t}\Pi(t) \\
&= \int_0^\infty dt \cos(\omega t)\mathrm{Im}\Pi(t) + \int_0^\infty dt \sin(\omega t)\mathrm{Re}\Pi(t) \\
&= \frac{1}{2}\int_{-\infty}^\infty dt \cos(\omega t)\mathrm{Im}\Pi(t) + \frac{1}{2}\int_{-\infty}^\infty dt \sin(\omega t)\mathrm{Re}\Pi(t)
\end{aligned} \tag{5.164}$$

where we used the symmetry properties of real and imaginary parts of $\Pi(t)$, $\cos(\omega t)$ and $\sin(\omega t)$ to extend the integration limits. Comparing against the imaginary part of the Fourier transform of unrestricted $\Pi(t)$, we get

$$\mathrm{Im}\mathcal{F}[\Pi](\omega) = \int_{-\infty}^\infty dt \cos(\omega t)\mathrm{Im}\Pi(t) + \int_{-\infty}^\infty dt \sin(\omega t)\mathrm{Re}\Pi(t), \tag{5.165}$$

we get

$$\mathrm{Im}\mathcal{F}[\Pi^R](\omega) = \frac{1}{2}\mathcal{F}[\Pi](\omega). \tag{5.166}$$

(continued)

Then plugging Eqs. (5.162) and (5.166) in Eq. (5.160), we get the desired fluctuation-dissipation theorem for the retarded functions that was is in the main text in Eq. (5.147):

$$\boxed{\frac{\mathrm{Im}\chi^R(\omega)}{\omega} = \frac{1-e^{-\beta\omega}}{\omega}\mathrm{Im}\mathcal{F}\left[\Pi^R\right](\omega)}. \tag{5.167}$$

5.4.4 Current-Current Correlations

The core quantity governing charge transport in our formalism is the current-current correlation function $\langle I(0)I(t)\rangle$. It appears as the key integrand in Eq. (5.150), determining the dynamical conductivity and the DC conductivity through Eq. (5.151). We are interested in evaluating the current-current correlation $\langle I(0)I(t)\rangle$ for the three chains ($\kappa = \{1/2, 1, 2\}$).

For physical times $t > 0$ (relevant to response functions), only the forward branch deformations of the Keldysh contour contributes (Fig. 2.1). This simplifies the analysis by switching off all backward contour deformations $\vec{\eta} = 0$ in Eq. (5.84), leaving us with the effective action

$$\begin{aligned}\frac{S_I}{|D|^2} &= S_I|_{\vec{\lambda}=0} + \frac{1}{2}\sum_{n>m}\left[\lambda_n\bar{\lambda}_m F(\tau_n-\tau_m) + \bar{\lambda}_n\lambda_m F(\tau_n-\tau_m)\right] \\ &\quad - \imath\sum_m\left[\bar{\lambda}_m+\lambda_m\right]\left(\int_{-\infty}^0 \mathrm{d}t Y(t)\right) + \frac{1}{2}\sum_n \lambda_n\bar{\lambda}_n X(0)\end{aligned} \tag{5.168}$$

Here, $\vec{\lambda} = \left(\{\lambda_n\}, \{\bar{\lambda}_m\}\right)$ denotes forward-branch deformations at times τ_n. The indices $\{n, m\}$ label deformation points along the forward contour. For all calculations in this work, two deformations suffice ($n, m \leq 2$). Higher-order moments (if needed) generalize straightforwardly.

The total action $S = S_0 + S_I$ partitions into: (1) Free action S_0: independent of $\vec{\lambda}, \vec{\eta}$ (contains no coupling terms), & (2) interacting action S_I: solely responsible for deformation dependence. Consequently $\partial_{\lambda_n} S = \partial_{\lambda_n} S_I$ (since $\partial_{\lambda_n} S_0 = 0$). The partition function satisfies $\mathcal{Z}|_{\vec{\lambda},\vec{\eta}=0} = 1$ (normalization at zero deformation).

Example of Energy Expectation

We now present a calculation for energy expectation as an example. To illustrate, we compute $\langle \mathcal{H}_{\to}(t) \rangle$—required for initial conditions like $\sigma(t = 0)$ in Eq. (5.140). Using the simplified action in Eq. (5.168). Functional derivative is given by

$$\langle \mathcal{H}_{\to}(t) \rangle = i\partial_{\lambda_1} \mathcal{Z}|_{\vec{\lambda}=0} . \tag{5.169}$$

Partition function expansion gives

$$\mathcal{Z} = \int \mathcal{D}\mathcal{G}\mathcal{D}\Sigma e^{-NLS(\vec{\lambda})[\mathcal{G},\Sigma]} \tag{5.170}$$

Action derivative is calculated as

$$\partial_{\lambda_1} \mathcal{Z}|_{\vec{\lambda}=0} = \int \mathcal{D}\mathcal{G}\mathcal{D}\Sigma \left(- NL \left[\partial_{\lambda_1} S_I(\vec{\lambda}) \right] \Big|_{\vec{\lambda}=0} \right) e^{-NLS(0)[\mathcal{G},\Sigma]}. \tag{5.171}$$

Then we use the fact that undeformed $\mathcal{Z}|_{\vec{\lambda}=0} = 1$ and use the linear terms in λ_m in Eq. (5.168) to get

$$\partial_{\lambda_1} S_I|_{\vec{\lambda}=0} = -\imath|D|^2 \int_{-\infty}^{0} dt' Y\left(t'\right). \tag{5.172}$$

Therefore, the final expectation value is given by

$$\boxed{\langle \mathcal{H}_{\to}(t) \rangle = -NL|D|^2 \int_{-\infty}^{0} dt' Y\left(t'\right)} . \tag{5.173}$$

This feeds directly into $\langle \mathcal{H}_{\text{trans}} \rangle$ (Eq. (5.90)), enabling computations like Eq. (5.140)—determining the initial condition for $\sigma(0)$ completely.

Then proceeding along the similar lines to compute multi-operator expectations like $\left\langle \mathcal{H}_{\to}(0)\mathcal{H}_{\to}^{\dagger}(t) \right\rangle$ for $t > 0$, we use functional derivatives of the partition function:

$$\left\langle \mathcal{H}_{\to}(0)\mathcal{H}_{\to}^{\dagger}(t) \right\rangle = -\partial_{\lambda_1} \partial_{\bar{\lambda}_2} \mathcal{Z} \quad (t > 0). \tag{5.174}$$

NOTE: There is a critical constraint: this expression only gives time-ordered correlators where operators appear chronologically from right to left. Attempting to compute $\langle \mathcal{H}^{\dagger}_{\rightarrow}(t)\mathcal{H}_{\rightarrow}(0)\rangle$ by swapping derivatives $\left(-\partial_{\bar{\lambda}_2}\partial_{\lambda_1}\mathcal{Z}\right)$ fails because

- Functional derivatives commute, but operators generally satisfy $\left[\mathcal{H}_{\rightarrow}(0), \mathcal{H}^{\dagger}_{\rightarrow}(t)\right] \neq 0$.
- Resolution: We enforce temporal ordering by associating: λ_1 with time $\tau_1 = 0$, $\bar{\lambda}_2$ with time $\tau_2 = t > 0$, ensuring $\tau_2 > \tau_1$ matches the operator sequence.

Then using the definition of current from Eq. (5.122), we get

$$\begin{aligned}\langle I(0)I(t)\rangle &= \frac{1}{4}\left\langle\left(ir\mathcal{H}_{\rightarrow}(0) - ir\mathcal{H}^{\dagger}_{\rightarrow}(0)\right)\left(ir\mathcal{H}_{\rightarrow}(t) - ir\mathcal{H}^{\dagger}_{\rightarrow}(t)\right)\right\rangle \\ &= \frac{r^2}{4}\left[\left\langle\mathcal{H}_{\rightarrow}(0)\mathcal{H}^{\dagger}_{\rightarrow}(t)\right\rangle + \left\langle\mathcal{H}^{\dagger}_{\rightarrow}(0)\mathcal{H}_{\rightarrow}(t)\right\rangle - \left\langle\mathcal{H}^{\dagger}_{\rightarrow}(0)\mathcal{H}^{\dagger}_{\rightarrow}(t)\right\rangle\right. \\ &\quad \left. - \langle\mathcal{H}_{\rightarrow}(0)\mathcal{H}_{\rightarrow}(t)\rangle\right].\end{aligned} \tag{5.175}$$

Here $\mathcal{H}_{\rightarrow}$ is defined in Eq. (5.89) which explicitly factors out transport couplings $|D|^2$, isolating dynamics in the correlators. Using the derivative expressions, we compactly write

$$\langle I(0)I(t)\rangle = -\frac{r^2}{4}\left(\partial_{\lambda_1}\partial_{\bar{\lambda}_2} + \partial_{\bar{\lambda}_1}\partial_{\lambda_2} - \partial_{\bar{\lambda}_1}\partial_{\bar{\lambda}_2} - \partial_{\lambda_1}\partial_{\lambda_2}\right)\mathcal{Z}\bigg|_{\vec{\lambda}=0}. \tag{5.176}$$

The evaluation of these derivatives are along the same lines as done above in the box for the energy expectation value. The calculations are long but straightforward and is provided in depth in Appendix D of Ref. [10] which we encourage the reader to consult if needed. The final result for $r/2$-body hopping (for us, $r = \kappa q$)

$$\boxed{\langle I(0)I(t)\rangle = +\frac{r^2}{4}NL|D|^2F(t)}. \tag{5.177}$$

We now specialize to our three chains where $r = \kappa q$ and $\kappa = \{1/2, 1, 2\}$. We have already evaluated F in Eq. (5.85) which we plug in here to get

$$\boxed{\langle I(0)I(t)\rangle_{\kappa} = +\frac{\kappa}{8}NL|\mathcal{D}|^2e^{\kappa g^+(t)}}, \tag{5.178}$$

where we used the definition of $|\mathcal{D}|^2 \equiv 2|D|^2e^{-2\kappa q\mathcal{Q}}$ with charge density $\mathcal{Q} = \mathcal{O}(1/\sqrt{q})$, ensures $|\mathcal{D}| = \mathcal{O}\left(q^0\right)$ at large q. Substituting into the conductivity integral (Eq. (5.150)):

$$\boxed{\mathcal{W}_{\kappa}(t) = \int_{-\infty}^{t} d\tau\frac{|\mathcal{D}|^2}{4}e^{\kappa g^+(\tau)}}. \tag{5.179}$$

This function $\mathcal{W}_\kappa(t)$ directly determines the dynamical conductivity and the DC conductivity via Eq. (5.151).

Therefore we have been successfully able to utilize the Keldysh contour deformations in calculating the transport energy expectation value in Eq. (5.173), the current-current correlation for any SYK-like setup in Eq. (5.177), as well as for our three chains in Eq. (5.178) where the conductivities can be evaluated by evaluating the integral in Eq. (5.179) which is plugged in Eq. (5.151). This all boils down to solving for the Green's function $g^+(t)$ that appears in Eq. (5.179) for various values of κ—a feat that we have already performed in Sect. 5.4.2. We utilize those solutions in the next section to evaluate the integral and calculate the DC resistivities of the three chains (simply given by the inverse of the DC conductivities).

5.4.5 DC Resistivity

Recall that the chain can be mapped to a dot via the mapping in Eq. (5.100). This allows us to use the exact expressions for $g^+(t)$[13] to leading-order in $1/q$, derived for $\kappa = 1/2$, 1 and 2 cases in Eqs. (5.107), (5.111) and (5.114), respectively and the closure relation given in Eqs. (5.109), (5.112) and (5.116), respectively. These solutions enable explicit determination of the function $g^+(t)$, which directly yields the current-current correlation function in Eq. (5.178) via Eq. (5.179) in closed analytical form across all temperatures. So, all we have to do is to evaluate the integral in Eq. (5.179) for all three cases of κ corresponding to their $g^+(t)$.

Evaluating the integrals is a bit involved, however, this has been discussed in detail and shown explicitly in Section VII of Ref. [10] where in Section VII.A, a general framework is provided for the integral of interest. We briefly mention the technique here, however we recommend the reader to refer to the aforementioned section in Ref. [10]. We introduce the special function $\Upsilon_{s,\gamma}(\theta)$, a cornerstone for evaluating key integrals in our analysis. Defined via polylogarithms $\mathrm{Li}_s(z)$,

$$\Upsilon_{s,\gamma}(\theta) \equiv i\left[\mathrm{Li}_s\left(-e^{\theta+\iota\gamma}\right) - \mathrm{Li}_s\left(-e^{\theta-\iota\gamma}\right)\right], \tag{5.180}$$

it exhibits a small-γ expansion $\Upsilon_{s,\gamma}(\theta) = -2\mathrm{Li}_{s-1}\left(-e^{\theta}\right)\gamma + \mathcal{O}\left(\gamma^3\right)$. The core properties of $\mathrm{Li}_s(z)$ are

- Definitions & Recursion:
  - $\mathrm{Li}_0(z) = z/(1-z)$, $\mathrm{Li}_1(z) = -\ln(1-z)$, with recursion $\partial_\theta \mathrm{Li}_{s+1}\left(ae^{\theta}\right) = \mathrm{Li}_s\left(ae^{\theta}\right)$.
  - Series: $\mathrm{Li}_s(z) = \sum_{k=1}^{\infty} z^k/k^s$ for $|z| < 1$.
- Boundary Behaviors:
  - $\mathrm{Li}_s(0) = 0 \Rightarrow \Upsilon_{s,\gamma}(-\infty) = 0$.

[13] The differential equation for all three chains are given by in an unified manner $\ddot{g}^+(t) = -2|\mathcal{D}|^2 e^{\kappa g^+(t)} - 2\mathcal{J}^2 e^{g^+(t)}$.

– As $\mathrm{Re}(z) \to \infty$ and $s \geq 0$: $\mathrm{Li}_s(e^z) \to -z^s/s!$, leading to

$$\Upsilon_{s,\gamma}(\theta) \underset{\theta\to\infty}{\longrightarrow} \imath \frac{(\theta - \imath\gamma)^s - (\theta + \imath\gamma)^s}{s!}.$$

- Integration Identity: Anti-derivatives satisfy

$$\int d\theta \Upsilon_{s,\gamma}(\theta) = \Upsilon_{s+1,\gamma}(\theta),$$

enabling recursive evaluation of integrals.

We provide the explicit solutions which will be critical for computations of resistivities below:

- $s = 0$:

$$\Upsilon_{0,\gamma}(\theta) = \frac{\tan\gamma}{1 + \sqrt{1 + \tan^2\gamma}\cosh\theta}. \tag{5.181}$$

- $s = 1$:

$$\Upsilon_{1,\gamma}(\theta) = \gamma + 2\tan^{-1}[\tan(\gamma/2)\tanh(\theta/2)]. \tag{5.182}$$

The small-γ approximation yields $\Upsilon_{1,\gamma}(\theta) \approx [1 + \tanh(\theta/2)]\gamma + \mathcal{O}(\gamma^3)$.

The special function $\Upsilon_{s,\gamma}(\theta)$ systematizes complex integrals arising in correlation functions. Its recursive integrability (via $s \to s+1$) and exact solutions for $s = 0, 1$ provide analytical traction for asymptotic and numerical work. The boundary conditions ensure well-behaved limits in physical applications. For practical calculations, we recommend to focus on the integration identity and explicit $s = 0, 1$ forms—these are practical workhorses. The asymptotics anchor sanity checks for large-θ regimes.

We now evaluate the integrals across all temperature ranges for all three cases while explicitly deriving the relations for low-temperature limit which is what we are typically interested in condensed matter physics. As mentioned above, we refer the reader to Section VII of Ref. [10] for details of computation.

- $\kappa = \frac{1}{2}$: The integral in Eq. (5.179) gets evaluated to the following closed form that holds across all temperature regimes (recall the definitions of $\mathcal{J}$ and $\mathcal{D}$ from Eq. (5.99))

$$\begin{aligned}\mathcal{W}_{1/2}(t) &= \int_{-\infty}^{t} d\tau \frac{|\mathcal{D}|^2}{4} e^{g^+(\tau)/2} = \int_{-\infty}^{t} d\tau \frac{1}{4}\frac{d\theta}{d\tau}\frac{\pi\nu T}{\tan\gamma_{1/2}}\Upsilon_{0,\gamma_{1/2}}(\theta) \\ &= \frac{1}{4}\frac{\pi\nu T}{\tan\gamma_{1/2}}\int d\theta \Upsilon_{0,\gamma_{1/2}}(\theta) = \frac{1}{4}\frac{\pi\nu T}{\tan\gamma_{1/2}}\Upsilon_{1,\gamma_{1/2}}(\theta),\end{aligned} \tag{5.183}$$

where

$$\tan \gamma_{1/2} \equiv \pi \nu T \mathcal{J} |\mathcal{D}|^{-2}, \quad \theta \equiv \pi \nu T t + \imath \pi \nu/2 \tag{5.184}$$

and $\Upsilon_{0,\gamma_{1/2}}(\theta)$ and $\Upsilon_{1,\gamma_{1/2}}(\theta)$ are given in Eqs. (5.181) and (5.182), respectively. The subscript 1/2 on $\mathcal{W}$ and γ denote that we are solving for the $\kappa = \frac{1}{2}$ case.

We now specialize to the low-temperature limit where the closure relation for ν in Eq. (5.109) yields

$$\nu \xrightarrow{\text{low-temperature}} 2 - 2\alpha_{1/2} T + \mathcal{O}(T^2), \tag{5.185}$$

where (subscript 1/2 again denote $\kappa = \frac{1}{2}$ case)

$$\alpha_{1/2} \equiv \frac{2}{|\mathcal{D}|} \sqrt{2 + |\mathcal{J}/\mathcal{D}|^2}. \tag{5.186}$$

Furthermore, we use the small-$\gamma_{1/2}$ expansion (which implies low-temperature regimes) for $\Upsilon_{1,\gamma_{1/2}}(\theta)$ using the expression below Eq. (5.182) and use the limit $\lim_{x \to 0} x/\tan(x) = 1$ to get

$$\begin{aligned} \mathcal{W}_{1/2}(t) &= \frac{1}{4} \pi \nu T \left(1 + \tanh(\theta/2)\right) \\ &= \frac{1}{4} \pi \nu T \left(1 + \tanh(\pi \nu T t/2 + \imath \pi \nu/4)\right). \end{aligned} \tag{5.187}$$

Plugging in Eq. (5.151) to calculate the DC conductivity, we get

$$\sigma_{\text{DC}}^{(\kappa=1/2)} = \beta N \text{Re} \mathcal{W}_{1/2}(0)/4 = \frac{N \pi \nu}{16}, \tag{5.188}$$

where we used the identity $\tanh(\imath x) = \imath \tan(x)$. Taking the low temperature expansion for ν, we get

$$\sigma_{\text{DC}}^{(\kappa=1/2)} \xrightarrow{\text{low-}T} \frac{N\pi}{8} \left(1 - \alpha_{1/2} T\right). \tag{5.189}$$

Thus, the DC resistivity at low temperatures become

$$\boxed{\rho_{\text{DC}}^{(\kappa=1/2)} = \frac{8}{N\pi} \left(1 + \alpha_{1/2} T\right)} \tag{5.190}$$

which is linear-in-temperature at low temperatures—a signature of strange metallic phase of matter! (More on this below.)

Furthermore, we find a universal DC resistivity, independent of coupling constants $\mathcal{J}$ and $|\mathcal{D}|$, at zero-temperature (which also happens to be the residual resistivity)

$$\boxed{\rho_{\text{DC}}^{\text{min}} = \frac{8}{N\pi}}. \tag{5.191}$$

- $\kappa = 1$: Evaluating the integral leads to the following expression for Eq. (5.179) that holds across all temperatures

$$\mathcal{W}_1(t) = \frac{\pi \nu T}{4|\mathcal{J}/\mathcal{D}|^2 + 4} \left(1 + \imath \tan\left[\pi \nu (1/2 - \imath t/\beta)\right]\right), \tag{5.192}$$

giving us the following for DC conductivity:

$$\sigma_{\text{DC}}^{(\kappa=1)} = \beta N \text{Re}\mathcal{W}_1(0)/2 = \frac{N\pi\nu}{8(|\mathcal{J}/\mathcal{D}|^2 + 1)}. \tag{5.193}$$

Note that this holds across all temperatures.

Now, we go to the low-temperature limit where using the closure relation for ν from Eq. (5.112), we get

$$\nu \xrightarrow{\text{low-temperature}} 1 - \alpha_1 T + \mathcal{O}(T^2) \tag{5.194}$$

where (subscript 1 denotes $\kappa = 1$ case)

$$\alpha_1 \equiv \frac{2}{|\mathcal{D}|} \frac{1}{\sqrt{1 + |\mathcal{J}/\mathcal{D}|^2}}. \tag{5.195}$$

Thus, we have

$$\sigma_{\text{DC}}^{(\kappa=1)} \xrightarrow{\text{low-}T} \frac{N\pi}{8(|\mathcal{J}/\mathcal{D}|^2 + 1)} (1 - \alpha_1 T). \tag{5.196}$$

Inverting this at low temperatures gives

$$\boxed{\rho_{\text{DC}}^{(\kappa=1)} = \frac{8(|\mathcal{J}/\mathcal{D}|^2 + 1)}{N\pi} (1 + \alpha_1 T)}. \tag{5.197}$$

This is also linear-in-temperature, characteristic of a strange metal, the same as we obtained for the $\kappa = \frac{1}{2}$ case above in Eq. (5.190).

Furthermore, we see that the resistivity achieves its minimum value at $T = 0$ (which also happens to be the residual resistivity) for $\mathcal{J} = 0$ and $\forall\ \mathcal{D}$, given by

$$\boxed{\rho_{\text{DC}}^{\text{min}} = \frac{8}{N\pi}}. \tag{5.198}$$

whose value matches exactly with $\kappa = \frac{1}{2}$ case in Eq. (5.191).

- $\kappa = 2$: We get for Eq. (5.179) the following:

$$\mathcal{W}_2(t) = \frac{\pi \nu T}{8}\left[1 + \frac{\sinh(\theta)}{\cos\gamma_2 + \cosh\theta} - \frac{1}{\tan\gamma_2}\Upsilon_{1,\gamma_2}(\theta)\right] \tag{5.199}$$

where

$$\tan\gamma_2 \equiv \pi\nu T\frac{|\mathcal{D}|}{\sqrt{2}\mathcal{J}^2}, \qquad \theta \equiv \pi\nu T t + \imath\pi\nu/2 \tag{5.200}$$

and subscript 2 on $\mathcal{W}$ and γ denote the $\kappa = 2$ case. Note that $\cos\gamma_2 = \frac{1}{\sqrt{1+\tan^2\gamma_2}}$ and $\Upsilon_{1,\gamma_2}(\theta)$ is defined in Eq. (5.182). This relation holds true for all temperature ranges, which accordingly decides the DC conductivity via Eq. (5.151) for all temperature ranges.

Now, we go to the low-temperature regime where for we can expand $\Upsilon_{1,\gamma_2}(\theta)$ for small-γ_2 value (as done below Eq. (5.182)). This also helps simplify $\tan\gamma_2 \approx \gamma_2$. Furthermore the closure relation for ν in Eq. (5.116) admits the following low-temperature expansion:

$$\nu \xrightarrow{\text{low-temperature}} 2 - 2\alpha_2 T + \mathcal{O}(T^2) \tag{5.201}$$

where

$$\alpha_2 \equiv \frac{\sqrt{2}}{\mathcal{J}}\sqrt{2 + |\mathcal{D}/\mathcal{J}|^2}. \tag{5.202}$$

Therefore, we get the following DC conductivity at low-temperatures

$$\sigma_{\text{DC}}^{(\kappa=2)} = \frac{N\pi\nu}{8}\left(1 - \frac{\gamma_2}{\tan\gamma_2}\right) = \frac{N\pi\nu\gamma_2^2}{24} + \mathcal{O}(\gamma_2^4) = \frac{N\pi^3 T^2|\mathcal{D}|^2}{12\mathcal{J}^4} + \mathcal{O}(T^3). \tag{5.203}$$

Inverting, we get the following DC resistivity behavior

$$\boxed{\rho_{\text{DC}}^{(\kappa=2)} \sim T^{-2}}, \tag{5.204}$$

which is an insulating model, unlike $\kappa = 1/2, 1$ cases.

Furthermore, we note from the first equality in Eq. (5.203) that the highest universal DC conductivity happens when $\gamma_2/\tan(\gamma_2) = 0$ which implies $\gamma_2 = \pi(n - 1/2)$ for $n \in \mathbb{Z}$. This happens at finite (non-zero) temperatures (unlike $\kappa = 1/2, 1$ cases), so in this particular case, the universal maximum DC conductivity does not match with the residual conductivity (which goes to zero as $T \to 0$).

The universal value is

$$\sigma_{\text{DC}}^{\max} = \frac{N\pi}{8} \quad \Rightarrow \boxed{\rho_{\text{DC}}^{\min} = \frac{8}{N\pi}}. \tag{5.205}$$

The value matches the ones in $\kappa = 1/2, 1$ cases even though here it's obtained at a finite (non-zero) temperature.

NOTE: We would like to bring attention to a formal mathematical connection between $\kappa = 1/2$ and $\kappa = 2$ cases. Both are solved above in Sect. 5.4.2 where a mathematical symmetry exists between the $\kappa = 1/2$ and $\kappa = 2$ cases: the solution for $\kappa = 2$ is obtained from the $\kappa = 1/2$ solution via the transformation

$$(g^+, \mathcal{K}_{q/2}^2, \mathcal{J}_q^2)\Big|_{\kappa=1/2} \to 2\left(g^+, \mathcal{J}_q^2, \mathcal{K}_{2q}^2\right)\Big|_{\kappa=2}. \tag{5.206}$$

Crucially, this symmetry is purely algebraic and does not extend to the physical behavior. As our analysis demonstrates, the transport properties of these systems are fundamentally distinct:

- The $\kappa = 1/2$ system exhibits conducting behavior (finite DC conductivity) for the temperature all the way down to zero.
- The $\kappa = 2$ system behaves as an insulator (vanishing DC conductivity) for $T \to 0$.

We have derived closed-form expressions for DC resistivity $\rho_\kappa(T)$ in the thermodynamic limit, valid across all temperatures for the three SYK chain models ($\kappa = 1/2, 1, 2$). Having established the low-temperature behavior and providing explicit closed form expressions for DC resistivities for all three chains, we now analyze the full temperature dependence of $\rho^{(\kappa)}(T)$, unify microscopic insights from our solutions, and extract universal physical implications for transport in strongly correlated non-Fermi liquids (in particular strange metals).

5.4.6 DC Resistivities Across All Temperatures

Leveraging our derived DC conductivity expressions via Eq. (5.151) where $\mathcal{W}_\kappa$ is explicitly evaluated across all temperature ranges in Eq. (5.183) for $\kappa = 1/2$, Eq. (5.192) for $\kappa = 1$, and Eq. (5.199). Crucially, these results incorporate the exact closure relations for the variable ν (Eqs. (5.109), (5.112), (5.116) for $\kappa = 1/2, 1, 2$, respectively), which remain valid across all temperatures, including the $T \to 0$ limit.

The qualitative temperature dependence of resistivity remains unchanged by the ratio $|\mathcal{D}|/\mathcal{J}$, but larger values of this ratio systematically reduce the resistivity to the minimum resistivity $\rho_{\text{DC}}^{\min}$. In the limit of the fraction tending to ∞, $\rho_{\text{DC}}^{\min}$ universally approaches the value $8/(N\pi)$ for all three models ($\kappa = 1/2, 1, 2$). Crucially, for sufficiently small value of the ratio $|\mathcal{D}|/\mathcal{J}$, the resistivities at low temperatures fall below the Mott-loffe-Regel (MIR) bound:

$$\rho_{\text{DC}}^{\text{MIR}} = \frac{2\pi}{N} \quad \text{(for lattice spacing } a = 1, \text{ flavor charge } q_e = 1) , \tag{5.207}$$

signifying a true strange metal phase (linear-in-T resistivity without quasiparticles below the Mott-Ioffe-Regel (MIR) bound of DC resistivity). The observed linear-in-T resistivity—without quasiparticles and below the MIR bound—signatures a true strange metal phase. This contrasts with "bad metals", where resistivity exceeds the MIR limit due to incoherent transport. Therefore, we have strange metal ($\rho \sim T$ and quasiparticles are absent) below the MIR bound while bad metal (also $\rho \sim T$ without quasiparticles) above the MIR bound. There are no characteristic change in a phenomenological way between the two phases, rather this is a theoretical bound above which a loss of quasiparticle coherence happens. Achieving strange metallicity below the MIR bound is a hallmark of non-Fermi liquid physics, as it violates semiclassical transport expectations while maintaining Planckian dissipation. We do not wish to go in any further details, but rather refer the reader to Ref. [5] for a nice review on MIR bound, Ref. [6] for Planckian dissipation in metals, and Sections I, VII.E, VIII of Ref. [10] for MIR bounds in SYK chains.

Now we proceed to plot (for a given set of parameters) the DC resistivity in Fig. 5.1 for the three chains across all temperatures using the expressions above. A detailed implementation for obtaining the plots of normalized DC resistivities for all three chains across all temperature ranges is provided in Appendix J.

As evident from Fig. 5.1, we find that $\kappa = 1/2$ and $\kappa = 1$ cases have linear-in-temperature resistivity all the way down to $T \to 0$ where there exists a true strange metallic phase for temperature ranges where the resistivity is below the MIR bound, while acting as bad metals above the MIR bound. There are no characteristic phenomenological change across this crossover.

The $\kappa = 2$ chain exhibits a remarkable four-stage temperature evolution (Fig. 5.1):

1. Insulator: $\rho \sim T^{-2}$ as $T \to 0$.
2. Fermi-Liquid Crossover: Quadratic dependence $\rho \sim (T - T_{\text{min}})^2$ near T_{min} (resistivity minimum).
3. Strange Metal: Linear $\rho \sim T$ below ρ_{MIR}.
4. Bad Metal: Linear $\rho \sim T$ above ρ_{MIR} at high T.

This sequence resolves a key challenge in correlated systems: the smooth connection between Fermi-liquid, strange metal, and bad-metal regimes without abrupt crossovers—addressing the puzzle noted in Ref. [2] regarding the elusive continuity between low-T non-Fermi liquids and high-T incoherent transport.

5.4.6.1 Hints of Holography*

The resistivity of our insulating system exhibits a distinct power-law scaling $\rho(T) \propto T^{-2}$ at low temperatures. This contrasts sharply with conventional insulators, where

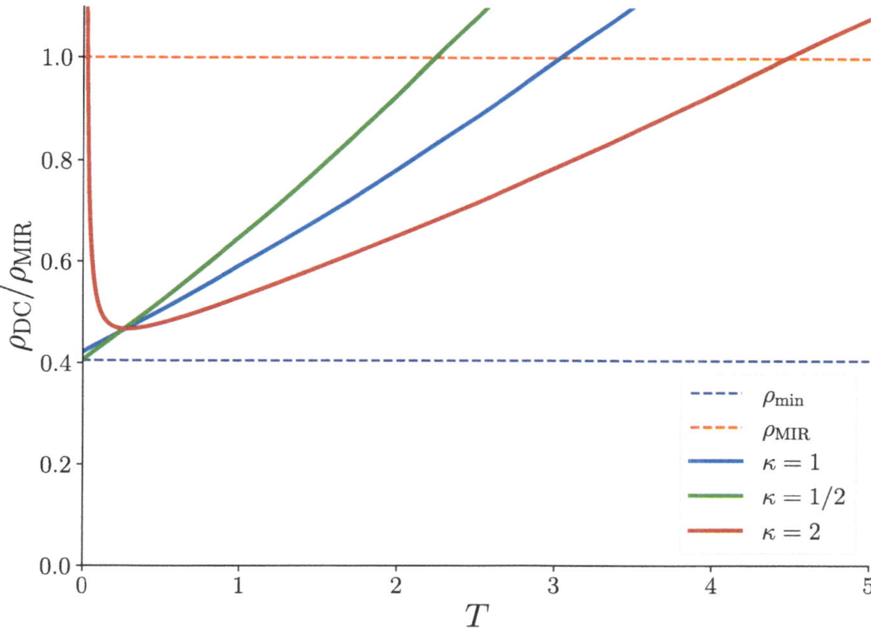

Fig. 5.1 Resistivity normalized to the Mott-Ioffe-Regel (MIR) bound versus temperature for the three universality classes ($\kappa = 1/2, 1, 2$) at fixed couplings $\mathcal{J} = 1$, $|\mathcal{D}| = 5$. The $\kappa = 1/2$ and $\kappa = 1$ chains exhibit robust linear-in-T resistivity characteristic of strange metals across all temperatures (including below the MIR bound) and they become "bad metals" at higher temperatures above the MIR bound (see the text for more details). In contrast, the $\kappa = 2$ system shows insulating behavior ($d\rho/dT < 0$) at low temperatures but crosses over to linear-in-T resistivity at higher T. A universal minimum resistivity $\rho_{\rm min} = 8/(N\pi)$ is achieved for all couplings in the $\kappa = 1/2$ case, while $\kappa = 1$ and $\kappa = 2$ reach this bound only at specific parameter values. Here, at $|\mathcal{D}| = 5$, $\mathcal{J} = 1$, the latter two systems operate above $\rho_{\rm min}$, while $\kappa = 1/2$ achieves the minimum value. The implementation is provided in Appendix J

resistivity follows $\rho(T) \sim e^{\Delta/T}$ due to thermal activation of carriers across an energy gap Δ.

Such T^{-2} scaling is characteristic of *holographic insulators*, which emerge from gauge/gravity duality frameworks. In our model, this behavior arises because the system flows to a conformal field theory (CFT) at low energies. The absence of an intrinsic energy scale in this critical state-a hallmark of scale invariance-likely enables the observed power-law transport. This is an active area of research and we refer the reader to the literature [1, 7, 12].

With this, we conclude the chapter by briefly summarizing what we did. This chapter investigated non-equilibrium dynamics and transport in SYK systems through quench protocols and Keldysh formalism. For isolated SYK dots, a quench to a single large-q Hamiltonian yields *instantaneous thermalization of Green's functions*, where time-translational invariance and the KMS condition emerge immediately post-quench, defining an effective temperature. In contrast, SYK chains exhibit finite thermalization times due to spatial charge transport. A proof by contradiction showed energy density constraints prevent instantaneous equilibrium, and charge dynamics follow a wave like equation, independent of on-site interactions. Keldysh contour deformations enable exact computation of energy densities and current-current correlations. This framework reveals $\langle I(0)I(t)\rangle_\kappa$ for three chain types ($\kappa = \{1/2, 1, 2\}$) whose transport properties and DC resistivities were studied in details, where we found the DC resistivity showing distinct universality classes for the three chains (summarized in Fig. 5.1).

References

1. Andrade, T., Krikun, A., Schalm, K., Zaanen, J.: Doping the holographic Mott insulator. Nat. Phys. **14**, 1049–1055 (2018)
2. Chowdhury, D., Georges, A., Parcollet, O., Sachdev, S.: Sachdev-Ye-Kitaev models and beyond: a window into non-Fermi liquids. Rev. Mod. Phys. **94**(3), 035004 (2022). arXiv:2109.05037 [cond-mat, physics:hep-th, physics:quant-ph]
3. Eberlein, A., Kasper, V., Sachdev, S., Steinberg, J.: Quantum quench of the Sachdev-Ye-Kitaev model. Phys. Rev. B **96**(20), 205123 (2017)
4. Gogolin, C., Eisert, J.: Equilibration, thermalisation, and the emergence of statistical mechanics in closed quantum systems. Rep. Prog. Phys. **79**(5), 056001 (2016). Publisher: IOP Publishing
5. Hartnoll, S.A.: Theory of universal incoherent metallic transport. Nat. Phys. **11**, 54–61 (2015)
6. Hartnoll, S.A., Mackenzie, A.P.: Colloquium: planckian dissipation in metals. Rev. Mod. Phys. **94**(4), 041002 (2022)
7. Horowitz, G.T., Santos, J.E., Tong, D.: Optical conductivity with holographic lattices. J. High Energy Phys. **2012**(7), 168 (2012)
8. Jaramillo, S.S., Jha, R., Kehrein, S.: Thermalization of a closed Sachdev-Ye-Kitaev system in the thermodynamic limit. Phys. Rev. B **111**(19), 195153 (2025)
9. Jha, R., Louw, J.C.: Dynamics and charge fluctuations in large-q Sachdev-Ye-Kitaev lattices. Phys. Rev. B **107**(23), 235114 (2023)
10. Jha, R., Kehrein, S., Louw, J.C.: Current correlations and conductivity in SYK-like systems: an analytical study. Phys. Rev. B **111**(4), 045111 (2025)
11. Louw, J.C., Kehrein, S.: Thermalization of many many-body interacting Sachdev-Ye-Kitaev models. Phys. Rev. B **105**(7), 075117 (2022)
12. Mefford, E., Horowitz, G.T.: Simple holographic insulator. Phys. Rev. D **90**(8), 084042 (2014)
13. Tsutsumi, M.: On solutions of Liouville's equation. J. Math. Anal. Appl. **76**(1), 116–123 (1980)

Conclusion and Outlook

6

Abstract

This concluding chapter synthesizes the Sachdev-Ye-Kitaev (SYK) paradigm as a transformative framework for strongly correlated quantum matter that unifies analytical tractability, maximal quantum chaos, and holographic duality. We summarize how the SYK model resolves pivotal challenges in non Fermi liquid physics, particularly the strange metal phase, while serving as a pedagogical laboratory for modern many body techniques including disorder averaging, replica symmetry, Schwinger-Dyson and Kadanoff-Baym equations, Keldysh formalism, and contour deformations. Key achievements across the preceding chapters are reviewed, including emergent conformal symmetry and Schwarzian actions in both Majorana and complex fermion variants, mean-field critical exponents matching those of charged Anti-de Sitter black holes and van der Waals fluids, saturation of the Maldacena-Shenker-Stanford quantum chaos bound at low temperatures, and transport signatures ranging from linear-in-temperature strange metal resistivity to insulating and bad metallic regimes in SYK chains. The chapter outlines two pivotal research frontiers: incorporating bosonic degrees of freedom via Yukawa SYK frameworks to study superconducting instabilities and higher-dimensional transport, and extending to open quantum dynamics where decoherence demands noise resilient diagnostics for entanglement and chaos. We emphasize the robustness of SYK universality under spatial extension and competing interactions, positioning the model as a computational microscope for decoding quantum chaos, criticality, and holographic duality in strongly correlated systems.

This review has synthesized the SYK paradigm as a transformative framework for strongly correlated quantum matter. By unifying analytical tractability, maximal chaos, and holographic duality, the SYK model resolves pivotal challenges in non-Fermi liquid physics, particularly the enigmatic strange metal phase, while

R. Jha, *Theory of Strongly Correlated Systems*, Lecture Notes in Physics 1051,
https://doi.org/10.1007/978-3-032-20910-8_6

offering a pedagogical laboratory for advanced many-body techniques. We conclude by summarizing key insights, methodological advances, and emergent research frontiers.

The SYK model's power stems from its simultaneous capture of strong correlations without quasiparticles, maximal quantum chaos, and emergent conformal symmetry, addressing the breakdown of Landau's Fermi liquid theory in strange metals. As demonstrated throughout this work, we uncovered the universal thermodynamics and chaotic dynamics at low temperatures where both Majorana (Chap. 2) and complex SYK variants (Chap. 3) exhibit emergent reparameterization symmetry, described by Schwarzian actions as well as uncovered non-Fermi liquid physics, in particular the strange metallic phase of matter, such as violation of equilibration rate being proportional to T^2, reproduce hallmark strange metal signatures: linear-in-T resistivity ($\rho \propto T$), among others. Finally, we highlighted the connection to the classical gravity dual via the gauge-gravity duality (holography) by highlighting the similarities of SYK thermodynamics with charged AdS black holes, chaotic bounds, while transport in SYK chain further revealing insulating phases ($\rho \propto T^{-2}$) akin to holographic insulators.

We started by introducing the Majorana variant of the SYK in Chap. 2 where we established the foundational methodology—disorder averaging, replica symmetry, and Schwinger-Dyson equations, among others—and solved the theory in the IR limit where a conformal symmetry arises and showed how the large-q expansion of the Green's function leads to great simplification without losing any physical features. Real-time Keldysh formalism established the stage to study thermalization and non-equilibrium properties where we solved for a simple quench from $\text{SYK}_q+\text{SYK}_2 \to \text{SYK}_2$ dot and found SYK_2 to thermalize instantaneously post-quench with respect to the Green's functions.

In Chap. 3, we extended this formalism to complex fermions with conserved $U(1)$ charge. The resulting Liouville effective action and large-q solutions laid the groundwork for studying quantum chaos, thermodynamic phase transitions, and transport phenomena in chains in chapters to follow.

Chapter 4 studied equilibrium properties such as thermodynamics and chaos: critical exponents ($\alpha = 0, \beta = 1/2, \gamma = 1, \delta = 3$) reveals a mean-field universality class for a continuous phase transition that occurs in the SYK_q dot—a phenomenon shared with AdS black holes and van der Waals fluids—while the quantum Lyapunov exponent λ_L saturates the MSS bound ($\lambda_L = 2\pi T$) at low temperatures. SYK chains were introduced and dealt in the most general setup, followed by providing explicit solutions under uniform equilibrium conditions. This set the stage to study various non-equilibrium and transport properties in the next chapter.

We found the non-equilibrium quenches in chains to not thermalize instantaneously with respect to the Green's function in Chap. 5 (unlike the isolated dots which thermalize instantaneously as showed next in the chapter), where we found a closed form, wave-like solution for the local charge density transport that is completely independent of the on-site interactions. We developed in detail the Keldysh contour deformation techniques to compute the DC resistivity. For three

SYK chains considered, linear-T resistivity emerged naturally, while one specific chain showed crossovers from insulating ($\rho \sim T^{-2}$) to Fermi liquid ($\rho \sim T^2$) to strange metallic ($\rho \sim T$) to bad metallic (no distinctive phenomenological difference from strange metals, except that linear-in-T resistivity violates the MIR resistivity bound) regimes.

The SYK model's solvability has been leveraged in this work to demystify many advanced many-body techniques such as

- Disorder Averaging & Effective Actions: Large-N techniques transform intractable interactions into bi-local ($\mathcal{G}$, Σ) field theories
- Keldysh Dynamics: Non-equilibrium quenches as well as Keldysh contour deformations enable exact solutions for thermalization dynamics and transport
- Universal Criticality: SYK critical exponents provide a tractable window into Landau-Ginzburg universality, spanning black holes, van der Waals fluids, and various other quantum matter.
- Quantum Chaos: In the large-N limit, the SYK model saturates the Maldacena-Shenker-Stanford (MSS) bound $\lambda_L \leq 2\pi T/\hbar$ with $\lambda_L = 2\pi T$ at low temperatures. This exact solution, obtained via analytical continuation of the real-time four-point function, establishes SYK as a rare example of a solvable, maximally chaotic quantum system. Crucially, this mirrors the chaos saturation of black holes in holographic duality, providing a microscopic mechanism for Planckian scrambling.

Like all great theoretical frameworks, SYK transforms the unknown into the unsettled. By providing exact solutions for strange metal phenomenology, maximal chaos, and holographic duality, it shifts research from whether these phenomena exist to how they manifest in broader contexts: higher dimensions, finite-size systems, experimental platforms, and importantly, reconciling SYK's abstract all-to-all non-local disordered interactions with physical microscopics, in particular identifying material realizations where disordered, highly non-local couplings emerge organically to bridge phenomenological effectiveness and fundamental reality. The analytical foundation established for closed fermionic SYK systems in this work invites two pivotal extensions:

1. Incorporating Bosonic Degrees of Freedom: Hybrid models like the Yukawa-SYK framework [2, 4, 5] unify fermions and bosons while preserving solvability. These systems retain maximal chaos [3] and closed-form Schwinger-Dyson equations, enabling exact studies of strange-metal conductivity scaling, non-Fermi liquid transport in higher dimensions, superconducting instabilities where spinful fermions lead to anomalous non-zero expectation values, thermalization in bosonic sectors, as well as the nature and the robustness of thermodynamic phase transitions and chaos in the presence of bosonic degrees of freedom.
2. Open Quantum Dynamics: Extending SYK physics to dissipative environments that causes decoherence (and accordingly represents more realistic situations) raises fundamental questions: How to quantify true entanglement when deco-

herence corrupts standard measures (e.g., von Neumann entropy)? Can SYK-inspired diagnostics (e.g., out-of-time-order correlators) detect chaos in noisy settings? What are the true diagnostics for criticality in such open systems? How are they characterized? The key hurdle involves defining noise-resilient metrics (e.g., negativity for entanglement [1]) to distinguish true quantum correlations from decoherence effects.

Within the pure SYK framework, key universal features—maximal quantum chaos, thermodynamic phase transitions and the associated mean-field critical exponents—exhibit remarkable robustness. As highlighted in the box above Sect. 4.4, these properties are robust under introduction of a spatial dimensionality (e.g., SYK chains) as well as Inclusion of new energy scales (e.g., competing interaction terms). Notably, low-temperature physics remains unaltered, suggesting that SYK universality transcends specific realizations. This invariance raises a fundamental question: Do arbitrary combinations of SYK Hamiltonians preserve these universal signatures (for instance, at large-q, $\mathcal{H} = \sum_{\kappa} \mathcal{K}_{\kappa q} \mathcal{H}_{\kappa q}$)?

The SYK paradigm transcends its origins as a toy model for holography. It provides a computational microscope for strong correlations, demystifying strange metals, black hole thermodynamics, and non-equilibrium quantum matter through analytical exactness for a quantum system in the thermodynamic limit. As mystery of Planckian metals and non-Fermi liquid physics get resolved more and more, SYK-inspired frameworks can play an indispensable role for decoding universality in quantum chaos and criticality.

References

1. Bhattacharyya, A., Hanif, T., Haque, S.S., Paul, A.: Decoherence, entanglement negativity, and circuit complexity for an open quantum system. Phys. Rev. D **107**(10), 106007 (2023)
2. Esterlis, I., Schmalian, J.: Cooper pairing of incoherent electrons: an electron-phonon version of the Sachdev-Ye-Kitaev model. Phys. Rev. B **100**(11), 115132 (2019)
3. Kim, J., Cao, X., Altman, E.: Low-rank Sachdev-Ye-Kitaev models. Phys. Rev. B **101**(12), 125112 (2020)
4. Patel, A.A., Sachdev, S.: Critical strange metal from fluctuating gauge fields in a solvable random model. Phys. Rev. B **98**(12), 125134 (2018)
5. Wang, Y., Chubukov, A.V.: Quantum phase transition in the Yukawa-SYK model. Phys. Rev. Res. **2**(3), 033084 (2020)

A Euclidean/Imaginary Time

In terms of the real coordinate time t, the definition of the imaginary (or Euclidean) time is given by the *Wick rotation*:

$$\imath t \to \tau \quad \Rightarrow \quad t \to -\imath\tau. \tag{A.1}$$

We now discuss the periodic properties in the imaginary-time formalism. The Wick rotated time τ is not periodic in itself, but the thermal averages calculated in τ turns out to be periodic in τ. For instance, consider a general time-dependent operator $\hat{A}(\tau)$ whose time evolution is given by (in Heisenberg picture)

$$\hat{A}(\tau) = e^{\mathcal{H}\tau}\hat{A}(0)e^{-\mathcal{H}\tau} \tag{A.2}$$

where the unitary time evolution is $e^{\imath\mathcal{H}t} \to e^{\mathcal{H}\tau}$ ($\mathcal{H}$ is the Hamiltonian). Then the thermal average is

$$A(\tau) = \langle\hat{A}(\tau)\rangle = \frac{1}{\mathcal{Z}}\mathrm{Tr}[e^{-\beta\mathcal{H}}\hat{A}(\tau)] \tag{A.3}$$

where $\mathcal{Z}$ is the partition function and β is the inverse temperature. Then if we substitute Eq. (A.2) in Eq. (A.3) and evaluate the averaged quantity for τ and $\tau + \beta$, we find

$$\boxed{A(\tau) = A(\tau + \beta)}. \tag{A.4}$$

Therefore there is a periodicity of β in τ.

Next we study the thermal 2-point (Green's) function which is defined as

$$\mathcal{G}_\beta = \frac{1}{\mathcal{Z}}\mathrm{Tr}\Big[e^{-\beta\mathcal{H}}\hat{K}(\tau, x)\hat{K}(0, 0)\Big] \tag{A.5}$$

R. Jha, *Theory of Strongly Correlated Systems*, Lecture Notes in Physics 1051,
https://doi.org/10.1007/978-3-032-20910-8

where $\hat{K}(\tau, x)$ is any arbitrary operator at (imaginary) time τ and position x. It can be bosonic or fermionic (such as the creation and annihilation operators). Using Eq. (A.2), the time evolution by β units in time gives

$$e^{\beta\mathcal{H}}\hat{K}(\tau, x)e^{-\beta\mathcal{H}} = \hat{K}(\tau + \beta, x). \tag{A.6}$$

Focusing on $\tau > 0$ and $\beta > 0$ (positive temperatures), we have

$$\begin{aligned}
\mathcal{G}_\beta(\tau + \beta, x) =& \frac{1}{\mathcal{Z}}\mathrm{Tr}\Big[e^{-\beta\mathcal{H}}\hat{K}(\tau + \beta, x)\hat{K}(0, 0)\Big] \\
=& \frac{1}{\mathcal{Z}}\mathrm{Tr}\Big[\underbrace{e^{-\beta\mathcal{H}}e^{\beta\mathcal{H}}}_{=1}\hat{K}(\tau, x)e^{-\beta\mathcal{H}}\hat{K}(0, 0)\Big] \\
=& \frac{1}{\mathcal{Z}}\mathrm{Tr}\Big[e^{-\beta\mathcal{H}}\hat{K}(0, 0)\hat{K}(\tau, x)\Big]
\end{aligned} \tag{A.7}$$

where cyclic property of trace is used in the last step $\mathrm{Tr}[\hat{A}\hat{B}\hat{C}] = \mathrm{Tr}[\hat{B}\hat{C}\hat{A}] = \mathrm{Tr}[\hat{C}\hat{A}\hat{B}]$. Since $\hat{K}$ can be bosonic or fermionic, we can write

$$\mathcal{G}_\beta(\tau + \beta, x) = \eta\frac{1}{\mathcal{Z}}\mathrm{Tr}\Big[e^{-\beta\mathcal{H}}\hat{K}(\tau, x)\hat{K}(0, 0)\Big]; \qquad \eta = \begin{cases} +1 & (\text{bosonic } \hat{K}) \\ -1 & (\text{fermionic } \hat{K}) \end{cases} \tag{A.8}$$

Therefore we obtain that the thermal Green's function is periodic (for bosonic operators) and anti-periodic (for fermionic operators), written in a unified as

$$\boxed{\mathcal{G}_\beta(\tau, x) = \eta\mathcal{G}_\beta(\tau \pm \beta, x); \qquad \eta = \begin{cases} +1 & (\text{bosonic } \hat{K}) \\ -1 & (\text{fermionic } \hat{K}) \end{cases}} \tag{A.9}$$

A Note on Gaussian Integrals and Matrices B

Matrices come in different forms. The purpose of this appendix is to introduce the basics of Gaussian-type integrals of real Grassmann variables. With the benefit of hindsight, we start with focusing on *real $2n \times 2n$ skew-symmetric (equivalently, anti-symmetric) matrices A* which is defined by

$$A^T = -A \qquad (T \text{ denotes transpose}). \tag{B.1}$$

Then as an eigenvalue problem, we consider $A\vec{x} = \lambda\vec{x}$ where λ are the eigenvalues of the real skew-symmetric matrix A and $\vec{x}$ are the corresponding eigenfunctions. Since A is real, any Hermitian conjugate simply yields $A^\dagger = A^T$. We use this by taking the Hermitian conjugate of $A\vec{x} = \lambda\vec{x}$ (recall that $(AB)^\dagger = B^\dagger A^\dagger$)

$$\begin{aligned} \vec{x}^\dagger A^T &= \lambda^\star \vec{x}^\dagger \\ \Rightarrow \vec{x}^\dagger(-A) &= \lambda^\star \vec{x}^\dagger \\ \Rightarrow \vec{x}^\dagger(-A)\vec{x} &= \lambda^\star \vec{x}^\dagger \vec{x} \qquad (\text{multiplying both sides from right by } \vec{x}) \\ \Rightarrow -\lambda\vec{x}^\dagger\vec{x} &= \lambda^\star \vec{x}^\dagger \vec{x}. \end{aligned} \tag{B.2}$$

Therefore, we get

$$\lambda^\star = -\lambda \qquad \Rightarrow \text{Re}[\lambda] = 0. \tag{B.3}$$

Furthermore, taking just the complex conjugation of $A\vec{x} = \lambda\vec{x}$ gives (recall A is real, so $A^\star = A$)

$$A\vec{x}^\star = \underbrace{\lambda^\star}_{=-\lambda} \vec{x}^\star \Rightarrow A\vec{x}^\star = -\lambda\vec{x}^\star. \tag{B.4}$$

R. Jha, *Theory of Strongly Correlated Systems*, Lecture Notes in Physics 1051,
https://doi.org/10.1007/978-3-032-20910-8

Hence, all eigenvalues are imaginary and they always appear in $\pm\lambda$ pairs. Let's label the $2n$ eigenvalues as (recall the dimension of A is $2n \times 2n$)

$$\begin{aligned}\lambda(A) =& \{\pm\imath\ell_1, \pm\imath\ell_2, \pm\imath\ell_3, \ldots, \pm\imath\ell_n\}, \quad (\ell_k \in \mathbb{R} \quad \forall k) \\ =& \{\lambda_1, \lambda_2, \lambda_3, \ldots, \lambda_n\},\end{aligned} \tag{B.5}$$

where each λ_i comes in pair of $\{\pm\imath\ell_i\}$.Then

$$\det[A] = \prod_{i=1}^{n} \lambda_i^2 = \prod_{i=1}^{n} \ell_i^2 = \Big(\prod_{i=1}^{n} \ell_i\Big)^2. \tag{B.6}$$

Using special orthogonal transformation ($SO(N$[1]$)$, we can put A in block-diagonal form D where they are connected via

$$D = OAO^T \Rightarrow A = O^T DO \tag{B.7}$$

where O satisfies $O^T O = OO^T = \mathbb{1}$ (identity). Therefore the block diagonal structure of D looks like

$$D = \text{diag}\Big[\begin{pmatrix} 0 & +\lambda_1 \\ -\lambda_1 & 0 \end{pmatrix}, \begin{pmatrix} 0 & +\lambda_2 \\ -\lambda_2 & 0 \end{pmatrix}, \begin{pmatrix} 0 & +\lambda_3 \\ -\lambda_3 & 0 \end{pmatrix}, \ldots, \begin{pmatrix} 0 & +\lambda_n \\ -\lambda_n & 0 \end{pmatrix}\Big]. \tag{B.8}$$

Having set the stage for the real skew-symmetric matrix, we now head toward the integration rules for real Grassmann variables $\{\chi_i\}$. By definition, they satisfy

$$\chi_i^\star = \chi_i, \qquad \{\chi_i, \chi_j\} = 0 \quad \forall \quad i, j \quad \Rightarrow \chi_i^2 = 0. \tag{B.9}$$

Integration is defined as a linear operation

$$\int d\chi(a + b\chi) = b \quad (a, b \in \mathbb{R}) \tag{B.10}$$

[1] We discussed $O(N)$ transformations in Sect. 2.4 which includes both rotations and reflections, therefore determinant can be ± 1. A subgroup of $O(N)$ is $SO(N)$ where only rotations are considered and reflections excluded. Accordingly, the determinant only takes the value of $+1$.

which is just like a differentiation of "normal" variables. That's why, integration of a constant function is zero. However, ordering matters:

$$\iint d\chi_1 d\chi_2 e^{-a\chi_1\chi_2} = \iint d\chi_1 d\chi_2 (1 - a\chi_1\chi_2) = \iint \underbrace{d\chi_1 d\chi_2}_{\longrightarrow}(1 + a\, \underbrace{\chi_2\chi_1}_{\longleftarrow})$$
$$= \int d\chi_1 a\chi_1 = a \tag{B.11}$$

where $a \in \mathbb{R}$. Therefore the pattern is: increasing order of index values in the measure must match with the decreasing order of index values of the integrand and then start integrating out from the largest index value to the smallest.

We finally provide a matrix representation of the Gaussian-type integration for real Grassmann variables. Let's start with a 2×2 real skew-symmetric matrix a given by

$$A = \begin{pmatrix} 0 & a \\ -a & 0 \end{pmatrix} \tag{B.12}$$

which allows us to write the Gaussian-type integral in Eq. (B.11) as

$$\iint d\chi_1 d\chi_2 e^{-\frac{1}{2}\chi_i A^{ij}\chi_j} = a \tag{B.13}$$

where repeated indices are summed over (Einstein convention), in this case from $i, j = 1$ to $i, j = 2$. We can generalize this Gaussian integral to an arbitrary $N \times N$ real skew-symmetric matrix A where indices i, j run from 1 to N

$$\int \mathcal{D}\chi_i e^{-\frac{1}{2}\chi_i A^{ij}\chi_j} \tag{B.14}$$

where the measure $\mathcal{D}\chi_i = \prod_{i=1}^{N} \chi_i$. We are interested in evaluating this integral in the remaining of this appendix.[2]

Since A is skew-symmetric, we can bring it to the block-diagonal form using Eqs. (B.7) and (B.8). Suppose we have $A = O^T D O$ ($O \in SO(N)$) where we

[2] The Gaussian form in Eq. (B.14) justifies that we chose to focus on the skew-symmetric matrices at the start of this appendix. Consider A to be any general, non-singular matrix ($\det[A] \neq 0$). Any general matrix can be written as $A = A_S + A_{SS}$ where the symmetric part $A_S = (A + A^T)/2$ and skew-symmetric part $A_{SS} = (A - A^T)/2$. Then the real Grassmann variables will contract with this general matrix and due to the anti-commuting nature of the Grassmann variables (which makes them anti-symmetric), the symmetric component of the matrix A_S will have vanishing contribution, leaving us with the skew-symmetric component A_{SS}.

impose the orthogonal transformation on the real Grassmann variables via

$$O_i{}^j \chi_j = \chi_i' \quad \text{(Einstein summation convention implied).} \tag{B.15}$$

Then the measure in Eq. (B.14) remains invariant under such orthogonal transformation.

Proof We have the measure $\mathcal{D}\chi_i' = \prod_{i=1}^{N} \chi_i'$ which gives us (again, Einstein summation convention is implied)

$$\begin{aligned}
\prod_{i=1}^{N} \chi_i' &= \frac{1}{N!}\epsilon^{i_1 i_2 \ldots i_N} \chi_{i_1}' \chi_{i_2}' \cdots \chi_{i_N}' \\
&= \frac{1}{N!}\epsilon^{i_1 i_2 \ldots i_N} O_{i_1}{}^{j_1} \chi_{j_1} O_{i_2}{}^{j_2} \chi_{j_2} \ldots O_{i_N}{}^{j_N} \chi_{j_N} \\
&= \frac{1}{N!} \det[O] \underbrace{\epsilon^{j_1 j_2 \ldots j_N} \chi_{j_1} \chi_{j_2} \cdots \chi_{j_N}}_{=N! \prod_{j=1}^{N} \chi_j} \qquad\qquad \text{(B.16)} \\
&= \det[O] \prod_{j=1}^{N} \chi_j = \prod_{j=1}^{N} \chi_j \quad (\because \det[O] = 1).
\end{aligned}$$

□

Here $\epsilon^{i_1 i_2 \ldots i_N}$ is the Levi-Civita symbol where we follow the convention that all even permutations of indices $\{i_1 i_2 \ldots i_N\}$ are $+1$ while odd permutations are -1 and any repeated indices imply a vanishing symbol. The Levi-Civita symbol $\epsilon^{i_1 i_2 \ldots i_N}$ is antisymmetric under permutation of any two indices. This means $\epsilon^{i_1 i_2 \ldots i_N} = -\epsilon^{i_2 i_1 \ldots i_N}$, etc. We used the following identity in the third equality

$$\epsilon^{i_1 i_2 \ldots i_N} O_{i_1}{}^{j_1} O_{i_2}{}^{j_2} \ldots O_{i_N}{}^{j_N} = \det(O) \epsilon^{j_1 j_2 \ldots j_N}. \tag{B.17}$$

Proof The determinant of matrix O is given by

$$\det(O) = \frac{1}{N!}\epsilon^{i_1 i_2 \ldots i_N} \epsilon_{j_1 j_2 \ldots j_N} O_{i_1}{}^{j_1} O_{i_2}{}^{j_2} \ldots O_{i_N}{}^{j_N}. \tag{B.18}$$

Now let's consider the left-hand side of the identity we wish to prove in Eq.(B.17). The left-hand side, after contractions, is completely anti-symmetric in indices $\{j_1, j_2, \ldots, j_N\}$. Therefore the only unique possibility for this is to have

$$\epsilon^{i_1 i_2 \ldots i_N} O_{i_1}{}^{j_1} O_{i_2}{}^{j_2} \ldots O_{i_N}{}^{j_N} \propto \epsilon^{j_1 j_2 \ldots j_N} \tag{B.19}$$

where let the proportionality constant be C. So we have

$$\epsilon^{i_1 i_2 \ldots i_N} O_{i_1}{}^{j_1} O_{i_2}{}^{j_2} \ldots O_{i_N}{}^{j_N} = C \cdot \epsilon^{j_1 j_2 \ldots j_N}. \tag{B.20}$$

We contract both sides by $\epsilon_{j_1 j_2 \ldots j_N}$ where the left-hand side is equal to $N!\det[O]$ (Eq. (B.18)) while the right-hand side is equal to $C \cdot N!$ because $\epsilon_{j_1 j_2 \ldots j_N} \epsilon^{j_1 j_2 \ldots j_N} = N!$. This follows because the only non-zero terms occur when $j_1, j_2, \ldots, j_N$ are distinct permutations of $1, 2, \ldots, N$. For each permutation, the product of the two Levi-Civita symbols is $(\text{sign(permutation)})^2 = 1$. There are $N!$ distinct permutations of N indices. Each permutation contributes $+1$ to the sum. Summing over all permutations gives $\epsilon_{j_1 j_2 \ldots j_N} \epsilon^{j_1 j_2 \ldots j_N} = \sum_{\text{permutations}} 1 = N!$. Hence we get

$$C = \det[O]$$

and this concludes our proof of Eq. (B.17). □

To give some concrete examples, let's consider $N = 2$: $\epsilon_{12}\epsilon^{12} + \epsilon_{21}\epsilon^{21} = 1 + 1 = 2! = 2$. Then for $N = 2$, we expand the left-hand side of Eq. (B.17), namely $\epsilon^{i_1 i_2} O_{i_1}{}^{j_1} O_{i_2}{}^{j_2} = \epsilon^{12} O_1{}^{j_1} O_2{}^{j_2} + \epsilon^{21} O_2{}^{j_1} O_1{}^{j_2}$. Using $\epsilon^{12} = +1$ and $\epsilon^{21} = -1$, we get

$$\text{Left-hand side} = O_1{}^{j_1} O_2{}^{j_2} - O_2{}^{j_1} O_1{}^{j_2} \quad (\text{anti-symmetric in } j_1 \text{ and } j_2).$$

Recognizing $\det[O] = O_1{}^1 O_2{}^2 - O_1{}^2 O_2{}^1$, we have for right-hand side of Eq. (B.17) $\det[O] e^{j_1 j_2}$. Therefore, we can verify $\epsilon^{i_1 i_2} O_{i_1}{}^{j_1} O_{i_2}{}^{j_2} = \det[O] \epsilon^{j_1 j_2}$ by considering

- For $j_1 = 1, j_2 = 2$: Left-hand side = $O_1{}^1 O_2{}^2 - O_2{}^1 O_1{}^2 = \det[O]$ and right-hand side = $\det[O]\epsilon^{12} = \det[O] \cdot 1 = O_1{}^1 O_2{}^2 - O_1{}^2 O_2{}^1$ = left-hand side.
- For $j_1 = 2, j_2 = 1$: Left-hand side = $O_1{}^2 O_2{}^1 - O_2{}^2 O_1{}^1 = -\det[O]$ and right-hand side = $\det[O]\epsilon^{12} = \det[O] \cdot (-1) = O_1{}^2 O_2{}^1 - O_1{}^1 O_2{}^2$ = left-hand side.
- For $j_1 = j_2$: This holds trivially ($0 = 0$).

Note that a particular element of matrix O, such as $O_i{}^j$ for a particular value of i and j is just a number and can be moved across other numbers (e.g., $O_1{}^1 O_2{}^2 = O_2{}^2 O_1{}^1$) without worrying about any sign issue.

Moving forward, we have established that the measure in Gaussian integral in Eq. (B.14) is invariant under orthogonal transformations of Eq. (B.15). Then we re-express Eq. (B.14) as

$$\int \mathcal{D}\chi_i e^{-\frac{1}{2}\chi^T A \chi} = \int \mathcal{D}\chi_i \exp\Big(-\frac{1}{2} \underbrace{\chi^T O^T}_{\chi'^T} D \underbrace{O\chi}_{\chi'}\Big) = \int \mathcal{D}\chi'_i e^{-\frac{1}{2}\chi'^T D \chi'} \tag{B.21}$$

where we substituted for the measure $\mathcal{D}\chi_i = \mathcal{D}\chi_i'$ as proved above. The dimension of the real skew-symmetric matrix A is $N \times N$ where let's have $N = 2n$. Then D is given by Eq. (B.8). Let's simplify further (ignoring $'$ for convenience)

$$
\begin{aligned}
&\int \mathcal{D}\chi_i e^{-\frac{1}{2}\chi_i D^{ij}\chi_j} \\
&\quad= \int \mathcal{D}\chi_i \exp\Big\{-\frac{1}{2}\Big(\lambda_1\chi_1\chi_2 - \lambda_1\chi_2\chi_1 + \lambda_2\chi_3\chi_4 - \lambda_2\chi_4\chi_3 \\
&\qquad\quad + \ldots + \lambda_n\chi_{2n-1}\chi_{2n} - \lambda_n\chi_{2n}\chi_{2n-1}\Big)\Big\} \\
&= \int \mathcal{D}\chi_i \exp\Big\{\lambda_1\chi_2\chi_1 + \lambda_2\chi_4\chi_3 + \lambda_3\chi_6\chi_5 + \ldots + \lambda_n\chi_{2n}\chi_{2n-1}\Big\} \\
&\quad= \int \mathcal{D}\chi_i \exp\Big\{\sum_{i=1}^{n}\lambda_i\chi_{2i}\chi_{2i-1}\Big\} \\
&\quad= \int \mathcal{D}\chi_i \prod_{i=1}^{n}\exp\Big\{\lambda_i\chi_{2i}\chi_{2i-1}\Big\} = \int \mathcal{D}\chi_i \prod_{i=1}^{n}\Big(1 + \lambda_i\chi_{2i}\chi_{2i-1}\Big)
\end{aligned}
\tag{B.22}
$$

Here, the first term is just a constant whose integration will vanish. So we are left with

$$
\begin{aligned}
\Rightarrow \int \mathcal{D}\chi_i e^{-\frac{1}{2}\chi_i D^{ij}\chi_j} &= \int \mathcal{D}\chi_i\Big(\prod_{i=1}^{n}\lambda_i\Big)(\chi_2\chi_1)(\chi_4\chi_3)(\chi_6\chi_5)\ldots(\chi_{2n}\chi_{2n-1}) \\
&= \Big(\prod_{i=1}^{n}\lambda_i\Big)\int \mathcal{D}\chi_i(\chi_2\chi_1)(\chi_4\chi_3)(\chi_6\chi_5)\ldots(\chi_{2n}\chi_{2n-1}).
\end{aligned}
\tag{B.23}
$$

Each χ_i are Grassmann variables (physically, they can be thought of as spinless Majorana fermions) that satisfy anti-commutation relations, accordingly pairing them makes them act like a "bosonic" partner. So each pair $(\chi_2\chi_1)$, $(\chi_4\chi_3)$, $(\chi_6\chi_5)$, …, $(\chi_{2n}\chi_{2n-1})$ can be thought as a single "boson" and can be transferred through each other without any minus signs appearing. So we can re-arrange the integrand as

$$
\begin{aligned}
&(\chi_2\chi_1)(\chi_4\chi_3)(\chi_6\chi_5)\ldots(\chi_{2n}\chi_{2n-1}) \\
&\qquad=(\chi_{2n}\chi_{2n-1})(\chi_{2n-2}\chi_{2n-3})(\chi_{2n-4}\chi_{2n-5})\ldots(\chi_6\chi_5)(\chi_4\chi_3)(\chi_2\chi_1)
\end{aligned}
$$

where the pattern is clear that the indices are arranged in descending order. The reason this matters is because the ordering plays a big role in integration of Grassmann variables (see Eq. (B.11) and the paragraph below) where the measure is arranged in increasing order of index values, namely $\mathcal{D}\chi_i = d\chi_1 d\chi_2 d\chi_3 d\chi_4 \ldots d\chi_{2n-2} d\chi_{2n-1} d\chi_{2n}$. Then the increasing order of measure matches the decreasing order of integrand and we start integrating out from the largest index value to the smallest to get (recall that integration is like differentiation of "normal" variables, therefore the integral will yield unity)

$$\Rightarrow \int \mathcal{D}\chi_i e^{-\frac{1}{2}\chi_i D^{ij}\chi_j} = \Big(\prod_{i=1}^{n} \lambda_i\Big) \tag{B.24}$$

where we use Eq. (B.6) to identify and obtain the final result for a finite-dimensional Gaussian-type integration for real Grassmann variables

$$\boxed{\int \mathcal{D}\chi_i \exp\Big\{-\frac{1}{2}\sum_{i,j=1}^{N} \chi_i A_{ij}\chi_j\Big\} = \sqrt{\det[A]}}\,. \tag{B.25}$$

We can immediately generalize this to infinite-dimensional Grassmann algebra $\chi(\tau)$ where $\{\chi(\tau), \chi(\tau')\} = 0$ for $\tau, \tau' \in \mathbb{R}$ as follows:

$$\boxed{\int \mathcal{D}\chi \exp\Big\{-\frac{1}{2}\iint d\tau d\tau' \chi(\tau) A(\tau, \tau')\chi(\tau')\Big\} = \sqrt{\det[A]}} \tag{B.26}$$

where we have taken the limit $n \to \infty$ (recall the dimension of A as $N \times N$ and $N = 2n$). The measure is given by $\mathcal{D}\chi = \lim_{n\to\infty} \mathcal{D}\chi_i = \lim_{n\to\infty} \prod_{i=1}^{2n} d\chi_i$ which integrates over entire infinite-dimensional Grassmann algebra.

Derivation of the Schwarzian Identity

C

We have to prove the Schwarzian identity (Eq. (2.114)), namely

$$\left\{f(g(\tau)), \tau\right\} = \left(g'(\tau)\right)^2 \{f, g\} + \{g, \tau\}. \tag{C.1}$$

Proof We start with the left-hand side where we use the definition of the Schwarzian derivative from Eq. (2.110) which we reproduce here for convenience

$$\{g, \tau\} \equiv \frac{g'''(\tau)}{g'(\tau)} - \frac{3}{2}\left(\frac{g''(\tau)}{g'(\tau)}\right)^2 \quad \text{(Schwarzian derivative)}.$$

Then the left-hand side simplifies to

$$\text{Left-hand side } = \frac{\frac{d^3 f(g(\tau))}{d\tau^3}}{\frac{df(g(\tau))}{d\tau}} - \frac{3}{2}\left(\frac{\frac{d^2 f(g(\tau))}{d\tau^2}}{\frac{df(g(\tau))}{d\tau}}\right)^2. \tag{C.2}$$

Let

$$\dot{f} \equiv \frac{df}{dg}, \quad g' \equiv \frac{dg}{d\tau}, \tag{C.3}$$

then

$$\frac{df(g(\tau))}{d\tau} = \frac{df}{dg}\frac{dg}{d\tau} = \dot{f}g'. \tag{C.4}$$

R. Jha, *Theory of Strongly Correlated Systems*, Lecture Notes in Physics 1051,
https://doi.org/10.1007/978-3-032-20910-8

Therefore,

$$
\begin{aligned}
\frac{d^2 f(g(\tau))}{d\tau^2} &= \frac{d}{d\tau}\left(\frac{df}{dg}\frac{dg}{d\tau}\right) \\
&= \left(\frac{d}{d\tau}\left(\frac{df}{dg}\right)\right)\frac{dg}{d\tau} + \frac{df}{dg}\frac{dg^2}{d\tau^2} = \left(\frac{d}{dg}\left(\frac{df}{d\tau}\right)\right)\frac{dg}{d\tau} + \frac{df}{dg}\frac{dg^2}{d\tau^2} \\
&= \frac{d}{dg}\left(\frac{df}{dg}\frac{dg}{d\tau}\right)\frac{dg}{d\tau} + \frac{df}{dg}\frac{dg^2}{d\tau^2} \\
&= \frac{d^f}{dg^2}\left(\frac{dg}{d\tau}\right)^2 + \frac{df}{dg}\Big(\underbrace{\frac{d}{dg}\left(\frac{dg}{d\tau}\right)}_{=0}\Big) + \frac{df}{dg}\frac{dg^2}{d\tau^2} \\
&= \ddot{f}(g')^2 + \dot{f}g''
\end{aligned}
\tag{C.5}
$$

where we used in the second line the fact that partial derivatives commute. Also, we used the fact that g and g' are independent variables, so $\frac{d}{dg}\frac{dg}{d\tau} = 0$.

Finally,

$$
\begin{aligned}
\frac{d^3 f(g(\tau))}{d\tau^3} &= \frac{d}{d\tau}\left(\frac{d^2 f}{dg^2}\left(\frac{dg}{d\tau}\right)^2 + \frac{df}{dg}\frac{d^2 g}{d\tau^2}\right) \\
&= \frac{d}{d\tau}\left(\frac{d^2 f}{dg^2}\right)\left(\frac{dg}{d\tau}\right)^2 + 2\frac{dg}{d\tau}\frac{d^2 g}{d\tau^2}\frac{d^2 f}{dg^2} + \frac{d}{d\tau}\left(\frac{df}{dg}\right)\frac{d^2 g}{d\tau^2} + \frac{df}{dg}\frac{d^3 g}{d\tau^3} \\
&= \frac{d^3 f}{dg^3}\left(\frac{dg}{d\tau}\right)^3 + 3\frac{dg}{d\tau}\frac{d^2 g}{d\tau^2}\frac{d^2 f}{dg^2} + \frac{df}{dg}\frac{d^3 g}{d\tau^3} \\
&= (g')^3\,\dddot{f} + 3g'g''\ddot{f} + \dot{f}g'''.
\end{aligned}
\tag{C.6}
$$

Accordingly, we have evaluated all three derivatives required to calculate the left-hand side in Eq. (C.2) which simplifies to

$$
\begin{aligned}
\text{Left-hand side} &= \{f(g(\tau)), \tau\} = \frac{\frac{d^3 f(g(\tau))}{d\tau^3}}{\frac{df(g(\tau))}{d\tau}} - \frac{3}{2}\left(\frac{\frac{d^2 f(g(\tau))}{d\tau^2}}{\frac{df(g(\tau))}{d\tau}}\right)^2 \\
&= \frac{(g')^3\,\dddot{f} + 3g'g''\ddot{f} + \dot{f}g'''}{\dot{f}g'} - \frac{3}{2}\Big(\frac{\ddot{f}(g')^2 + \dot{f}g''}{\dot{f}g'}\Big)^2 \\
&= (g')^2\frac{\dddot{f}}{\dot{f}} \cancel{+3g''\frac{\ddot{f}}{\dot{f}}} + \frac{g'''}{g'} - \frac{3}{2}(g')^2\left(\frac{\ddot{f}}{\dot{f}}\right)^2 \cancel{-3g''\frac{\ddot{f}}{\dot{f}}} - \frac{3}{2}\left(\frac{g''}{g'}\right)^2
\end{aligned}
$$

$$
\begin{aligned}
&= (g')^2 \left[\frac{\dddot{f}}{\dot{f}} - \frac{3}{2} \left(\frac{\ddot{f}}{\dot{f}} \right)^2 \right] + \left[\frac{g'''}{g'} - \frac{3}{2} \left(\frac{g''}{g'} \right)^2 \right] \\
&= \left(g'(\tau) \right)^2 \{f, g\} + \{g, \tau\} = \text{Right-hand side.}
\end{aligned}
\tag{C.7}
$$

□

A Note on the Langreth Rules

D

Langreth rules are a cornerstone of non-equilibrium many-body theory, providing a systematic method to convert contour-ordered equations (defined on the Keldysh contour $\mathcal{C}$) into real-time components. They decompose contour convolutions into real-time integrals. For the sake of deriving the Langreth rules, consider a general (out-of-equilibrium) Keldysh integral

$$Z(t_1, t_2) = \int_{\mathcal{C}} dt_3 X(t_1, t_3) Y(t_3, t_2) \quad \text{(D.1)}$$

where Z, X and Y are some arbitrary bi-temporal Keldysh functions. The integration is done over the Keldysh contour $\mathcal{C}$.

We begin with the case where time t_1 resides on the forward branch $\mathcal{C}_+$and t_2 on the backward branch $\mathcal{C}_-$ (Fig. 2.1), having omitted the imaginary-time segment via Bogoliubov's weakening of correlations. By definition, the left-hand side corresponds to the lesser function $Z^<(t_1, t_2)$ (Eq. (2.190)). Applying the same definition to the right-hand side gives

$$\Rightarrow Z^<(t_1, t_2) = \int_{\mathcal{C}_+} dt_3 X(t_1, t_3) Y^<(t_3, t_2) + \int_{\mathcal{C}_-} dt_3 X^<(t_1, t_3) Y(t_3, t_2). \quad \text{(D.2)}$$

To resolve the contour integrals, we deform $\mathcal{C}$ using forward-backward loops terminating at their maximum real-time values (Fig. D.1), a method established in Ref. [1] and detailed in Ref. [3]. This deformation leaves integrals invariant.

R. Jha, *Theory of Strongly Correlated Systems*, Lecture Notes in Physics 1051,
https://doi.org/10.1007/978-3-032-20910-8

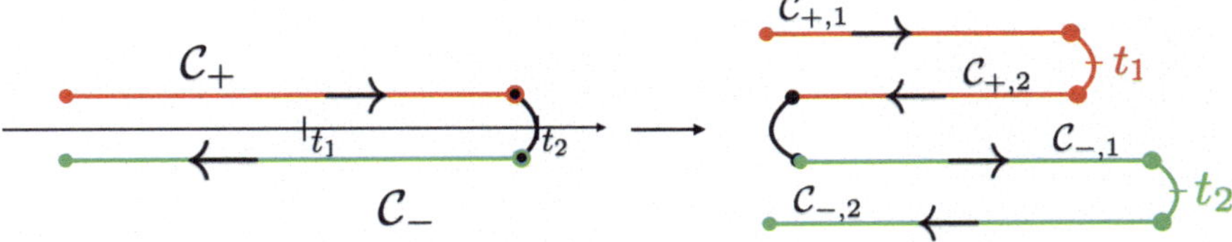

Fig. D.1 Schematic representing deformation of the Keldysh contour in Fig. 2.1

Splitting $\mathcal{C}_+ = \mathcal{C}_{+,1} \cup \mathcal{C}_{+,2}$, the first term simplifies as

$$\begin{aligned}\int_{\mathcal{C}_+} dt_3 X(t_1, t_3) Y^<(t_3, t_2) &= \int_{\mathcal{C}_{+,1}} dt_3 X(t_1, t_3) Y^<(t_3, t_2) \\ &\quad + \int_{\mathcal{C}_{+,2}} dt_3 X(t_1, t_3) Y^<(t_3, t_2) \\ &= \int_{-\infty}^{t_1} dt_3 X^>(t_1, t_3) Y^<(t_3, t_2) \\ &\quad + \int_{t_1}^{-\infty} dt_3 X^<(t_1, t_3) Y^<(t_3, t_2) \\ &= \int_{-\infty}^{+\infty} dt_3 \Theta(t_1 - t_3) \left[X^>(t_1, t_3) \right. \\ &\quad \left. - X^<(t_1, t_3)\right] Y^<(t_3, t_2) \\ &= \int_{-\infty}^{+\infty} dt_3 X^R(t_1, t_3) Y^<(t_3, t_2).\end{aligned} \tag{D.3}$$

The final step uses the retarded function definition (Eq. (2.193)), where $X^R \equiv X^> - X^<$.

Similarly, splitting $\mathcal{C}_- = \mathcal{C}_{-,1} \cup \mathcal{C}_{-,2}$, the second term reduces to

$$\begin{aligned}\int_{\mathcal{C}_-} dt_3 X^<(t_1, t_3) Y(t_3, t_2) &= \int_{\mathcal{C}_{-,1}} dt_3 X^<(t_1, t_3) Y(t_3, t_2) \\ &\quad + \int_{\mathcal{C}_{-,2}} dt_3 X^<(t_1, t_3) Y(t_3, t_2) \\ &= \int_{-\infty}^{t_2} dt_3 X^<(t_1, t_3) Y^<(t_3, t_2) \\ &\quad + \int_{t_2}^{-\infty} dt_3 X^<(t_1, t_3) Y^>(t_3, t_2)\end{aligned} \tag{D.4}$$

$$= \int_{-\infty}^{\infty} dt_3 \Theta(t_2 - t_3) X^<(t_1, t_3) \times \left[Y^<(t_3, t_2) - Y^>(t_3, t_2) \right]$$
$$= \int_{-\infty}^{\infty} dt_3 X^<(t_1, t_3) Y^A(t_3, t_2).$$

Here, the advanced function $Y^A \equiv Y^< - Y^>$ (Eq. (2.193)) emerges.

Combining results, Eq. (D.2) yields the Langreth rule for the lesser component:

$$\boxed{Z^<(t_1, t_2) = \int_{-\infty}^{\infty} dt_3 [X^R(t_1, t_3) Y^<(t_3, t_2) + X^<(t_1, t_3) Y^A(t_3, t_2)]}. \tag{D.5}$$

For the complementary case where t_1 lies on the backward branch $\mathcal{C}_-$and t_2 on the forward branch $\mathcal{C}_+$, an analogous derivation yields the Langreth rule for the greater component

$$\boxed{Z^>(t_1, t_2) = \int_{-\infty}^{\infty} dt_3 [X^R(t_1, t_3) Y^>(t_3, t_2) + X^>(t_1, t_3) Y^A(t_3, t_2)]} \tag{D.6}$$

Equations (D.5) and (D.6) constitute the Langreth rules employed in Sect. 2.9.3 to derive the Kadanoff-Baym equations from the Schwinger-Dyson equations.

Finally, we require Langreth rules for products (as opposed to convolutions), which appear in self-energy terms of the Schwinger-Dyson equations. For a simple product on the Keldysh contour, we have

$$C(t, t') = A(t, t') B(t, t') \tag{D.7}$$

whose real-time components are given by the Langreth rules

$$\boxed{\begin{aligned} C^{\gtrless}(t, t') &= A^{\gtrless}(t, t') B^{\gtrless}(t, t') \\ C^R(t, t') &= A^<\left(t, t'\right) B^R\left(t, t'\right) + A^R\left(t, t'\right) B^<\left(t, t'\right) + A^R\left(t, t'\right) B^R\left(t, t'\right) \end{aligned}}. \tag{D.8}$$

We refer the reader to Ref. [5] for a detailed dive-in.

Coherence Temperature Scale

E

Understanding renormalization group (RG) fixed points is fundamental to describing universal low-energy physics in quantum many-body systems. A fixed point in the RG flow represents a special set of coupling constants $\{J_i\}$ where the theory becomes invariant under scale transformations. This invariance manifests in two key ways:

1. Absence of RG Flow: At the fixed point, the couplings cease to evolve as the energy scale Λ is reduced, mathematically expressed by $\frac{dJ_i}{d\ln\Lambda} = 0$ for all couplings J_i.
2. Scale Invariance: Physical observables, particularly correlation functions, exhibit characteristic power-law behavior. This absence of any intrinsic energy scale signifies the system's self-similarity across different length or energy scales.

The Sachdev-Ye-Kitaev (SYK_q) model provides a remarkable example of this physics. At low temperatures (deep infrared, IR), the dominant SYK_q interaction term drives the system towards a conformally invariant fixed point. This critical point is vividly reflected in the long-time behavior of the fermionic Green's function $\mathcal{G}(\tau)$ and the self-energy $\Sigma(\tau)$, which both obey power-law scaling (as derived in this work in detail)—a direct consequence of the emergent scale invariance and the lack of a characteristic energy scale in the IR.

For cases of mixed Hamiltonian such as in Eq. (2.235) ($\mathcal{H} = \mathrm{SYK}_q + \mathrm{SYK}_2$), the RG flow of the SYK model is fundamentally governed by the competition between different terms in its Hamiltonian. At high energies (ultraviolet, UV), various contributions (like SYK_q, SYK_2, or higher-q interactions) are significant. However, as we flow down in energy towards the IR, the term with lesser number of fermions start to dominate. This is particularly crucial for this model because SYK_q is a model without stable quasiparticles while SYK_2 is a model with stable quasiparticles. Accordingly, this reflects a competition between the two terms. The *coherence temperature* $T_{\mathrm{coherence}}$ sets the scale for the crossover where SYK_2

R. Jha, *Theory of Strongly Correlated Systems*, Lecture Notes in Physics 1051,
https://doi.org/10.1007/978-3-032-20910-8

becomes comparable to SYK_q which is required for Planckian dynamics [2] to be exhibited (since SYK_q alone does not admit Planckian dynamics). Below $T_{coherence}$, SYK_2 dominates driving the system toward a Fermi liquid-like state; above it, SYK_q takes over driving to strange metal behavior.

The coherence temperature is derived by equating the self-energies of SYK_q and SYK_2 at the crossover scale where both terms are equally important. At this scale:

- The system is still near the SYK_q fixed point, so the Green's function retains its SYK_q scaling.
- The SYK_2 term is treated as a perturbation using the SYK_q Green's function because the calculation assumes proximity to the SYK_q fixed point and slightly above the coherence temperature where the SYK_q is still dominant.

E.1 SYK_q Fixed Point

The SYK_q model is governed by random $q/2$-body interactions. At low temperatures, it flows to a conformally invariant fixed point which we explain below:

1. Schwinger-Dyson Equations: The Green's function $\mathcal{G}(\tau)$ and self-energy $\Sigma(\tau)$ satisfy:

$$\Sigma(\tau) = J_q^2[\mathcal{G}(\tau)]^{q-1}, \quad \mathcal{G}(i\omega_n)^{-1} = i\omega_n - \Sigma(i\omega_n) \tag{E.1}$$

   where ω_n are Matsubara frequencies. In the IR limit ($\omega_n \to 0$), the $i\omega_n$ term is negligible, leading to $\mathcal{G}^{-1} \approx -\Sigma$.
2. Conformal Ansatz: Assume $\mathcal{G}(\tau) \propto \frac{\operatorname{sgn}(\tau)}{|J_q\tau|^{2\Delta_q}}$. Plugging this into the Schwinger-Dyson equations and solving for the scaling dimension Δ_q, we find $\Delta_q = \frac{1}{q}$. The fixed-point Green's function becomes:

$$\mathcal{G}(\tau) \sim \frac{1}{(J_q\tau)^{2/q}} \tag{E.2}$$

E.2 Treating SYK_2 as a Perturbation

Self-energy contributions are

- SYK_q self-energy: $\Sigma_q(\tau) \sim J_q^2[\mathcal{G}(\tau)]^{q-1} \sim J_q^{2/q}\tau^{-2(q-1)/q}$.
- SYK_2 self-energy: $\Sigma_2(\tau) \sim J_2^2\mathcal{G}(\tau) \sim J_2^2 J_q^{-2/q}\tau^{-2/q}$.

The coherence temperature $T_{\text{coherence}}$ is defined when $\Sigma_q \sim \Sigma_2$:

$$\Rightarrow J_q^{2/q} \tau^{-2(q-1)/q} \sim J_2^2 J_q^{-2/q} \tau^{-2/q} \tag{E.3}$$

At temperature T, the imaginary-time scale $\tau \sim 1/T$. Substituting $\tau \sim 1/T$ and solving for T gives

$$\boxed{T_{\text{coherence}} = \left(\frac{J_2^q}{J_q^2}\right)^{\frac{1}{q-2}}} \tag{E.4}$$

As examples, we show five cases: $q = 4$, $q = 8$, $q = 16$, $q = 32$ and the large-q limit to get

- $q = 4$: The coherence temperature is given by $T_{\text{coherence}} = \frac{J_2^2}{J_q}$ which matches with the known result from Ref. [4].
- $q = 8$: $T_{\text{coherence}} = \frac{J_2^{4/3}}{J_q^{1/3}}$, $q = 16$: $T_{\text{coherence}} = \frac{J_2^{8/7}}{J_q^{1/7}}$, $q = 32$: $T_{\text{coherence}} = \frac{J_2^{16/15}}{J_q^{1/15}}$
- Large-q ($q \to \infty$): $T_{\text{coherence}} = J_2$ which is independent of J_q.

Matsubara Frequencies

F

In quantum many-body systems at finite temperature $T = 1/\beta$, real-time dynamics become analytically intractable due to thermal fluctuations. To circumvent this, we use the Matsubara formalism (or imaginary-time formalism), which exploits a profound connection between thermal equilibrium and Euclidean (imaginary-time) path integrals. This approach:

- Replaces oscillatory real-time evolution $e^{-\imath\mathcal{H}t}$ with exponentially decaying evolution $e^{-\mathcal{H}\tau}$ (Wick's rotation: $\tau = \imath t$), ensuring convergence ($\mathcal{H}$ is the Hamiltonian).
- Encodes the Boltzmann factor $e^{-\beta\mathcal{H}}$ via periodicity in imaginary time, linking quantum statistics to boundary conditions.

As we already showed in Appendix A, any a generic bosonic or fermionic operator $\mathcal{O}(\tau)$ in the Heisenberg picture whose expectation value is given by

$$\langle\mathcal{O}(\tau)\rangle = \frac{1}{\mathcal{Z}}\,\mathrm{Tr}\left(e^{-\beta\mathcal{H}}\mathcal{O}(\tau)\right), \quad \mathcal{Z} = \mathrm{Tr}\left(e^{-\beta\mathcal{H}}\right), \tag{F.1}$$

has the boundary conditions imposed by the Kubo-Martin-Schwinger (KMS) condition

- Bosons: $\mathcal{O}(\tau)$ must be β-periodic:

$$\langle\mathcal{O}(\tau)\rangle = \langle\mathcal{O}(\tau + \beta)\rangle.$$

- Fermions: $\mathcal{O}(\tau)$ must be β-anti-periodic:

$$\langle\mathcal{O}(\tau)\rangle = -\langle\mathcal{O}(\tau + \beta)\rangle.$$

R. Jha, *Theory of Strongly Correlated Systems*, Lecture Notes in Physics 1051,
https://doi.org/10.1007/978-3-032-20910-8

This arises because inserting $e^{-\beta\mathcal{H}}$ cyclically permutes operators under the trace, with a sign change for fermions due to anticommutativity.

Let's consider any two-point function $G(\tau) = \langle T_\tau \mathcal{O}(\tau)\mathcal{O}^\dagger(0)\rangle$ (the Green's function is a special case of this where, for instance, the operator $\mathcal{O}$ is a fermionic field), whose Fourier transform is given by

$$G(\imath\omega_n) = \int_0^\beta d\tau e^{\imath\omega_n\tau} G(\tau), \quad G(\tau) = \frac{1}{\beta}\sum_{\omega_n} e^{-\imath\omega_n\tau} G(\imath\omega_n), \tag{F.2}$$

where ω_n are Matsubara frequencies (bosonic or fermionic). Using this, one can see how the partial derivative transforms under Fourier transform: $\partial_\tau G(\tau) \to -i\omega_n G(i\omega_n)$ (one needs to use integration by parts to come to this result). In chapters above, we have simply used ω in the imaginary-time formalism without the subscript n but here we wish to make things explicit, so we keep it. Boundary condition: $G(\tau+\beta) = \pm G(\tau)$ (+: bosons, −: fermions) implies

$$e^{-\imath\omega_n(\tau+\beta)} = \pm e^{-\imath\omega_n\tau} \Longrightarrow e^{-\imath\omega_n\beta} = \pm 1. \tag{F.3}$$

Solving gives

- Bosons: $e^{-\imath\omega_n\beta} = 1 \Longrightarrow \boxed{\omega_n = \frac{2n\pi}{\beta}} \quad (n \in \mathbb{Z}).$
- Fermions: $e^{-\imath\omega_n\beta} = -1 \Longrightarrow \boxed{\omega_n = \frac{(2n+1)\pi}{\beta}} \quad (n \in \mathbb{Z}).$

The sign in the boundary condition directly reflects the symmetry as in for bosons (integer spin): Symmetric wavefunctions $\to$ periodic boundary conditions $\to$ even harmonics $\omega_n = 2n\pi T$, while for fermions (half-integer spin): Antisymmetric wavefunctions $\to$ anti-periodic boundary conditions $\to$ odd harmonics $\omega_n = (2n+1)\pi T$.

There is a complementary picture to the Matsubara frequencies, namely in the pole structure of Fermi-Dirac/Bose-Einstein distributions under analytic continuation. Consider the thermal distribution functions:

- Bose-Einstein: $n_B(\epsilon) = \frac{1}{\epsilon^{2x}-1}$
- Fermi-Dirac: $n_F(\epsilon) = \frac{1}{\epsilon^{\epsilon t}+1}$

Their poles in the complex ϵ-plane occur where denominators vanish

- Bosons:

$$e^{\beta\epsilon} - 1 = 0 \Longrightarrow \epsilon = \imath\omega_n = \imath\frac{2n\pi}{\beta}, \quad n \in \mathbb{Z}$$

- Fermions:

$$e^{\beta\epsilon} + 1 = 0 \Longrightarrow \epsilon = \imath\omega_n = \imath\frac{(2n+1)\pi}{\beta}, \quad n \in \mathbb{Z}$$

These are precisely the Matsubara frequencies for fermions and bosons.

As concluding remarks, Matsubara frequencies are discrete and pure imaginary, reflecting compactification of time to $[0, \beta]$. The spacing $\Delta\omega = \frac{2\pi}{\beta} = 2\pi T$ for both bosons and fermions ensures resolution of thermal fluctuations. Physical response functions (e.g., retarded correlators/Green's functions $G^{\mathrm{ret}}(\omega)$) are obtained by analytically continuing

$$G^{\mathrm{ret}}(\omega) = \lim_{\eta \to 0^+} G\left(i\omega_n \to \omega + i\eta\right). \tag{F.4}$$

This maps the discrete set $\{i\omega_n\}$ to the complex plane, exposing poles and branch cuts encoding excitations.

Mathematica® Implementation for Critical Exponents

G

Here we provide the Mathematical implementation required to evaluate the thermodynamic expressions for complex SYK model in Sect. 4.2.5. We start by evaluating the expressions for equation of state m and grand potential f in terms of reduced variables (Eqs. (4.74) and (4.75)). We have used Mathematica 13.2 for the following codes. The system requirements to run this version of Mathematica can be found at https://support.wolfram.com/62016.

```
ClearAll["Global`*"];
(*Critical point values*)
Qc=Sqrt[3/2];          (* \tilde{\mathcal{Q}}_c*)
Tc=2 Jq Exp[-3/2];         (* \tilde{T}_c*)
muc=12 Qc Exp[-3/2];    (* \tilde{\mu}_c*)
Omegac=-14 Exp[-3/2];   (* \tilde{\varOmega}_c*)

(*Reduced variables:Q=Qc(1+\[Rho]),T=Tc(1+t)*)
Qval=Qc (1+\[Rho]);
Tval=Tc (1+t);

(*Chemical potential \[Mu]*)
Print["\[Mu]: "]
\[Mu]=4*Qval*(Tval+Jq Exp[-Qval^2]) //FullSimplify(* \tilde{\mu}=4\tilde{\mathcal{Q}} (\tilde{T}+e^{-\tilde{\mathcal{Q}}^2)*)

        (*Grand potential \[CapitalOmega]*)
        Print["\[CapitalOmega]: "]
        \[CapitalOmega]=-2 Jq Exp[-Qval^2] (1+2 Qval^2)-2 Qval^2 Tval //FullSimplify

        (*Reduced variables:m,t (\[Rho] is already defined)*)
        Print["m: "]
        m=\[Mu]/muc-1 //FullSimplify

        (*Shifted and rescaled grand potential f*)
        Print["f: "]
```


R. Jha, *Theory of Strongly Correlated Systems*, Lecture Notes in Physics 1051,
https://doi.org/10.1007/978-3-032-20910-8

```
    f=(\[CapitalOmega]-Omegac)/(muc Qc)-t/3+m //FullSimplify

    (*Expand f in series around \[Rho]=0 and t=0 up to O(\[Rho]^5)*)
    fSeries=Series[f,{\[Rho],0,4},{t,0,1}]//Normal//Expand//
  FullSimplify;
    gSeries=Series[m,{\[Rho],0,3},{t,0,1}]//Normal//Expand//
  FullSimplify;
    (*Extract the leading terms:-\[Rho]^2 (t+(3\[Rho]/2)^2)/3*)
    Print["Computed f_series: ",fSeries];
    Print["Computed m_series: ",gSeries];
```

Listing G.1 Mathematica code for thermodynamic properties of complex SYK model

Next we evaluate expressions in the vicinity of the critical point that gave us the critical exponents in Sect. 4.2.5.

```
$Assumptions={h\[Element]Reals,t\[Element]Reals,\[Rho]\[Element]Reals};
Print["Critical value of \[CapitalOmega]: "]
\[CapitalOmega]c=-2 Jq Exp[-Qc^2] (1+2 Qc^2)-2 Qc^2 Tc (*Verifying
    critical value of Omega, used above*)
Print["Critical value of \[Mu]: "]
\[Mu]c=4*Qc*(Tc+Jq Exp[-Qc^2]) (*Verifying critical value of mu, used
    above*)
(*Continuing with the analysis in terms of the mixing field h*)
Print["Inverting \[Rho] in terms of h and t: "]
sol=Reduce[\[Rho]^3+2 t \[Rho]/3-h==0,\[Rho],Cubics->True]
fExpr=-\[Rho]^2 (t+(3\[Rho]/2)^2)/3;
fExpanded=FullSimplify[fExpr/.{\[Rho] ->(2 2^(1/3) t)/(3 (-27 h+Sqrt[729
     h^2+32 t^3])^(1/3))-(-27 h+Sqrt[729 h^2+32 t^3])^(1/3)/(3 2^(1/3))
    }];
Print["Expansion of f for small h: "]
fSmallH=Series[fExpanded,{h,0,2}, Assumptions->{t<0, h>0}]//Normal //
    FullSimplify
Print["Expansion of f for small t: "]
fSmallT=Series[fExpanded,{t,0,0}, Assumptions->{t<0, h>0}]//Normal//
    FullSimplify
Print["Expansion of \[Rho] for small h: "]
\[Rho]SmallH=Series[((2 2^(1/3) t)/(3 (-27 h+Sqrt[729 h^2+32 t^3])^(1/3)
    )-(-27 h+Sqrt[729 h^2+32 t^3])^(1/3)/(3 2^(1/3))),{h,0,1},
    Assumptions->{t<0, h>0}]//Normal//FullSimplify
Print["Expansion of \[Rho] for small t: "]
\[Rho]SmallT=Series[((2 2^(1/3) t)/(3 (-27 h+Sqrt[729 h^2+32 t^3])^(1/3)
    )-(-27 h+Sqrt[729 h^2+32 t^3])^(1/3)/(3 2^(1/3))),{t,0,0},
    Assumptions->{t<0, h>0}]//Normal//FullSimplify
```

Listing G.2 Mathematica code for expansion around the critical point to get critical exponents for complex SYK model

Mathematica® Implementation for Complex SYK Model

Here we provide the Mathematical implementation required to evaluate the Liouville equation for the complex SYK model in Sect. 5.1.2. We have used Mathematica 13.2 for the following codes. The system requirements to run this version of Mathematica can be found at https://support.wolfram.com/62016.

All equations are provided in Sect. 5.1.2, however we reproduce them here for convenience.

H.1 Verification of Most General Form of Liouville Equation

The Liouville equation we are interested in is given by

$$\partial_{t_1}\partial_{t_2} g^+(t_1, t_2) = 2\mathcal{J}_q^2 e^{g_+(t_1,t_2)}. \tag{H.1}$$

The most general form of solution can be written as [6]:

$$e^{g(t_1,t_2)} = \frac{-\dot{u}\,(t_1)\,\dot{u}^*\,(t_2)}{\mathcal{J}^2\,[u\,(t_1) - u^*\,(t_2)]^2}, \tag{H.2}$$

where we use $g^+(t_1, t_2)^\star = g^+(t_2, t_1)$. We verify here using the following Mathematica code:

```
ClearAll["Global‘*"];

(*Define the expression for exp(g(t1,t2)) without the negative sign*)
expG[t1_,
t2_] := -(D[u[t1], t1]*
Conjugate[D[u[t2], t2]])/(J^2*(u[t1] - Conjugate[u[t2]])^2);

(*Express g(t1,t2)=Log[expG(t1,t2)]*)
g[t1_, t2_] := Log[expG[t1, t2]];
```


R. Jha, *Theory of Strongly Correlated Systems*, Lecture Notes in Physics 1051,
https://doi.org/10.1007/978-3-032-20910-8

```
(*Ensure derivatives of conjugates are handled correctly*)
Unprotect[Conjugate];
Conjugate /: D[Conjugate[f_[t_]], t_] := Conjugate[D[f[t], t]];
Protect[Conjugate];

(*Compute mixed partial derivative*)
mixedPartial = D[g[t1, t2], t1, t2];

(*Right-hand side of Liouville's equation*)
rhs = 2 J^2*expG[t1, t2];

(*Simplify the difference with assumptions*)
$Assumptions = {t1 \[Element] Reals, t2 \[Element] Reals,
        J \[Element] Reals};
difference = FullSimplify[mixedPartial - rhs];

(*Verify if the solution satisfies the equation*)
If[difference == 0,
Print["The solution satisfies the Liouville equation."],
Print["The solution does NOT satisfy the equation. Difference: ",
difference]]
```

Listing H.1 Mathematica code for verifying Liouville solution

The output one gets is

```
If[(2 Derivative[1][u][
t1] (Conjugate[Derivative[1][u][t2]] -
Conjugate[Derivative[1][u[t2]]] Derivative[1][u][
t2]))/(Conjugate[u[t2]] - u[t1])^2 == 0,
Print["The solution satisfies the Liouville equation."],
Print["The solution does NOT satisfy the equation. Difference: ",
difference]]
```

Listing H.2 Output of the Mathematica code for verifying Liouville solution

Mathematica cannot differentiate the commuting properties of taking complex conjugate and taking derivative. Since we have

$$\left(\frac{du(t_2)}{dt_2}\right)^{\star} = \frac{du(t_2)^{\star}}{dt_2}, \tag{H.3}$$

accordingly the if condition is satisfied in the output and the ansatz satisfies the Liouville equation.

H.2 Verifying the Ansatz for u

We take the ansatz for u in the above equation as $(a, b, c, d, \sigma \in \mathbb{R})$

$$u(t) = \frac{a e^{\imath \pi \nu/2} e^{\sigma t} + \imath b}{c e^{\tau \pi \nu/2} e^{\sigma t} + \imath d}, \quad \nu \in [-1, 1] \tag{H.4}$$

then we use the following Mathematica code for substituting back in Eq. (H.2):

```
ClearAll["Global`*"];

(*Define the function u(t)*)
u[t_] := (a Exp[I \[Pi] v/2] Exp[\[Sigma] t] +
I b)/(c Exp[I \[Pi] v/2] Exp[\[Sigma] t] + I d);

(*Define parameters as real-valued*)
$Assumptions = {a \[Element] Reals, b \[Element] Reals,
        c \[Element] Reals, d \[Element] Reals, \[Sigma] \[Element]
    Reals,
        v \[Element] Reals, J \[Element] Reals, t1 \[Element] Reals,
        t2 \[Element] Reals, t1 >= 0, t2 >= 0};

(*Compute derivative of u(t)*)
uDot[t_] = D[u[t], t] // FullSimplify;

(*Compute complex conjugate of u(t)*)
uConj[t_] = Conjugate[u[t]] // FullSimplify;

(*Construct the expression for e^{g(t1,t2)}*)
expG = -(uDot[t1] Conjugate[uDot[t2]])/(J^2 (u[t1] - uConj[t2])^2);

(*Simplify the expression*)
result = FullSimplify[expG, Assumptions -> $Assumptions];

(*Display the result*)
result
```

Listing H.3 Mathematica code for ansatz verification for u

The output reads as

```
(E^(I \[Pi] v + (t1 + t2) \[Sigma]) \[Sigma]^2)/((E^(t2 \[Sigma]) +
E^(I \[Pi] v + t1 \[Sigma]))^2 J^2)
```

Listing H.4 Output of the Mathematica code for ansatz verification for u

Written in LaTeXform, we have

$$\frac{\sigma^2 e^{\sigma(\mathrm{t}_1+\mathrm{t}_2)+i\pi\nu}}{\mathcal{J}_q^2 \left(e^{\sigma \mathrm{t}_2} + e^{\sigma \mathrm{t}_1 + i\pi\nu}\right)^2}. \tag{H.5}$$

where there is a perfect cancellation of a, b, c and d. This expression is exactly the same as

$$e^{g(t_1,t_2)} = \frac{(\sigma/2)^2}{\mathcal{J}_q^2 \cos^2\left(\pi v/2 - \sigma \imath \left(t_1 - t_2\right)/2\right)}, \quad \sigma \geq 0. \tag{H.6}$$

The following Mathematica code verifies this:

```
ClearAll["Global`*"];
$Assumptions = {\[Sigma] > 0, v \[Element] Reals, t1 \[Element] Reals,
        t2 \[Element] Reals, J \[Element] Reals};

(*Given expression from u(t) substitution*)
expr1 = (\[Sigma]^2 Exp[\[Sigma] (t1 + t2) +
I \[Pi] v])/(J^2 (Exp[\[Sigma] t2] +
Exp[\[Sigma] t1 + I \[Pi] v])^2);

(*Expected expression*)
expr2 = (\[Sigma]/2)^2/(J^2 Cos[\[Pi] v/2 - I \[Sigma] (t1 - t2)/2]^2);

(*Show expr1 simplifies to expr2*)
simplifiedExpr1 = FullSimplify[expr1, Assumptions -> $Assumptions];
Print["Expression from substitution: ", simplifiedExpr1];
Print["Expected expression: ", expr2];

(*Verify equality*)
diff = simplifiedExpr1 - expr2;
If[FullSimplify[diff == 0, Assumptions -> $Assumptions],
Print["The expressions are equal."],
Print["The expressions are NOT equal. Difference: ", diff]]
```

Listing H.5 Mathematica code for equivalence of Liouville solutions

where the output is

```
Expression from substitution: (E^(I \[Pi] v+(t1+t2) \[Sigma]) \[Sigma
     ]^2)/((E^(t2 \[Sigma])+E^(I \[Pi] v+t1 \[Sigma]))^2 J^2)

Expected expression: (\[Sigma]^2 Sec[(\[Pi] v)/2-1/2 I (t1-t2) \[Sigma
     ]]^2)/(4 J^2)

The expressions are equal.
```

Listing H.6 Output of the Mathematica code for equivalence of Liouville solutions

Therefore, Eq. (H.6) shows that the most general solution $e^{g^{+}(t_1,t_2)}$ satisfies the time-translational.

H.3 KMS Relation for g^+ and g^-

We know that the KMS relation $g(t) = g(-t - \imath 2\pi\nu/\sigma)$ is satisfied for systems in equilibrium where we can read-off the temperature via the relation

$$\beta = \frac{2\pi\nu}{\sigma}. \tag{H.7}$$

We know that $g^{\gtrless}(t)$ inherits this property. Using the definition of $g^{\pm}(t) \equiv \frac{g^{>}(t)\pm g^{<}(-t)}{2}$, we show using Mathematica that g^+ inherits the KMS relation which allows us to extract the temperature of the system, while g^- does not.

```
ClearAll["Global‘*"];
$Assumptions = {\[Sigma] > 0,
        v \[Element] Reals, -1 <= v <= 1, \[Tau] \[Element]
        Reals, \[Beta] = (2 \[Pi] v)/\[Sigma]};

(*Define g^+(\[Tau])=g(\[Tau])*)
gPlus[t_] :=
Log[(\[Sigma]/2)^2/(J^2*Cos[\[Pi] v/2 - I \[Sigma] t/2]^2)];

(*Shifted argument:-\[Tau]-i\[Beta]*)
shiftedgPlus = gPlus[-t - I \[Beta]];

(*Check equality*)
If[FullSimplify[gPlus[t] == shiftedgPlus],
Print["g^+(t) satisfies KMS: g^+(t) = g^+(-t - i\[Beta])"],
Print["Verification failed"]]
```

Listing H.7 Mathematica code for KMS relation for $g^+(t)$

whose output is

```
g^+(t) satisfies KMS: g^+(t) = g^+(-t - i\[Beta])
```

Listing H.8 Output of the Mathematica code for KMS relation for $g^+(t)$

The following code shows that $g^-(t)$ does not inherit the KMS relation:

```
ClearAll["Global‘*"];
$Assumptions = {\[Kappa] \[Element] Reals, \[Beta] > 0,
        t \[Element] Reals};

(*Define g^-(t)=i\[Kappa]t*)
gMinus[t_] := I \[Kappa] t;

(*Shifted argument:-t-i\[Beta]*)
shiftedgMinus = gMinus[-t - I \[Beta]];

(*Check equality*)
If[FullSimplify[gMinus[\[Tau]] != shiftedgMinus],
```

```
Print["g^-(t) does NOT satisfy KMS: g^-(t) \[NotEqual] g^-(-t - i\
\[Beta])"], Print["Verification failed"]]
```

Listing H.9 Mathematica code for KMS relation for $g^{-}(t)$

whose output is

```
If[\[Kappa] (t + I \[Beta] + \[Tau]) != 0,
Print["g^-(t) does NOT satisfy KMS: g^-(t) \[NotEqual] g^-(-t - i\
\[Beta])"], Print["Verification failed"]]
```

Listing H.10 Output of the Mathematica code for KMS relation for $g^{-}(t)$

where the if condition is not true, accordingly g^{-} does not satisfy the KMS relation, only g^{+} does.

We finally provide the Mathematica code to show that Eq. (H.6) satisfies the KMS relation, showing that the system is in equilibrium (whose temperature can be extracted using the KMS relation in Eq. (H.7)) with respect to the Green's function.

```
ClearAll["Global`*"];
(*Define assumptions*)
$Assumptions = {\[Sigma] > 0, v \[Element] Reals,
        t \[Element] Reals, -1 <= v <= 1  (*Physical constraint for v*)
    };

(*Define the expression for e^{g(t)}*)
expG[t_] := (\[Sigma]/2)^2/(J^2*Cos[\[Pi] v/2 - I \[Sigma] t/2]^2);

(*Define the shifted argument \[Tau]=-t-i 2\[Pi]v/\[Sigma]*)
shiftedExpG = expG[-t - I (2 \[Pi] v)/\[Sigma]];

(*Simplify both expressions*)
expGT = FullSimplify[expG[t]];
expGShifted = FullSimplify[shiftedExpG];

(*Check if expG[t]==expG[shifted argument]*)
If[Simplify[expGT == expGShifted],
Print["e^{g(t)} satisfies the KMS condition: e^{g(t)} = e^{g(-t - i \
        2\[Pi]v/\[Sigma])}"],
Print["Verification failed. Difference: ",
Simplify[expGT - expGShifted]]];

(*Optional:Check g(t) equality (principal branch)*)
g[t_] := Log[expG[t]];
gShifted := g[-t - I (2 \[Pi] v)/\[Sigma]];
difference =
FullSimplify[g[t] - gShifted, Assumptions -> $Assumptions];

(*Since difference might be 2\[Pi]i n,check if it's an integer \
multiple of 2\[Pi]i*)
If[Simplify[difference/(2 \[Pi] I) \[Element] Integers],
Print["g(t) = g(-t - i 2\[Pi]v/\[Sigma]) + 2\[Pi]i n holds for \
```

```
integer n."],
Print["g(t) equality verification failed. Difference: ", difference]]
```

Listing H.11 Mathematica code for KMS relation satisfied by the Louville solution $e^{g^+(t_1,t_2)}$ evaluated above in Eq. (H.6)

whose output gives the confirmation:

```
e^{g(t)} satisfies the KMS condition: e^{g(t)} = e^{g(-t - i 2\[Pi]v/\[
    Sigma])}

g(t) = g(-t - i 2\[Pi]v/\[Sigma]) + 2\[Pi]i n holds for integer n.
```

Listing H.12 Output of the Mathematica code for KMS relation satisfied by the Louville solution $e^{g^+(t_1,t_2)}$ evaluated above in Eq. (H.6)

We also showed that KMS relation holds true for arbitrary integer n when $2\pi n$ is added to the KMS relation.

Initial Condition for Conductivity

I

The initial condition for conductivity $\sigma(0)$ is taken from the expression in Eq. (5.138) which gives

$$\sigma(0) = -\imath \frac{\langle [X(0), I(0)] \rangle}{L}, \tag{I.1}$$

where X is the polarization operator, defined in Eq. (5.128) and I is the total current operator, defined in Eq. (5.122). We reproduce those equations for convenience:

$$\begin{aligned} X &= \sum_{j=1}^{L} jN\mathcal{Q}_j, \quad \mathcal{Q}_i = \frac{1}{N}\sum_{\alpha=1}^{N}\left(c^\dagger_{i,\alpha}c_{i,\alpha} - \frac{1}{2}\right), \\ I &= \frac{\imath r}{2}(\mathcal{H}_\rightarrow - \mathcal{H}^\dagger_\rightarrow), \quad \mathcal{H}_\rightarrow = \sum_{i=1}^{L}\mathcal{H}_{i\rightarrow i+1}. \end{aligned} \tag{I.2}$$

The bond Hamiltonian and $\mathcal{H}_{\text{trans}}$ are defined in Eqs. (5.118) and (5.90) respectively, which we reproduce here

$$\mathcal{H}^{i,i+1} = \mathcal{H}_{i\rightarrow i+1} + \mathcal{H}^\dagger_{i\rightarrow i+1}, \quad \mathcal{H}_{\text{trans}} = \mathcal{H}_\rightarrow + \mathcal{H}^\dagger_\rightarrow. \tag{I.3}$$

The Galitskii-Migdal relations provide

$$[N\mathcal{Q}_i, \mathcal{H}_{i\rightarrow i+1}] = -\frac{r}{2}\mathcal{H}_{i\rightarrow i+1}, \quad [N\mathcal{Q}_i, \mathcal{H}_{i-1\rightarrow i}] = \frac{r}{2}\mathcal{H}_{i-1\rightarrow i}. \tag{I.4}$$

R. Jha, *Theory of Strongly Correlated Systems*, Lecture Notes in Physics 1051,
https://doi.org/10.1007/978-3-032-20910-8

For the global current I, we have

$$\begin{aligned}[N\mathcal{Q}_i, I_i] &= \imath r[N\mathcal{Q}_i, \mathcal{H}_{i-1\to i} - \mathcal{H}^\dagger_{i-1\to i}]/2 \\ &= \imath r[N\mathcal{Q}_i, \mathcal{H}_{i-1\to i}]/2 - \imath r[N\mathcal{Q}_i, \mathcal{H}^\dagger_{i-1\to i}]/2 \\ &= \imath\frac{r^2}{4}\mathcal{H}_{i-1\to i} + \imath\frac{r^2}{4}\mathcal{H}^\dagger_{i-1\to i} = \imath\frac{r^2}{4}\mathcal{H}^{i-1,i}.\end{aligned} \tag{I.5}$$

Therefore, we get for the full commutator at equal-time (in this case $t = 0$) $[X, I]$

$$[X, I] = \sum_{j=1}^{L} j\,[N\mathcal{Q}_j, I] = \imath\frac{r^2}{4}\sum_{j=1}^{L} j\mathcal{H}^{j-1,j}. \tag{I.6}$$

Using index shift and open-boundary conditions ($\mathcal{H}^{0,1} = 0, \mathcal{H}^{L,L+1} = 0$):

$$\sum_{j=1}^{L} j\mathcal{H}^{j-1,j} = \sum_{j=1}^{L} \mathcal{H}^{j-1,j} = \mathcal{H}_{\text{trans}}, \tag{I.7}$$

where $\mathcal{H}_{\text{trans}} = \sum_{i=1}^{L}\mathcal{H}^{i,i+1}$. Thus

$$[X, I] = \imath\frac{r^2}{4}\mathcal{H}_{\text{trans}}. \tag{I.8}$$

Finally, we have the initial condition for the conductivity

$$\sigma(0) = -\imath\frac{\langle[X, I]\rangle}{L} = \frac{r^2}{4}\frac{\langle\mathcal{H}_{\text{trans}}\rangle}{L}, \tag{I.9}$$

which matches with the first equality in the main text in Eq. (5.140).

We can further evaluate the expression for $\langle\mathcal{H}_{\text{trans}}\rangle$ by using the definition of $\mathcal{H}_{\text{trans}}$ from above. We get

$$\langle\mathcal{H}_{\text{trans}}\rangle = \langle\mathcal{H}_{\to}\rangle + \langle\mathcal{H}^\dagger_{\to}\rangle = -2NL|D|^2\int_{-\infty}^{0} \mathrm{d}t\,Y(t), \tag{I.10}$$

where we used Eq. (5.173) for $\langle\mathcal{H}_{\to}\rangle$ and its complex conjugate (whose explicit expressions happens to be the same as for $\langle\mathcal{H}_{\to}\rangle$). However, we know that $Y = \mathrm{Im}[F]$ from Eq. (5.85) and F is related to current-current correlation via Eq. (5.177) which we reproduce here

$$\begin{aligned}\langle I(0)I(t)\rangle &= \frac{r^2}{4}NL|D|^2F(t) \quad \Rightarrow \mathrm{Im}\langle I(0)I(t)\rangle \\ &= \frac{r^2}{4}NL|D|^2\mathrm{Im}F(t) = \frac{r^2}{4}NL|D|^2Y(t).\end{aligned} \tag{I.11}$$

Thus, we replace $Y(t) = \frac{4}{r^2NL|D|^2}\mathrm{Im}\langle I(0)I(t)\rangle$ in Eq. (I.10), which in turn is plugged in Eq. (I.9) to get

$$\sigma(0) = \frac{r^2}{4}\frac{\langle\mathcal{H}_{\mathrm{trans}}\rangle}{L} = -\frac{2}{L}\mathrm{Im}\int_{-\infty}^{0} d\tau\langle I(0)I(\tau)\rangle. \tag{I.12}$$

where we used the fact that taking imaginary component commutes with performing integration (as long as the integral exists, i.e., converges)). This matches with the second equality in Eq. (5.140) and this concludes our proof.

DC Resistivities Across All Temperatures

J

We have seen in Sect. 5.4 the transport properties of three SYK chains where we found the relation in Eq. (5.151). The integral that appears there, namely $\mathcal{W}_\kappa$ ($\kappa = \{1/2, 1, 2\}$ for the three chains considered) is defined in Eq. (5.150). As we can see, $\mathcal{W}_\kappa$ depends on the current-current correlation $\langle I(0)I(t)\rangle$. We simplified the expressions reach Eq. (5.179) where we evaluated the integral for all three chains across all temperature ranges in Sect. 5.4.5. We also provided the low-temperature limit of DC resistivity and found Eqs. (5.190), (5.197) and (5.204). We further found that all three models admit a universal minimum resistivity, given by $\frac{8}{N\pi}$ when the coupling ratio $|\mathcal{D}|/\mathcal{J}$ becomes very large. The minimum resistivity for $\kappa = 1/2$ and $\kappa = 1$ cases coincide with their residual resistivity (i.e., DC resistivity at zero temperature) while the minimum resistivity occurs at a finite (non-zero) temperature for $\kappa = 2$ case.

We obtain exact temperature-dependent DC resistivity by leveraging closed-form solutions of the integrals $\mathcal{W}_\kappa$ (Eqs. (5.183), (5.192), (5.199)) that remain valid across all temperature regimes. The key innovation lies in our treatment of the scaling variable ν - rather than employing low-temperature approximations of the closure relations (Eqs. (5.109), (5.112), (5.116)), we implement a robust numerical procedure to determine $\nu(T)$ at arbitrary temperatures.

We start by noticing the three closure relations that we reproduce here for convenience:

$$\begin{aligned}
\pi\nu &= \sqrt{\left(\beta\mathcal{J}_q\right)^2 + \left(\frac{\left(\beta\mathcal{K}_{q/2}\right)^2}{\pi\nu}\right)^2}\cos(\pi\nu/2) + \frac{\left(\beta\mathcal{K}_{q/2}\right)^2}{\pi\nu},\\
\pi\nu &= \beta\sqrt{\mathcal{J}_q^2 + \mathcal{K}_q^2}\cos(\pi\nu/2), \qquad\qquad (J.1)\\
\pi\nu &= \sqrt{2\left(\beta\mathcal{K}_{2q}\right)^2 + \left(\frac{2\left(\beta\mathcal{J}_q\right)^2}{\pi\nu}\right)^2}\cos(\pi\nu/2) + \frac{2\left(\beta\mathcal{J}_q\right)^2}{\pi\nu}.
\end{aligned}$$

R. Jha, *Theory of Strongly Correlated Systems*, Lecture Notes in Physics 1051,
https://doi.org/10.1007/978-3-032-20910-8

We can absorb β in the coupling coefficients and redefine $\tilde{\mathcal{K}}_{\kappa q} = \beta\mathcal{K}_{\kappa q}$ and $\tilde{\mathcal{J}}_q = \beta\mathcal{J}_q$. We chose any particular parameter values, for instance $\mathcal{J}_q = 1$ and $\mathcal{K}_{\kappa q} = 5$, then for any given temperature we invert these relations numerically and plug in the exact expressions for $\mathcal{W}_\kappa$ obtained in Eqs. (5.183), (5.192), and (5.199). That in turn will give us the DC conductivity via the relation in Eq. (5.151), namely $\sigma_{\mathrm{DC}}^{(\kappa)} = \frac{\beta\kappa N}{2}\mathrm{Re}\mathcal{W}_\kappa(0)$. Taking inverse leads us to DC resistivity across all temperatures.

We start by providing examples for how to invert the closure relations. The following is the Mathematical implementation of inverting the relations at $T = 1$ for coupling strengths $\mathcal{J}_q = 1$ and $\mathcal{K}_{\kappa q} = 5$, for instance.

```
(*Common parameters*)J0 = 1.0;
D0 = 5.0;
rhoMIR = 2*Pi;

SolveKappa1[T_] := Module[{beta, J, D, v0, vSol, rho}, beta = 1/
T;
J = beta*J0;
D = beta*D0;
(*Initial guess*)v0 = 1 - (2/D)*(1/Sqrt[1 + (Abs[J]/D)^2]);
v0 = Clip[v0, {0.01, 0.99}];
(*Solve equation*)
vSol = v /.
FindRoot[(J^2 + D^2)*Cos[Pi*v/2]^2 - (Pi*v)^2 == 0, {v, v0,
0.01, 0.99}];
(*Compute resistivity*)rho = (8*((Abs[J]/D)^2 + 1))/(Pi*vSol);
rho/rhoMIR]

SolveKappaHalf[T_] := Module[{beta, J, D, v0, vSol, rho}, beta =
1/T;
J = beta*J0;
D = beta*D0;
(*Initial guess*)v0 = 2 - (4/D)*Sqrt[2 + (Abs[J]/D)^2];
v0 = Clip[v0, {0.1, 1.9}];
(*Solve equation*)
vSol = v /.
FindRoot[
Sqrt[J^2 + (D^4/(Pi^2 v^2))]*Cos[Pi*v/2] + D^2/(Pi*v) - Pi*v ==
0, {v, v0, 0.01, 1.99}];
(*Compute resistivity*)rho = 16/(Pi*vSol);
rho/rhoMIR]

SolveKappa2[T_] :=
Module[{beta, J, D, v0, vSol, tanGamma, gamma, rho}, beta = 1/T;
J = beta*J0;
D = beta*D0;
(*Initial guess*)v0 = 1 - (Sqrt[2]/Abs[J])*Sqrt[2 + (D/J)^2];
v0 = Clip[v0, {0.1, 0.9}];
(*Solve equation*)
vSol = v /.
FindRoot[
```

```
    Sqrt[2*D^2 + (J^4/(Pi^2 v^2))]*Cos[Pi*v] + J^2/(Pi*v) - 2*Pi*v
==
    0, {v, v0, 0.01, 0.99}];
    (*Compute resistivity with gamma term*)
    tanGamma = (2*Pi*vSol*Sqrt[2]*Abs[D])/J^2;
    gamma = ArcTan[tanGamma];
    rho = (4/(Pi*vSol))*(1/(1 - gamma/tanGamma));
    rho/rhoMIR]

    (*Test all cases at T=1.0*)
    Print["\[Kappa]=1 at T=1.0: \[Rho]/\[Rho]_MIR = ", SolveKappa1
[1.0]]
    Print["\[Kappa]=1/2 at T=1.0: \[Rho]/\[Rho]_MIR = ",
    SolveKappaHalf[1.0]]
    Print["\[Kappa]=2 at T=1.0: \[Rho]/\[Rho]_MIR = ", SolveKappa2
[1.0]]
```

Listing J.1 Mathematica example for inverting closure relation

The output reads as

```
\[Kappa]=1 at T=1.0: \[Rho]/\[Rho]_MIR = 0.592629

\[Kappa]=1/2 at T=1.0: \[Rho]/\[Rho]_MIR = 0.649347

\[Kappa]=2 at T=1.0: \[Rho]/\[Rho]_MIR = 0.530365
```

Listing J.2 Output of the Mathematica example for inverting closure relation

We also provide the Python script for the same example of inversion as above at $T = 1$ and coupling strengths $\mathcal{J}_q = 1$, $\mathcal{K}_{\kappa q} = 5$:

```
import numpy as np
from scipy.optimize import fsolve

# Constants
J0, D0 = 1.0, 5.0
rho_MIR = 2 * np.pi

def solve_kappa1(T):
"""Solve for Kappa = 1 at temperature T"""
beta = 1.0 / T
J = beta * J0
D = beta * D0

# Initial guess
v0_guess = 1 - (2 / D) * (1 / np.sqrt(1 + (np.abs(J) / D)**2))
v0_guess = np.clip(v0_guess, 0.01, 0.99)  # Clamp to physical bounds

# Define equation
def equation(v):
return (J**2 + D**2) * np.cos(np.pi * v / 2)**2 - (np.pi * v)**2
```

```
# Solve numerically
v_sol = fsolve(equation, v0_guess)[0]
v_sol = np.clip(v_sol, 0.01, 0.99)

# Compute resistivity
rho = (8 * ((np.abs(J)/D)**2 + 1)) / (np.pi * v_sol)
return rho / rho_MIR

def solve_kappa_half(T):
"""Solve for Kappa = 1/2 at temperature T"""
beta = 1.0 / T
J = beta * J0
D = beta * D0

# Initial guess
v0_guess = 2 - (4 / D) * np.sqrt(2 + (np.abs(J)/D)**2)
v0_guess = np.clip(v0_guess, 0.1, 1.9)

# Define equation
def equation(v):
term1 = np.sqrt(J**2 + (D**4 / (np.pi**2 * v**2)))
term2 = np.cos(np.pi * v / 2)
return term1 * term2 + (D**2 / (np.pi * v)) - np.pi * v

# Solve numerically
v_sol = fsolve(equation, v0_guess)[0]
v_sol = np.clip(v_sol, 0.01, 1.99)

# Compute resistivity
rho = 16 / (np.pi * v_sol)
return rho / rho_MIR

def solve_kappa2(T):
"""Solve for Kappa = 2 at temperature T"""
beta = 1.0 / T
J = beta * J0
D = beta * D0

# Initial guess
v0_guess = 1 - (np.sqrt(2) / np.abs(J)) * np.sqrt(2 + (D/J)**2)
v0_guess = np.clip(v0_guess, 0.1, 0.9)

# Define equation
def equation(v):
term1 = np.sqrt(2*D**2 + (J**4 / (np.pi**2 * v**2)))
term2 = np.cos(np.pi * v)
return term1 * term2 + (J**2 / (np.pi * v)) - 2*np.pi*v

# Solve numerically
v_sol = fsolve(equation, v0_guess)[0]
v_sol = np.clip(v_sol, 0.01, 0.99)
```

```
# Compute resistivity with gamma term
tan_gamma = (2 * np.pi * v_sol * np.sqrt(2) * np.abs(D)) / (J**2)
gamma = np.arctan(tan_gamma)
rho = 4/(np.pi * v_sol) * 1/(1 - gamma/tan_gamma)
return rho / rho_MIR

# Test all cases at T = 1.0
T_val = 1.0
print(f"Kappa=1 at T={T_val}:  Rho/Rho_MIR = {solve_kappa1(T_val):.6f}")
print(f"Kappa=1/2 at T={T_val}:  Rho/Rho_MIR = {solve_kappa_half(T_val)
    :.6f}")
print(f"Kappa=2 at T={T_val}:  Rho/Rho_MIR = {solve_kappa2(T_val):.6f}")
```

Listing J.3 Python example for inverting closure relation

whose output is (it matches with the Mathematica output as it should)

```
        Kappa=1 at T=1.0: Rho/Rho_MIR = 0.592629
        Kappa=1/2 at T=1.0: Rho/Rho_MIR = 0.649347
        Kappa=2 at T=1.0: Rho/Rho_MIR = 0.530365
```

Listing J.4 Output of the Python example for inverting closure relation

The full code plotting the DC resistivities for all three chains across all temperature ranges in Fig. 5.1 is given by

```
import numpy as np
import matplotlib.pyplot as plt
from scipy.optimize import fsolve

rho_MIR = 2 * np.pi
plt.rc('text', usetex=True)
plt.rc('font', family='serif')
plt.rc('axes', titlesize=20)
plt.rc('axes', labelsize=20)
plt.rc('xtick', labelsize=14)
plt.rc('ytick', labelsize=20)
plt.rc('legend', fontsize=20)

# Define the combined function f(v, J, D, kappa)
def f(v, J, D, kappa):
if kappa == 1:
return (J**2 + D**2) * np.cos(np.pi * v / 2)**2 - (np.pi * v)**2
elif kappa == 0.5:
return np.sqrt(J**2 + (D**2 / (np.pi * v))**2) * np.cos(np.pi * v / 2) +
     (D**2 / (np.pi * v)) - np.pi * v
elif kappa == 2:
return np.sqrt((np.sqrt(2) * D)**2 + ((np.sqrt(2) * J)**2 / (2 * np.pi *
     v))**2) * np.cos(np.pi * v) + ((np.sqrt(2) * J)**2 / (2 * np.pi * v
    )) - 2 * np.pi * v
else:
raise ValueError("Unsupported kappa value")
```

```
# Initial guess functions
def v0(J, D, kappa):
if kappa == 1:
return 1 - (2 / D) * (1 / np.sqrt(1 + abs(J / D)**2))
elif kappa == 0.5:
return 2 - 2 * (2 / D) * np.sqrt(2 + abs(J / D)**2)
elif kappa == 2:
return 1 - (np.sqrt(2) / J) * np.sqrt(2 + abs(D / J)**2)
else:
raise ValueError("Unsupported kappa value")

# Root finding function
def find_root(f, J, D, kappa, v_0, v_i, v_f):
if v_0 < v_i or v_0 > v_f:
raise ValueError("Initial guess v_0 is out of bounds.")

root = fsolve(lambda v: f(v, J, D, kappa), v_0)[0]

if root < v_i or root > v_f:
raise ValueError("Root found is out of bounds.")

return root

def ff(J_0, D_0):
# Define beta from 10 (T=0.1) to 0.2 (T=5.0)
beta_values = np.linspace(200, 0.2, 2000)
T_values = 1 / beta_values

results = {'kappa_1': [], 'kappa_0_5': [], 'kappa_2': []}
kappa_params = {
        1: {'label': 'kappa_1', 'v_i': 0, 'v_f': 1},
        0.5: {'label': 'kappa_0_5', 'v_i': 0, 'v_f': 2},
        2: {'label': 'kappa_2', 'v_i': 0, 'v_f': 1}
}

for kappa, params in kappa_params.items():
label = params['label']
v_i, v_f = params['v_i'], params['v_f']
v_0 = v0(beta_values[0] * J_0, beta_values[0] * D_0, kappa)

for beta in beta_values:
J, D = beta * J_0, beta * D_0
try:
v = find_root(f, J, D, kappa, v_0, v_i, v_f)
results[label].append((beta, v))
v_0 = v  # Update initial guess for next beta
except ValueError:
continue  # Skip problematic points

# Calculate resistivity (rho) for each kappa
rho_results = {'kappa_1': [], 'kappa_0_5': [], 'kappa_2': []}
for label in results:
```

```
for beta, v in results[label]:
T = 1 / beta
J, D = beta * J_0, beta * D_0

if label == 'kappa_1':
rho = (8 * (abs(J / D)**2 + 1)) / (np.pi * v)
elif label == 'kappa_0_5':
rho = 16 / (np.pi * v)
elif label == 'kappa_2':
tan_gamma = np.pi * 2 * v * abs(np.sqrt(2) * D )/ J**2
rho = 4 * (2 / (np.pi * 2 * v)) * (1 - np.arctan(tan_gamma) / tan_gamma)
    **-1

rho_results[label].append((T, rho))

return rho_results


# Calculate and plot results
rho_results = ff(J_0=1, D_0=5)

# Plotting begins
plt.figure(figsize=(8, 6), dpi=120)

# Plot horizontal reference lines with labels for the legend
plt.axhline(y=4/np.pi**2, color='b', linestyle='--', label=r'$\rho_{\
    mathrm{min}}$')  # rho_min
plt.axhline(y=1, color='r', linestyle='--', label=r'$\rho_{\mathrm{MIR}}
    $')           # rho_MIR

ax = plt.gca()
ax.spines['top'].set_visible(False)
ax.spines['right'].set_visible(False)
ax.yaxis.set_ticks_position('left')
ax.xaxis.set_ticks_position('bottom')

plt.ylim(0, 1.1)
plt.xlim(0, 5)
plt.xlabel(r'$T$', fontsize=22)
plt.ylabel(r'$\rho_{\mathrm{DC}}/\rho_{\mathrm{MIR}}$', fontsize=22)

# Define distinct colors for each kappa
colors = {
        'kappa_1': '#1f77b4',   # Vivid blue
        'kappa_0_5': '#2ca02c', # Forest green
        'kappa_2': '#d62728'    # Crimson red
}

# Plot the kappa curves
for label, data in rho_results.items():
if data:  # Ensure data exists
data_arr = np.array(data)
T_vals = data_arr[:, 0]
```

```
rho_vals = data_arr[:, 1] / rho_MIR

if label == 'kappa_1':
kappa_label = r'$\kappa=1$'
elif label == 'kappa_0_5':
kappa_label = r'$\kappa=1/2$'
elif label == 'kappa_2':
kappa_label = r'$\kappa=2$'

plt.plot(T_vals, rho_vals, label=kappa_label,
color=colors[label], linewidth=2.5)

# Set regular y-ticks without special labels
plt.yticks([0, 0.2, 0.4, 0.6, 0.8, 1.0], fontsize=16)
plt.xticks([0, 1, 2, 3, 4, 5], fontsize=16)

# Position legend in bottom right with two columns
plt.legend(fontsize=16, loc='lower right', framealpha=0.3, ncol=1,
edgecolor='gray', fancybox=True)

plt.tight_layout()
plt.savefig("rho_vs_T_D=5.pdf", format="pdf", bbox_inches='tight')
plt.show()
```

Listing J.5 Python code for reproducing Fig. 5.1 which shows the temperature dependence of the normalized DC resistivity for different chains. See the figure caption and the discussion of the figure in the main text for physical interpretation and parameter details

References

1. Danielewicz, P.: Operator expectation values, self-energies, cutting rules, and higher-order processes in quantum many-body theory. Ann. Phys. **197**(1), 154–201 (1990)
2. Hartnoll, S.A., Mackenzie, A.P.: Colloquium: planckian dissipation in metals. Rev. Mod. Phys. **94**(4), 041002 (2022)
3. Hyrkäs, M.J., Karlsson, D., van Leeuwen, R.: Contour calculus for many-particle functions. J. Phys. A: Math. Theor. **52**(21), 215303 (2019)
4. Song, X.Y., Jian, C.M., Balents, L.: Strongly Correlated metal built from Sachdev-Ye-Kitaev models. Phys. Rev. Lett. **119**(21), 216601 (2017). Publisher: American Physical Society
5. Stefanucci, G., van Leeuwen, R.: Nonequilibrium Many-Body Theory of Quantum Systems: A Modern Introduction. Cambridge University Press, Cambridge (2013)
6. Tsutsumi, M.: On solutions of Liouville's equation. J. Math. Anal. Appl. **76**(1), 116–123 (1980)

Index

R. Jha, *Theory of Strongly Correlated Systems*, Lecture Notes in Physics 1051,
https://doi.org/10.1007/978-3-032-20910-8

The manufacturer's authorised representative in the EU is Springer Nature Customer Service Centre GmbH, Europaplatz 3, 69115 Heidelberg, Germany. If you have any concerns regarding our products, please contact ProductSafety@springernature.com

Printed and bound by CPI Group (UK) Ltd, Croydon, CR0 4YY
07/07/2026
02160921-0002